Advances in Industrial Control

Advances in Industrial Control is a series of monographs and contributed titles focusing on the applications of advanced and novel control methods within applied settings. This series has worldwide distribution to engineers, researchers and libraries.

The series promotes the exchange of information between academia and industry, to which end the books all demonstrate some theoretical aspect of an advanced or new control method and show how it can be applied either in a pilot plant or in some real industrial situation. The books are distinguished by the combination of the type of theory used and the type of application exemplified. Note that "industrial" here has a very broad interpretation; it applies not merely to the processes employed in industrial plants but to systems such as avionics and automotive brakes and drivetrain. This series complements the theoretical and more mathematical approach of Communications and Control Engineering.

Indexed by SCOPUS and Engineering Index.

Proposals for this series, composed of a proposal form downloaded from this page, a draft Contents, at least two sample chapters and an author cv (with a synopsis of the whole project, if possible) can be submitted to either of the:

Series Editors

Professor **Michael J. Grimble**
Department of Electronic and Electrical Engineering, Royal College Building, 204 George Street, Glasgow G1 1XW, United Kingdom
e-mail: m.j.grimble@strath.ac.uk

Professor **Antonella Ferrara**
Department of Electrical, Computer and Biomedical Engineering, University of Pavia, Via Ferrata 1, 27100 Pavia, Italy
e-mail: antonella.ferrara@unipv.it

or the

In-house Editor

Mr. **Oliver Jackson**
Springer London, 4 Crinan Street, London, N1 9XW, United Kingdom
e-mail: oliver.jackson@springer.com

Proposals are peer-reviewed.

Publishing Ethics

Researchers should conduct their research from research proposal to publication in line with best practices and codes of conduct of relevant professional bodies and/or national and international regulatory bodies. For more details on individual ethics matters please see: https://www.springer.com/gp/authors-editors/journal-author/journal-author-helpdesk/publishing-ethics/14214

More information about this series at http://www.springer.com/series/1412

Zhe Wu · Panagiotis D. Christofides

Process Operational Safety and Cybersecurity

A Feedback Control Approach

Springer

Zhe Wu
Department of Chemical and Biomolecular Engineering
University of California
Los Angeles, CA, USA

Panagiotis D. Christofides
Department of Chemical and Biomolecular Engineering
University of California
Los Angeles, CA, USA

ISSN 1430-9491 ISSN 2193-1577 (electronic)
Advances in Industrial Control
ISBN 978-3-030-71185-6 ISBN 978-3-030-71183-2 (eBook)
https://doi.org/10.1007/978-3-030-71183-2

Mathematics Subject Classification (2010): 93C83, 49J15, 93C10

This Springer imprint is published by the registered company Springer Nature Switzerland AG
The registered company address is: Gewerbestrasse 11, 6330 Cham, Switzerland

Preface

Traditionally, the operational safety of chemical processes has been addressed through process design considerations and through a hierarchical, independent design of control and safety systems. By developing safety systems including alarms, emergency shutdown, and further emergency response systems to be activated when control systems fail to operate chemical processes in a normal operating region, process operational safety has been improved to prevent incidents that can lead to property damage, human injuries, and environmental impact. However, the consistent accidents throughout chemical process plant history (including several high-profile disasters in the last decade) have motivated researchers to design control systems that explicitly account for process operational safety considerations. In particular, a new design of control systems such as model predictive controllers (MPC) that incorporates safety considerations and can be coordinated with safety systems has the potential to significantly improve process operational safety and avoid unnecessary triggering of alarm systems. However, the rigorous design of safety-based control systems poses new challenges that cannot be addressed with traditional process control methods, including, for example, proving simultaneous closed-loop stability and safety. On the other hand, cybersecurity has become increasingly important in chemical process industries in recent years as cyber-attacks that have grown in sophistication and frequency have become another leading cause of process safety incidents. While the traditional methods of handling cyber-attacks in control systems still rely partly on human analysis and mainly fall into the area of fault diagnosis, the intelligence of cyber-attacks and their accessibility to control system information have recently motivated researchers to develop cyber-attack detection and resilient operation control strategies to address directly cybersecurity concerns.

The book covers several rigorous methods for the design of MPC systems that improve process operational safety and cybersecurity for chemical processes described by nonlinear dynamic models. Beginning with the motivation and organization of this book, a background on nonlinear systems analysis, Lyapunov-based control techniques, and MPC designs is first presented. Then, two MPC schemes that use a Safeness Index function and a control Lyapunov-barrier function, respectively, are presented with rigorous analysis provided on their closed-loop stability, operational safety, and recursive feasibility properties, followed by case studies of large-scale chemical processes under integrated process control and safety systems. Subsequently, the use of machine learning techniques to develop data-driven nonlinear dynamic process models to be used in the MPC schemes is presented with closed-loop stability and safety analysis as well as discussion on computational implementation issues. Next, the development of an integrated detection and control system for process cybersecurity is developed, in which several types of intelligent cyber-attacks, machine learning detection methods, and resilient control strategies are presented. The book closes with a two-tier control architecture that possesses inherent cybersecurity properties and could provide a blueprint for the design of cybersecure industrial process control systems. Throughout the book, the control methods are applied to numerical simulations of nonlinear chemical process examples and Aspen simulations of large-scale chemical process networks to demonstrate their effectiveness and performance.

The book requires some knowledge of nonlinear systems, nonlinear control theory, and nonlinear programming methods, and is intended for researchers, graduate students, and process control and safety engineers.

In conclusion, we would like to acknowledge Prof. Helen Durand, Prof. Fahad Albalawi, Dr. Anas Alanqar, Dr. Anh Tran, Dr. David Rincon, Dr. Zhihao Zhang, and Ms. Scarlett Chen, all at UCLA, who have contributed substantially to the research efforts and results included in this book. We would like to thank them for their hard work and contributions. We would also like to thank all the other people who contributed in some way to this project. In particular, we would like to thank our colleagues at UCLA, and the United States National Science Foundation and Department of Energy for financial support. Last but not the least, we would like to express our deepest gratitude to our families for their dedication, encouragement, and support over the course of this project. We dedicate this book to them.

Los Angeles, CA, USA

Zhe Wu
Panagiotis D. Christofides

Contents

List of Figures

List of Tables

Chapter 1
Introduction

1.1 Motivation

Process operational safety has been a long-standing research problem in optimal operation and control of dynamic systems and processes. The traditional approach to process operational safety is to employ a hierarchical approach as shown in Fig. 1.1. Specifically, a complete control and safety system used in industries includes basic process control systems (BPCSs), alarm systems, emergency shutdown systems (ESSs), and safety relief devices. Ideally, BPCS regulates process variables to their set-points, while the layers of the safety system should not be activated regularly. When the BPCS fails to maintain the process variables within acceptable ranges due to, for example, equipment faults or unusually large process disturbances, alarms are triggered that alert operators so that actions can be taken to prevent further unsafe deviations. If the process variables subsequently further exceed allowable values, the ESS is triggered, which takes automatic and extreme actions such as forcing a valve to its fully open position to bring the process to a safer state of operation. Safety relief devices such as relief valves are used on vessels that can become highly pressurized quickly to prevent an explosion. Containments are used to prevent hazardous materials from entering the environment or injuring workers when the other layers of the safety hierarchy fail to prevent the release of the materials. The emergency response plan is used in severe cases that cannot be mitigated by any other layers. The layers are independent of each other and of the control system (i.e., they have separate sensors, computing elements, and actuators) to allow redundancy and improve safety [119]. Design decisions for the location and sizing of the safety systems are aided through qualitative and quantitative studies (e.g., hazards and operability (HAZOP) studies, fault trees, event trees, what-if or worst-case scenarios, security indices, and layers of protection analysis (LOPA)) of the damage that may result from an accident (including life losses, capital equipment loss, and damage to the environment) which is evaluated to determine whether it is within an acceptable level of risk [55, 119, 125, 199].

Z. Wu and P. D. Christofides, *Process Operational Safety and Cybersecurity*, Advances in Industrial Control, https://doi.org/10.1007/978-3-030-71183-2_1

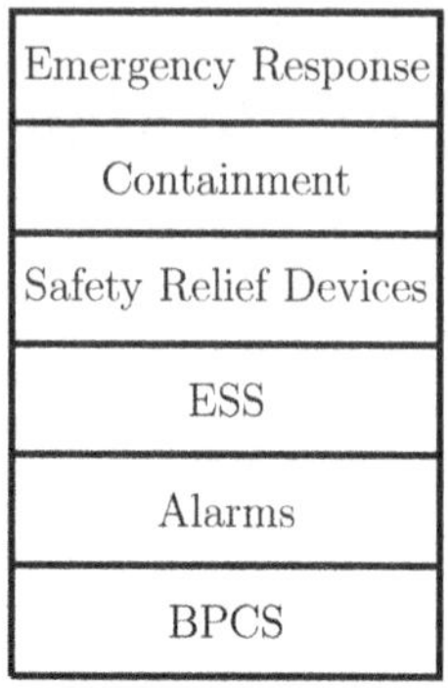

Fig. 1.1 Control/safety system layers [119]

Though safety systems and feedback control systems are critical to safe plant operation, they act fully independently in the hierarchical multilevel system of Fig. 1.1 and are not integrated to yield cooperative actions to ensure both operational safety and economic performance. This has resulted in staggering profit losses for the chemical process industries; for example, it was reported that the 20 major accidents in the hydrocarbon industry from 1974 to 2015 cost over \$15 billion, with the total accumulated value of the 100 largest losses at more than \$33 billion (estimates in 2015 dollars) [121]. It is clear from these numbers that it is necessary to coordinate the actions of process safety and control systems from both the ethical perspective of saving lives and property, and also from an economics standpoint for the chemical process industry. One potential solution is to incorporate safety considerations and safety system actions within optimization-based control schemes, e.g., model predictive control (MPC). While MPC has been widely used in real-time operation of industrial chemical plants to optimize chemical process performance accounting for closed-loop stability and control actuator constraints [66, 124, 130, 133, 160, 165], current MPC designs do not account for process safety considerations and actions and thus may lead to process operation in certain regions of the state space from which migration to an unsafe state may quickly occur. Therefore, a systematic methodology needs to be developed with rigorous analysis of process stability, operational safety, and recursive feasibility to coordinate MPC systems and safety systems to ensure operational safety while achieving desired operation performance.

In addition to process operational safety, cybersecurity has become crucially important in recent years due to increasing risks of cyber-attacks with the development of modern communication in industrial process controls and operations. Since both process safety and cybersecurity aim to prevent or mitigate events involving a loss of control of safety- and security-critical systems, the layers of protection analysis for safety systems can also be employed in the development of a defense-in-depth strategy for cyber-defense systems, where cybersecurity is incorporated into control network designs. Industrial control systems or supervisory control and data acquisition (SCADA) systems are generally large-scale, geographically dispersed, and life-critical systems in which embedded sensors, actuators, and controller networks are utilized to sense and control the physical devices [59]. The unsafe process operation due to the failure of cybersecurity can lead to catastrophic consequences in chemical process industries, causing environmental damage, capital loss, and human injuries. Cyber-attacks are essentially a series of computer actions that are designed to compromise the integrity, stability, and safety of control systems [58, 64, 152, 230]. Among cyber-attacks, targeted attacks are designed with the aim of modifying the control actions applied to an industrial process (for example, the Stuxnet worm was designed to attack the SCADA system by modifying the data sent to Programmable Logic Controllers [43]). Additionally, since targeted attacks are designed to be process and controller behavior aware and can have access to process operation information such as process state measurement, operating region, and control algorithms, they are stealthy and difficult to detect using conventional detection methods. Nevertheless, as the development of most of the existing detection methods still depends partly on human analysis, intelligent cyber-attacks that are process-aware and stealthy pose great challenges to the development of efficient detection methods with high detection accuracy for modern industrial control system where cyber- and physical components closely interact. Therefore, designing advanced detection systems and integrating them with MPC to handle cyber-attacks in safety-critical systems is a new frontier in control systems that will significantly improve the security of chemical production.

1.2 Background

Chemical process safety has traditionally been addressed through process design decisions (e.g., designing the process to be inherently safe in terms of its chemistry and physics [68, 77]) and control and safety system design decisions (e.g., adding sensors for critical process variables that trigger an alarm when a measurement outside of the desired range is obtained [119]). Inherently safer designs are achieved through four primary principles: minimize (reduce the quantity of hazardous substances used and stored by a process), substitute (utilize less hazardous process chemicals), moderate (dilute chemicals or change operating conditions), and simplify (choose designs with less complexity and less potential to create hazardous conditions when faults or errors occur) [71, 92]. However, it is not possible to eliminate all hazards at a plant, so a safety system, comprised of several independent layers, should be added (Fig. 1.1). While the hierarchical approach that utilizes control and safety systems independently for process safety has been successfully deployed in chemical process industries, the accidents throughout chemical plant history [96, 98, 117] have led some researchers to suggest that the philosophy used in the design of the control and safety system layers (i.e., designing barriers against specific unsafe scenarios using the safety system) is quite limited, particularly as economic considerations drive more optimized and integrated system designs [70, 75, 112, 140], and that a systems approach coordinating directly the actions of control and safety systems and analyzing closed-loop process operational safety should instead be used [7, 27, 54, 84, 109,

116, 195]. One step toward this systems approach is by incorporating safety considerations and safety system actions within the BPCS. However, the single-input/single-output controllers (e.g., proportional–integral–derivative controller (PID controller)) traditionally used within the BPCS cannot account for factors that are important to process safety such as multivariable interactions and state/input constraints. On the other hand, advanced model-based control methodologies such as model predictive control (MPC) can account for these factors and thus can be integrated with safety considerations [109, 124, 130, 160]. A large number of works in the MPC literature have addressed the robustness, performance, and closed-loop stability of MPC (e.g., [42, 62, 76, 82, 124, 128, 133, 146, 233] and the references therein), but have not considered explicit safety considerations and safety system actions in their formulations.

Several works have looked at coordinating control with safety considerations. For example, safety in the sense of fault/abnormality diagnosis and monitoring has been addressed, e.g., [53, 65, 197], as well as integrating fault tolerance within process control, e.g., [12, 35, 89, 105, 131, 229]; however, these methods do not address system-wide safety considerations and safety system actions in control. Furthermore, the coordination of control and safety systems through a system-wide safety metric (while operating the systems independently) has not been performed, though this has the potential to significantly reduce unnecessary triggering of the safety system and to help in the design of triggers and appropriate actions for automated elements of the ESS and relief systems. Thresholds on a recently developed state-based Safeness Index [8] may be incorporated as triggers for safety system activation that allow the safety system to be aware of system-level safety considerations; the same metric, with different thresholds, can be utilized in MPC design to provide some coordination between the designs. This can be particularly beneficial for mitigating alarm overloading [39, 69, 204], which is the triggering of too many alarms at once, either because of poor alarm design creating frequent alarms that require no operator actions, or too many correct alarms sounding at once triggered by the same root cause. The number of alarms that sound at a chemical process plant each day can be over seven times the recommended number [61, 172], making it difficult for operators to adequately address the alarms, which can lead to environment and plant damage, danger to lives [181, 184], and reduced operator confidence in the alarm system [204]. Industry [172] and academia [14, 20, 38, 44, 134, 137, 186, 203, 204] have addressed alarm issues with techniques based on, for example, models, statistical analysis, and metrics. Despite these efforts, the integration of operational safety considerations such as safeness metrics that characterize the safeness of chemical processes based on the values of the process states, as well as safety system actions (like on/off behavior of relief valves) within control system designs, has received limited attention.

Additionally, industrial process control systems rely heavily on information and communication technologies for automated operations. Particularly, industrial control systems integrate computers, data communications networks, and physical process components to seamlessly combine hardware and software resources for reliable operation and robust control. In more recent years, Internet communication and wireless networks are starting to replace or complement existing wired point-to-point communications in traditional large-scale process operations as well [49]. As these new developments bring efficiency to the existing system by enabling transmission of signals to remote locations without adding or altering the current hardwire infrastructure, heightened concern for security also arises [28]. Each device and communication channel in the control system network expand the possible attack surface that cyber-attacks can exploit, thereby increasing the vulnerability of the industrial cyber-physical system. Due to the connectivity and interaction between physical and cyber-components in these processes, a different strategy from the traditional information technology (IT) approaches is required for operational cybersecurity. Therefore, the design and implementation of cyber-defense in industrial control systems remain an ongoing scientific and practical issue. Moreover, with the increasing sophistication of attacks, they may lead to negative consequences beyond critical asset damage and the net economic loss of the system. Since the attackers may have full access to technical details of the process control system and production processes in the plant, process safety and operational integrity may also be compromised. In recent years, a number of industrial cyber-attacks have caused detrimental physical damage, for example, the Stuxnet worm compromising Iran's nuclear centrifuges, the 2014 cyber-attack attacking a German steel mill, and the 2015 cyber-attack compromising information systems of three energy distribution companies in Ukraine [94]. In light of conducting hazard analysis as part of standard process safety practice, there have been recent calls to incorporate cybersecurity-integrated hazard evaluations, where cyber-vulnerabilities in the production units are assessed and understood, and countermeasures are outlined to reduce these cyber-risks. However, at this stage, no systematic approach has been developed to actively monitor, detect, and mitigate the impact of these intrusions using the data network on the digital platform. Considering this gap, developing detection algorithms and mitigation measures from within the control system is fundamental to addressing the problem.

Recent IT developments such as enhancement of firewalls for guarding network security have given an edge to enterprise cybersecurity. As a huge amount of operational and instrumentation data is generated, collected and archived for process monitoring, control, and troubleshooting in production plants, safeguarding methodologies such as big data analytics may also be used to secure device measurements for safe process operation. With the rapid development of computing power and digital technologies, the potential application of these data goes beyond fault detection and preventative maintenance. One example usage of these process operational data is to detect and predict cyber-attacks in the industrial control systems. In recent years, cybersecurity and cyber-defense have garnered increasing research interests with the rise of virtualization and big data [26, 57, 99], where machine learning techniques that can learn the system pattern from big data provide a powerful tool to analyze industrial process data for the development of cyber-attack detection algorithms. In fact, machine learning has increasingly gained more popularity in classical engineering fields in addition to computer science and engineering [11, 30, 159, 161, 166, 177, 196, 211], and has shown promising potential for use in the detection of cyber-attacks. For example, [136] proposed a model-based fault diagnostic method for fault diagnosis and classification in electric drives, and [208] used hidden Markov models for automated fault detection and diagnosis of heating, ventilation, and air conditioning (HVAC) systems. Additionally, in [78], various machine learning classification methods were used to distinguish cyber-attacks on power systems from process disturbances, and in [86], a behavior-based intrusion detection algorithm was developed to identify the types of attacks. Moreover, an extensive literature review of machine learning methods deployed for attack detection are presented in [40, 147, 173, 192, 209, 236]. While the feasibility of data science and machine learning algorithms in anomaly management has been demonstrated in these recent literature contributions, the development of a protective safeguard through the integration of online machine-learning-based detection algorithms and existing advanced control techniques such as MPC to the multi-layer cyber-defense system that is of significant importance to next-generation smart manufacturing is still in its infancy.

1.3 Operational Safety and Cybersecurity of Chemical Processes

A chemical process example is presented in this section to provide the motivation for developing novel control algorithms that account for operational safety and cybersecurity. In the first case study, the chemical process is operated in an off steady-state manner under economic model predictive control (EMPC) to optimize process economic performance. While the formal definition of EMPC will not be presented until the subsequent chapters, we can think of EMPC as a predictive control scheme that optimizes operating strategy in real time to dynamically operate chemical processes in a bounded operating region in order to maximize process economic benefits accounting for various economic factors such as time-varying material and energy pricing. However, in the case that the economically optimal regions include unsafe operating conditions, the time-varying operation of EMPC without accounting for safety region constraints may lead to unsafe operations when attempting to maximize process economic profits. The second case study considers the same chemical process and demonstrates the impact of cyber-attacks that compromise one of the sensor measurements. Specifically, the system is normally operated at a pre-specified steady-state (either originally at the steady-state or forced to the steady-state from another operating condition) under feedback-based tracking model predictive control (MPC) with secure sensor measurements of process variables, e.g., temperature and species concentration; however, it will be demonstrated that process stability is no longer guaranteed in the sense that the system may deviate from the steady-state and even leave the normal operating region when sensor measurements are tampered by cyber-attacks. The two case studies indicate the importance of having advanced control systems that account for process operational safety and cybersecurity, and have motivated much of the work contained in this book. The chemical process example and the two case studies are provided below.

1.3.1 Continuously Stirred Tank Reactor

Continuously stirred tank reactor (CSTR) with a second-order reaction is a well-established chemical engineering example that demonstrates performance improvement through a time-varying operation. Specifically, we consider a non-isothermal CSTR in which an elementary second-order reaction takes place that produces the desired product B from the reactant A. As shown in Fig. 1.2, a feedstock stream with the reactant concentration C_{A0}, temperature T_0, and volumetric flow rate F is fed to the reactor. It is assumed that the CSTR contents are well-mixed, and the reactor has a static liquid holdup. The CSTR is equipped with an outer heating/cooling jacket to provide/remove heat to/from the reactor at a heat rate Q. Additionally, the constant fluid density and heat capacity are denoted by ρ_L and C_p, respectively. The following system of ordinary differential equations (ODEs) are developed by applying first-principles and the Arrhenius equation to describe the evolution of the CSTR reactant concentration and temperature:

$$\frac{\mathrm{d}C_A}{\mathrm{d}t} = \frac{F}{V}(C_{A0} - C_A) - k_0 e^{\frac{-E}{RT}} C_A^2 \tag{1.1a}$$

$$\frac{\mathrm{d}T}{\mathrm{d}t} = \frac{F}{V}(T_0 - T) + \frac{-\Delta H}{\rho_L C_p} k_0 e^{\frac{-E}{RT}} C_A^2 + \frac{Q}{\rho_L C_p V} \tag{1.1b}$$

where t is the time, C_A and T are the concentration of A and the temperature of the reactor contents in the reactor. k_0 is the rate constant, V is the volume of the liquid

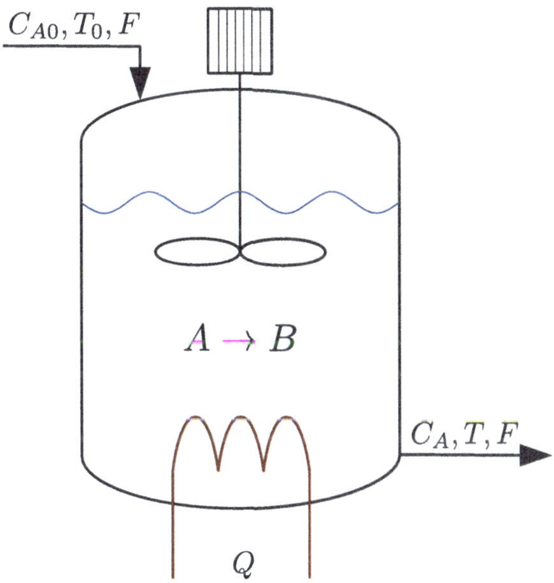

Fig. 1.2 Schematic of a CSTR with an irreversible, second-order reaction that converts the reactant A to the desired product B

Table 1.1 Process parameters of the CSTR

$T_0 = 300$	K	$F = 5$	$\frac{m^3}{h}$
$V = 1.0$	m^3	$E = 5 \times 10^4$	$\frac{kJ}{kmol}$
$k_0 = 8.46 \times 10^6$	$\frac{m^3}{kmol h}$	$\Delta H = -1.15 \times 10^4$	$\frac{kJ}{kmol}$
$C_p = 0.231$	$\frac{kJ}{kg K}$	$R = 8.314$	$\frac{kJ}{kmol K}$
$\rho_L = 1000$	$\frac{kg}{m^3}$	$C_{A0s} = 4$	$\frac{kmol}{m^3}$
$Q_s = 0$	$\frac{kJ}{h}$		

holdup in the reactor, R is the gas constant, ΔH is the enthalpy of reaction, and E is the reaction activation energy. The process parameter values are listed in Table 1.1.

The CSTR has an open-loop asymptotically stable steady-state at $[C_{As}\ T_s] = [1.22\ \frac{\text{kmol}}{\text{m}^3}\ 438\ K]$ and an unstable steady-state at $[C_{As}\ T_s] = [1.95\ \frac{\text{kmol}}{\text{m}^3}\ 402K]$ which correspond to the steady-state input $[C_{A0s}\ Q_s] = [4\ \frac{\text{kmol}}{\text{m}^3}\ 0\ \frac{\text{kJ}}{\text{h}}]$. In this example, the following production rate of B represents the operating profit of the reactor:

$$r_B = k_0 e^{-\frac{E}{RT}} C_A^2. \tag{1.2}$$

The two manipulated inputs are the heat input/removal rate Q and the concentration C_{A0} of the reactant A in the feed stream. Considering the physical bounds on C_{A0} and Q, the input constraints of the manipulated inputs are defined as follows: $|C_{A0} - C_{A0s}| \leq 3.5\ \frac{\text{kmol}}{\text{m}^3}$ and $|Q - Q_s| \leq 5 \times 10^5\ \frac{\text{kJ}}{\text{h}}$.

1.3.2 Case Study: Process Operational Safety in EMPC

We first demonstrate the operational safety issue during the time-varying operation of the CSTR system of Eq. 1.1 under EMPC. The EMPC is designed to maximize process operating profits while maintaining the process states C_A and T within a bounded operating region around the stable steady-state at $(C_{As},\ T_s) = (1.22\ \frac{\text{kmol}}{\text{m}^3},\ 438\ \text{K})$. Considering that thermal runaway may occur in CSTR systems when a temperature increase changes the process conditions in a way that leads to a further increase in temperature, operating conditions of high temperature should be avoided in the dynamic operation. Additionally, to ensure that the operating profits of the CSTR system are maximized while the consumption of reactant A (i.e., inlet concentration C_{A0}) does not exceed its steady-state value, i.e., C_{A0s}, over the entire operating period, the following material constraint is employed in the optimization problem of EMPC:

$$\frac{1}{t_p}\int_0^{t_p} (C_{A0}(t) - C_{A0s})\,\mathrm{d}t = C_{A0s}, \tag{1.3}$$

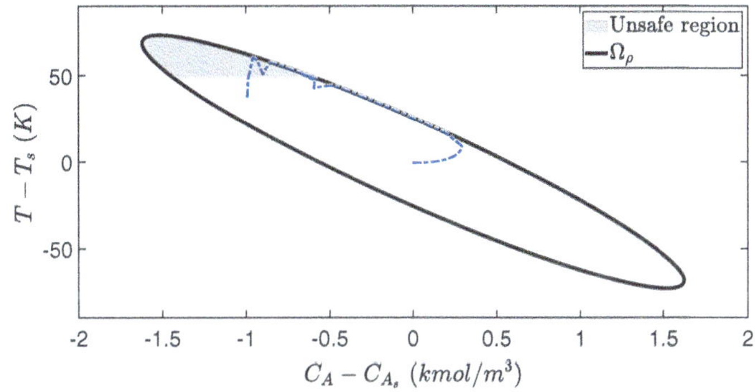

Fig. 1.3 State trajectory in closed-loop simulation of the CSTR under EMPC when the initial condition is at the steady-state, i.e., $(C_A(0) - C_{As}, T(0) - T_s)$=(0 $\frac{\text{kmol}}{\text{m}^3}$, $0K$)

where t_p is the length of the operation. It is straightforward to show that without material constraint, the system will attempt to maximize the production rate by using the maximum amount of material for all times, which is not desired in an economical viewpoint as our goal is to determine the optimal strategy to distribute the material to the reactor.

The closed-loop simulation result under EMPC is shown in Fig. 1.3. It is demonstrated that the CSTR system is operated in an optimal time-varying manner where the process state, i.e., $(C_A - C_{As}, T - T_s)$, starts from the steady-state and first reaches the boundary of the operating region Ω_ρ which is a bounded set around the steady-state, then enters the unsafe region where temperature $T - T_s$ is greater than 47 K, and is finally driven back into the safe region due to the material constraint. Figure 1.4b shows the accumulated operating profit profiles over the operating period $t_p = 1$ h, i.e., $\int_{t=0}^{t=t_p} r_B(\tau)\mathrm{d}\tau$, under EMPC and under the steady-state operation (i.e., the CSTR is operated at the steady-state (C_{As} T_s) for all times), from which it is demonstrated that the time-varying operation of EMPC outperforms the steady-state operation in terms of economic performance. Additionally, the production rate profile for all the operating conditions within the operating region Ω_ρ is shown in Fig. 1.4a. It is observed that the optimal operating profit (i.e., the maximum value for the production rate of Eq. 1.2) is achieved near the right boundary of Ω_ρ, which explains why the state trajectory (blue dashed) in Fig. 1.3 stays at the boundary of Ω_ρ for the majority of the operating time. From this case study, it is demonstrated that the EMPC scheme targeting process economic performance only is not able to achieve operational safety and economic optimality simultaneously, and therefore, a new EMPC design needs to be developed to incorporate safety considerations in its decision-making to ensure operational safety while achieving desired economic performance.

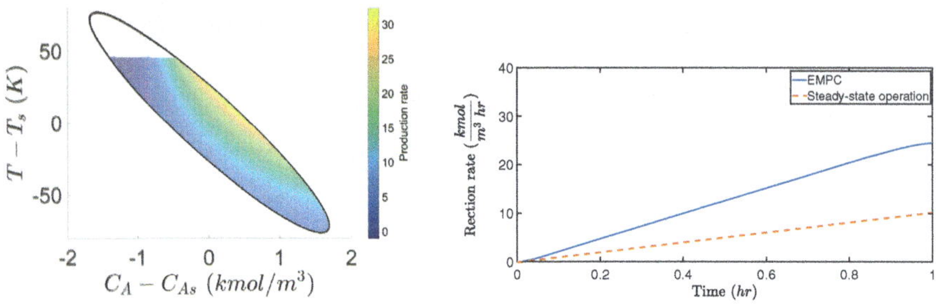

(a) (b)

Fig. 1.4 **a** Production rate profile $k_0 e^{-E/RT} C_A^2 (\frac{kmol}{m^3\ h})$ within the safe operating region of the CSTR, and **b** accumulated operating profits for the closed-loop CSTR under EMPC and steady-state operation, respectively

1.3.3 Case Study: Cybersecurity in Tracking MPC

Consider the same CSTR system under a tracking MPC that aims to track the system state to an unstable steady-state $(C_{As}\ T_s)$=($1.95\ \frac{\text{kmol}}{\text{m}^3}$, $402K$). The intrinsic unstable nature of the steady-state implies that without an appropriate controller, the CSTR system is not able to stably operate at the unstable steady-state; in other words, the steady-state inputs C_{A0s} and Q_s can neither stabilize the system at the steady-state if starting from another operating condition nor maintain the system at the original steady-state under small perturbations. Therefore, a stabilizing controller such as proportional–integral–derivative controller or tracking MPC is required to operate the system at the unstable steady-state. We assume that the temperature sensor measurement for MPC is vulnerable to cyber-attacks in the sense that the measurement value that will be sent to the controller can be manipulated by attackers. Additionally, intelligent cyber-attacks are assumed to be process and controller behavior aware by having access to process information such as the CSTR operating region and existing alarms configured on the input and output ranges (in this particular example, alarms are triggered when the process state leaves the operating region). In this case, the controller that takes falsified temperature measurements will compute unreasonable control actions that may destabilize the system and lead to unsafe operations by driving the process state off steady-state and ultimately out of the operating region. Figure 1.5 shows the closed-loop simulation results for the nominal CSTR system (i.e., under no attack) and the system under cyber-attacks. Specifically, the temperature sensor measurement is intruded by an intelligent cyber-attack that induces maximum disruption by setting the temperature value at its lower bound within the operating region since time $t = 0.03$ h. This type of cyber-attack is termed a min-max cyber-attack, and will be formally defined in Chap. 7. As the temperature measurements (red trajectory) are maintained within the operating region Ω_ρ for all times as shown in Fig. 1.5, the min-max cyber-attack cannot be detected by conventional detection methods designed based on the boundary values. In Fig. 1.5, it is shown

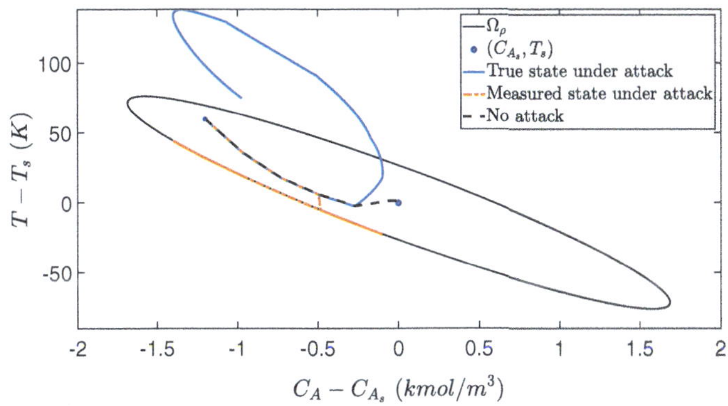

Fig. 1.5 Closed-loop state trajectories for the CSTR under tracking MPC when the temperature sensor is under no attack, and under a min-max attack, respectively

that starting from the initial condition $(C_A - C_{As}, T - T_s)$=$(-1.2\ \frac{\text{kmol}}{\text{m}^3}$, 60 K), the closed-loop state trajectory (black, dashed) is able to converge to the steady-state (C_{As}, T_s) under tracking MPC if no cyber-attack occurs. However, it is shown that without any detection system, the state trajectory (blue, solid) with the same initial condition initially moves toward the origin following the same path under no attack. Then it starts deviating from the direction toward the origin and quickly leaves the operating region Ω_ρ due to incorrect control actions computed based on falsified temperature measurement (red, dash-dotted) shown in Fig 1.6a. The system finally enters an unsafe region of extremely high temperature without being detected from sensor measurements, and therefore, alarm and emergency shutdown systems based on other process variables are employed to prevent further unsafe deviations. Although the above cyber-attack only attacks one sensor, i.e., temperature sensor, it cannot be easily detected by control engineers by reading sensor measurements since the compromised values are bounded in the operating region at all times. Moreover, the cyber-attacks that are designed for industrial control systems will be more complicated in the sense that they can attack sensor networks in a coupled way, which makes it barely possible to detect from the human analysis. Therefore, the example motivates the inquiry and theoretical developments of efficient data-based detection methods and resilient control strategies in the context of MPC systems that can eliminate the impact of cyber-attacks upon timely detection.

1.4 Objectives and Organization of the Book

This book develops a feedback control approach to process operational safety and cybersecurity, and illustrates the applications of the proposed control methods using chemical process examples. Specifically, the objectives of this book are summarized as follows:

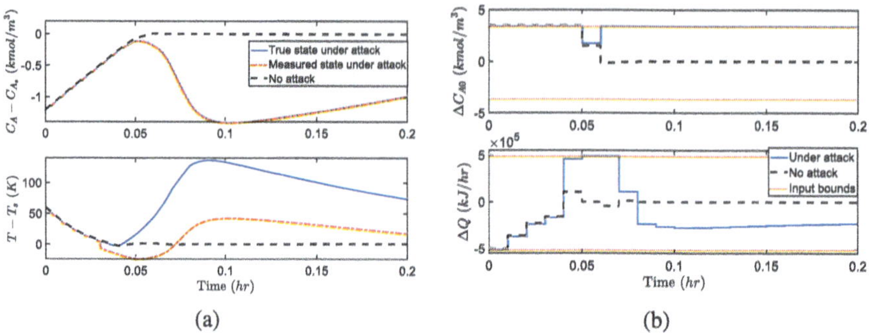

Fig. 1.6 **a** State and **b** input profiles for the CSTR under tracking MPC when the temperature sensor is under no attack, and under a min-max attack, respectively

1. To develop model predictive control and economic model predictive control methods that ensure simultaneous process stability and operational safety by incorporating system-level safety constraints.
2. To develop model predictive control systems using a new function termed control Lyapunov-barrier function to achieve guaranteed stability and safety properties and allow for recursive feasibility of MPCs.
3. To integrate the design and activation of safety systems with control system behavior to reduce safety system activation and eliminate unnecessary process shutdown.
4. To present a framework for integrating machine-learning-based detection systems with model predictive control methods to handle cyber-attacks in industrial control systems.
5. To illustrate the applications of the developed control methods that account for process operational safety and cybersecurity to benchmark chemical reactor examples and large-scale chemical process networks.

The book is organized as follows. In Chap. 2, a formal definition of the notation is provided. Some definitions and preliminary results on stability and stabilization of nonlinear systems using Lyapunov's method and on tracking MPC and economic MPC are given.

In Chap. 3, the concept of operational safety in process control is introduced, followed by a notion termed Safeness Index that characterizes the "safeness" of a process operation. Lyapunov-based MPC and EMPC schemes that incorporate Safeness Index-based constraints are developed to maintain the process state in the safe operating region and optimize process performance simultaneously. Stability and safety analysis of closed-loop systems are provided, and a benchmark chemical reactor example is used to illustrate the effectiveness of the proposed Safeness Index-based MPC methods.

In Chap. 4, control Lyapunov-barrier function (CLBF) and CLBF-based MPC schemes are developed to provide another approach to integrating operational safety within control systems that optimize process performance. Rigorous theoretical results of closed-loop stability, process operational safety, and recursive feasibility of MPCs are developed. The methodologies are applied to chemical process examples with different types of unsafe operating regions to demonstrate their effectiveness.

In Chap. 5, various methods and case studies of large-scale chemical processes using Aspen Plus and Apsen Plus Dynamics simulators are provided to demonstrate the integration of safety considerations into controller design. The dynamic interactions between feedback control and safety systems are first presented through the applications to a continuous stirred tank reactor (CSTR) and a high-pressure flash drum separator with both classical and model-based controllers. Subsequently, the Safeness Index-based MPC developed in Chap. 3 is applied to a flash drum and an ammonia production process to improve process operational safety.

In Chap. 6, issues relating to model development and real-time implementation of CLBF-based MPC schemes are addressed. The concept of recurrent neural networks (RNN) and a general framework to develop RNN models for nonlinear dynamic systems are first introduced. The CLBF-based MPC and EMPC schemes using RNN models for predicting system dynamics are developed, with sufficient conditions under which closed-loop stability and operational safety are derived. Parallel computing and online learning of machine learning models are developed to improve computational efficiency and model accuracy, respectively, in the real-time implementation of RNN-based controllers. The methods are applied to the benchmark chemical reactor example.

In Chap. 7, machine-learning-based detection systems and resilient control schemes are developed to detect and mitigate the impact of stealthy cyber-attacks in MPC and EMPC systems. The construction method of data-based machine learning detectors that can detect multiple classes of intelligent cyber-attacks is first presented. Several cyber-attack resilient control strategies are subsequently developed to contain and eliminate the impact of cyber-attacks by reconfiguring the control system. The application to a benchmark multivariable nonlinear process example is presented to evaluate the ability of the integrated detection and mitigation scheme.

In Chap. 8, a detector-integrated two-tier control architecture that can detect various types of cyber-attacks, and guarantee stability and operational safety of a closed-loop system upon detection, is presented. Specifically, cybersecure explicit feedback controllers in the lower-tier system that stabilize a nonlinear multivariable process at the steady-state are coupled with a model predictive controller that uses networked sensor measurements to improve closed-loop performance in the upper-tier system. Under cyber-attacks that target networked sensor measurement, the two-tier control architecture with machine learning detector ensures cybersecurity and system stability by turning off the upper-tier control and using the lower-tier control only upon detection. The use of safety systems together with the detector–controller architecture to handle cyber-attacks that can drive the process state to unsafe operating regions is also addressed. The detector-integrated two-tier control architecture is applied to a multivariable nonlinear process example.

Chapter 2
Background

This chapter provides a brief review of the concepts of stability for nonlinear systems that are used throughout this book, followed by model-based control schemes that ensure closed-loop stability. The first section presents a formal definition of the notation. The second section discusses the stability of nonlinear systems. The third section gives a brief overview of the control of nonlinear systems using Lyapunov-based control law and using model predictive control (MPC) schemes. For a more detailed overview of stability and control of nonlinear systems, the reader is referred to the textbooks [83, 90].

2.1 Notation

The set of real numbers is denoted by $\mathbf{R}$, and the set of nonnegative real numbers is denoted by $\mathbf{R}_+$. $\mathbf{R}^n$ is a real (Euclidean) space of n-dimension. The Euclidean norm of a vector is denoted by $|\cdot|$, and a weighted Euclidean norm of a vector is denoted by $|\cdot|_Q$ (i.e., $|x|_Q = \sqrt{x^T Q x}$ where Q is a positive definite matrix). The ceiling and floor functions, denoted as $\lceil a \rceil$ and $\lfloor a \rfloor$ for a scalar $a \in \mathbf{R}$, respectively, are the smallest integer not smaller than a and the largest integer not greater than a, respectively. x^T denotes the transpose of x. The variable $t \in \mathbf{R}_+$ is used to represent time. $\{t_k\}_{k \geq 0}$ denotes an infinite sequence, and $\{t_i\}_{i=0}^N$ denotes a finite sequence that describes the sequence: $t_0, t_1, \ldots, t_{N-1}, t_N$. $x(t) \in \mathbf{R}^n$ represents an n-dimensional time dependent vector.

The notation $L_f V(x)$ denotes the standard Lie derivative of function $V(x)$ with respect to the vector field f, i.e., $L_f V(x) := \frac{\partial V(x)}{\partial x} f$. A scalar continuous function $V : \mathbf{R}^n \to \mathbf{R}$ is proper if the set $\{x \in \mathbf{R}^n \mid V(x) \leq k\}$ is compact for all $k \in \mathbf{R}$, or equivalently, V is radially unbounded in the sense that $\lim_{|x| \to +\infty} V(x) = +\infty$ holds.

Z. Wu and P. D. Christofides, *Process Operational Safety and Cybersecurity*,
Advances in Industrial Control, https://doi.org/10.1007/978-3-030-71183-2_2

A function $V : \mathbf{R}^n \to \mathbf{R}_+$ is positive definite with respect to $x \in \mathbf{R}^n$ if $V(x) > 0$ for all $x \in \mathbf{R}^n$ except that $V(x) = 0$ if and only if $x = 0$. A function, $V : \mathbf{R}^n \to (-\infty, 0]$, is negative definite (with respect to the origin) if $-V$ is positive definite. The set Ω_ρ is used to represent a level set of a scalar-valued positive definite function V: $\Omega_\rho := \{x \in \mathbf{R}^n \mid V(x) \leq \rho\}$ where $\rho > 0$.

For given positive real numbers β and ε, $\mathscr{B}_\beta(\varepsilon) := \{x \in \mathbf{R}^n \mid |x - \varepsilon| < \beta\}$ is an open ball around ε with radius β. The relative complement of the set A in B is denoted by $A \backslash B := \{x \in A, x \notin B\}$. A function $f(\cdot)$ is of class $\mathscr{C}^1$ if it is continuously differentiable. A real-valued function $f(\cdot) : \mathbf{R}^n \to \mathbf{R}$ is called Lipschitz continuous if there exists a positive real constant k such that $|f(x) - f(y)| \leq k|x - y|$ holds for all $x, y \in \mathbf{R}^n$, and is called locally Lipschitz continuous if for each $y \in \mathbf{R}^n$, there exists an $L > 0$ such that f is Lipschitz continuous on the open ball $\mathscr{B}_L(y)$.

Given a set $\mathscr{D}$, the boundary, the closure, and the interior of $\mathscr{D}$ are denoted by $\partial\mathscr{D}$, $\overline{\mathscr{D}}$, and $\mathrm{Int}(\mathscr{D})$, respectively. A continuous function $\alpha : [0, a) \to \mathbf{R}_+$ is said to be of class $\mathscr{K}$ if it is strictly increasing and $\alpha(0) = 0$. $S(\Delta)$ denotes the family of piecewise constant functions with period $\Delta > 0$. If the vector-valued function $u(t) : [0, N\Delta) \to \mathbf{R}^m$ can be written in the form of $u(t) = \bar{u}_i$, for $t \in [i\Delta, (i + 1)\Delta)$, where $\bar{u}_i \in \mathbf{R}^m$, $i = 0, 1, \ldots, N - 1$, and N is a positive integer, then we say $u \in S(\Delta)$.

2.2 Stability of Nonlinear Systems

We first consider a class of unforced, time-invariant nonlinear systems described by the following system of first-order nonlinear ordinary differential equations (ODEs) to present stability properties:

$$\dot{x} = f(x) \tag{2.1}$$

where $x \in D \subset \mathbf{R}^n$, and $f : D \to \mathbf{R}^n$ is a smooth function of x on the domain $D \subset \mathbf{R}^n$. x is referred to as the *state* vector, which describes the current state of the system. The space $\mathbf{R}^n$ is referred to as *state-space*. The initial condition of system of Eq. 2.1 is given by $x_0 \in D$, i.e., $x(t_0) = x_0$ where the initial time is represented by $t_0 \in \mathbf{R}$.

We denote the solution of Eq. 2.1 as $x(t, x_0)$ for initial condition $x(t_0) = x_0$ and $t \geq t_0$. Without loss of generality, the initial time t_0 is taken to be zero ($t_0 = 0$) throughout this book. We also refer to the solution of Eq. 2.1 $x(t, x_0)$ as the state trajectory, and abbreviate it to $x(t)$ in the following texts. If every solution of Eq. 2.1 remains in a compact set $\mathbb{X} \subset D$ for all $t \geq t_0$, then a unique solution is ensured to exist for all $t \geq t_0$ [90].

Since $f(x)$ is a nonlinear function of x, the system of Eq. 2.1 may possess multiple isolated equilibrium points. Without loss of generality, it is assumed that $f(0) = 0$, and thus, the origin is an equilibrium point of the system of Eq. 2.1. For systems that do not have an equilibrium point at the origin, i.e., $f(0) \neq 0$, we can always rewrite the system of Eq. 2.1 in the form of $\dot{z} = f(z + x_s) =: g(z)$ such that the equilibrium point is $g(0) = 0$ and $z = 0$, by introducing deviation variables $z := x - x_s$ for the system $\dot{x} = f(x)$ with $f(x_s) = 0$ around an equilibrium point $x_s \neq 0$.

Now, we study the stability of the origin $x = 0$ for the system of Eq. 2.1. Specifically, the equilibrium point $x = 0$ is

- *stable* if, for each $\varepsilon > 0$, there is $\delta(\varepsilon) > 0$ such that

$$|x(0)| < \delta \Rightarrow |x(t)| < \varepsilon, \forall\, t \geq 0; \tag{2.2}$$

- *unstable* if it is not stable;
- *locally asymptotically stable* if it is stable and there exists a $\delta > 0$ such that

$$|x(0)| < \delta \Rightarrow \lim_{t\to\infty} |x(t)| = 0; \tag{2.3}$$

- *globally asymptotically stable* if it is stable and $\lim_{t\to\infty} |x(t)| = 0$ for all $x(0) \in \mathbf{R}^n$;
- *locally exponentially stable* if there exist positive real constants δ, c, and λ such that all solutions of Eq. 2.1 with $|x(0)| \leq \delta$ satisfy the inequality:

$$|x(t)| \leq c|x(0)|e^{-\lambda t} \quad \forall\, t \geq 0; \tag{2.4}$$

- *globally exponentially stable* if there exist positive real constants c and λ such that for all $x(0) \in \mathbf{R}^n$, all solutions of Eq. 2.1 satisfy the inequality:

$$|x(t)| \leq c|x(0)|e^{-\lambda t} \quad \forall\, t \geq 0. \tag{2.5}$$

Throughout this book, we will mainly discuss local stability properties for the nonlinear system of Eq. 2.1 around its equilibrium point $x = 0$, unless stated otherwise. Since Eq. 2.1 is a time-invariant system, all the stability properties above are *uniform*, that is, they do not depend on the initial time. Additionally, the concept of boundedness of the solution, a weaker notion of stability than asymptotic and exponential stability of the origin, is presented as follows. The solution of Eq. 2.1 is

- *bounded* if there exists a positive constant c and for every $a \in (0, c)$, there is $\beta(a) > 0$ such that

$$|x(0)| \leq a \Rightarrow |x(t)| \leq \beta, \forall\, t \geq 0. \tag{2.6}$$

Following the definition of boundedness of the solution, a set M is said to be a *positively invariant set* with respect to the system of Eq. 2.1 if $x(0) \in M \Rightarrow x(t) \in M, \ \forall\, t \geq 0$. We will also use the term *forward invariant set* to refer to a positively invariant set throughout this book.

2.2.1 Lyapunov's Direct Method

Lyapunov's direct method is often used to determine the stability of the equilibrium points for nonlinear systems by using a scalar-valued positive definite function that has negative (semi-)definite time-derivative along the state trajectory. Specifically, Lyapunov's stability theorem is presented as follows:

Theorem 2.1 (Lyapunov Stability Theorem, c.f. [90, Theorem 4.1]) *Let $x = 0$ be an equilibrium point for Eq. 2.1 and $D \subset \mathbf{R}^n$ be a domain containing the origin ($x = 0$). Let $V : D \to \mathbf{R}$ be a continuously differentiable positive definite function such that*

$$\dot{V}(x) = \frac{\partial V}{\partial x} f(x) \leq 0 \tag{2.7}$$

for all $x \in D$. Then, $x = 0$ is stable. If

$$\dot{V}(x) = \frac{\partial V}{\partial x} f(x) < 0 \tag{2.8}$$

for all $x \in D \setminus \{0\}$, then $x = 0$ is asymptotically stable.

A continuously differentiable positive definite function V satisfying the conditions in Theorem 2.1 is called a *Lyapunov function*. By considering the Lyapunov function as an abstract notion of the total energy of a physical system, the above theorem demonstrates that the system energy decays over time. Specifically, the condition $\dot{V} < 0, \forall x \in D \setminus \{0\}$, implies that the solution x of the system of Eq. 2.1 will cross the level surface $V(x) = c$ of a Lyapunov level set, i.e., $\Omega_c = \{x \in \mathbf{R}^n \mid V(x) \leq c\} \subset D$, $c > 0$, and stays inside the set Ω_c afterwards. Due to the fact that $\dot{V}(x) < 0$ holds for all $x \in D \setminus \{0\}$, the level surface that the state trajectory evolves along shrinks over time to the origin. However, if, instead, $\dot{V}(x) \leq 0$ holds for all $x \in D$, we cannot show that the trajectory will converge to the origin. In this case, the origin is still stable since the trajectory will be maintained inside $\mathscr{B}_\varepsilon(0)$, by requiring the initial condition x_0 to stay within a Lyapunov surface contained in that ball.

2.2.2 LaSalle's Invariance Principle

LaSalle's invariance principle gives a criterion for the asymptotic stability of the system of Eq. 2.1 in the case when $\dot{V}(x) \leq 0$ for all $x \in D$. Specifically, LaSalle's invariance principle states that if the state is initiated within any compact forward invariant subset of D, it will converge to the largest invariant set in D where $\dot{V}(x) = 0$.

Theorem 2.2 (LaSalle, c.f. [90, Theorem 4.4]) *Let $\Omega \subset D$ be a compact set that is positively invariant with respect to Eq. 2.1. If there exists a continuously differentiable function $V : D \to \mathbf{R}$ that satisfies $\dot{V}(x) \leq 0$ in Ω, then every solution in Ω will* *approach the set M as $t \to \infty$, where $E := \{x \in \Omega : \dot{V}(x) = 0\}$ is defined to be the set of states in Ω that meet $\dot{V}(x) = 0$ and M is the largest invariant set in E.*

Based on LaSalle's invariance principle, we can show asymptotic stability of the origin by establishing that the largest invariant set in E is the origin, i.e., $M = \{0\}$, in the following corollary.

Corollary 2.1 (c.f. [90, Corollary 4.1]) *Let $V : D \to \mathbf{R}$ be a continuously differentiable positive definite function on a domain D that contains the equilibrium point $x = 0$ of Eq. 2.1. If $\dot{V}(x) \leq 0$ is satisfied for all $x \in D$, and no solution other than the trivial solution $x \equiv 0$ can stay identically in S, where $S = \{x \in D : \dot{V}(x) = 0\}$, then the origin is asymptotically stable.*

2.3 Control of Nonlinear Systems

We now consider the class of forced nonlinear systems described by the following system of nonlinear ordinary differential equations:

$$\dot{x} = f(x, u, w) \tag{2.9}$$

where $x \in D \subseteq \mathbf{R}^n$ is the state vector, $u \in U \subset \mathbf{R}^m$ is the manipulated (control) input vector, and $w \in W$ is the unknown and un-modeled disturbance vector, where $W := \{w \in \mathbf{R}^l \mid |w| \leq \theta,\ \theta \geq 0\}$. The control action constraint is defined by $u \in U := \{u_{\min} \leq u \leq u_{\max}\} \subset \mathbf{R}^m$, where $u_{\min}$ and $u_{\max}$ are the lower and upper bounds for the input vector, respectively. Without loss of generality, it is assumed that $f : \mathbf{R}^n \times \mathbf{R}^m \times \mathbf{R}^l \to \mathbf{R}^n$ is a smooth vector function of its arguments with $f(0, 0, 0) = 0$, and thus, the origin is a steady-state of the nominal system of Eq. 2.9 with $w(t) \equiv 0$, (i.e., $(x_s^*, u_s^*) = (0, 0)$, where x_s^* and u_s^* denote the steady-state of Eq. 2.9).

2.3.1 Control Lyapunov Functions and Stabilization

To design a stabilizing feedback control law $u = \Phi(x)$ that renders the origin of the nominal closed-loop system of Eq. 2.9 with $w(t) \equiv 0$ globally asymptotically stable, one approach is to choose a Lyapunov function $V : \mathbf{R}^n \to \mathbf{R}_+$, and find a control law such that the time-derivative of V along the solutions of the closed-loop system $\dot{x} = f(x, \Phi(x), 0)$ satisfies the following inequality for all $x \in \mathbf{R}^n$:

$$\frac{\partial V(x)}{\partial x} f(x, \Phi(x), 0) \leq -W(x) \tag{2.10}$$

where $W : \mathbf{R}^n \rightarrow \mathbf{R}$ is a positive definite function. We can derive an explicit design of nonlinear controller $u = \Phi(x)$ based on the control Lyapunov function (CLF) of the following definition.

Definition 2.1 A control Lyapunov function for the system of Eq. 2.9 is a $\mathscr{C}^1$ and positive definite function $V : \mathbf{R}^n \rightarrow \mathbf{R}_+$ that satisfies

$$\inf_{u \in \mathbf{R}^m} \left\{ \frac{\partial V(x)}{\partial x} f(x, u, 0) \right\} < 0, \; \forall\, x \in \mathbf{R}^n, \;\; x \neq 0. \tag{2.11}$$

Equation 2.11 is a necessary and sufficient condition for showing the existence of a control law satisfying Eq. 2.10 [21].

For control-affine systems of the form:

$$\dot{x} = f(x) + g(x)u, \tag{2.12}$$

where $f : \mathbf{R}^n \rightarrow \mathbf{R}^n$, $g : \mathbf{R}^n \rightarrow \mathbf{R}^n \times \mathbf{R}^m$ are sufficiently smooth vector and matrix functions, and $f(0) = 0$, Eq. 2.11 is equivalent to the following condition using the Lie derivative notation:

$$L_f V(x) < 0, \;\; \forall x \in \{z \in \mathbf{R}^n \backslash \{0\} \mid L_g V(z) = 0\} \tag{2.13}$$

where $L_f V(x) := \frac{\partial V(x)}{\partial x} f(x)$ and $L_g V(x) := \frac{\partial V(x)}{\partial x} g(x)$.

If a CLF can be found for the system of Eq. 2.12, it implies that there exists a stabilizing feedback control law $\Phi(x)$ that renders the origin asymptotically stable in the sense that Eq. 2.13 holds for $u = \Phi(x)$, where $\Phi(x) \in U$. Additionally, if the CLF $V(x)$ satisfies the small control property [183], i.e., for every $\varepsilon > 0, \exists\, \delta > 0$, s.t. $\forall\, x \in \mathscr{B}_\delta(0)$, there exists u that satisfies $|u| < \varepsilon$ and $L_f V(x) + L_g V(x)u < 0$, then the controller $u = \Phi(x)$ is continuous at the origin and $\Phi(0) = 0$. The following equation gives an example of a continuous control law $k(x)$ that renders the origin asymptotically stable [110]:

$$k_i(x) = \begin{cases} -\dfrac{p + \sqrt{p^2 + |q|^4}}{|q|^2} q_i & \text{if } q \neq 0 \\ 0 & \text{if } q = 0 \end{cases} \tag{2.14}$$

where p denotes $L_f V(x)$, q_i denotes $L_{g_i} V(x)$, $q = [q_1 \ldots q_m]^T$, $f = [f_1 \cdots f_n]^T$, and $g_i = [g_{i1} \ldots g_{in}]^T$, $(i = 1, 2, \ldots, m)$. $k_i(x)$ of Eq. 2.14 represents the i_{th} component of the control law $u = k(x) \in \mathbf{R}^m$.

To further account for the saturation of the control action at the input bounds, the following controller $\Phi_i(x)$ is designed to represent the i_{th} component of the saturated control law that accounts for the input constraint $u \in U$.

$$\Phi_i(x) = \begin{cases} u_{\min} & \text{if } k_i(x) < u_{\min} \\ k_i(x) & \text{if } u_{\min} \le k_i(x) \le u_{\max} \\ u_{\max} & \text{if } k_i(x) > u_{\max}. \end{cases} \quad (2.15)$$

Therefore, the saturated Sontag's formula $u = \Phi(x) \in U$ of Eqs. 2.14–2.15 provides a universal nonlinear controller that asymptotically stabilizes the system of Eq. 2.9 at the origin while accounting for the input constraints $u \in U$. Additionally, there are other methods such as employing linear feedback control techniques, geometric control methods, and Lyapunov-based control techniques that can be used to design an explicit feedback control law $\Phi(x)$, under which the origin of Eq. 2.9 is rendered asymptotically stable. The interested reader may refer to [50, 101, 103, 104, 176].

Remark 2.1 It is noted that even though the controller is continuous at the origin, numerical implementation of Sontag's formula of Eq. 2.14 may lead to an oscillatory behavior around the origin due to numerical approximation in continuous simulations. In order to smoothen out the control action $k_i(x)$, a sufficiently small positive real number ε is often added to the denominator of Eq. 2.14. However, the addition of this parameter results in an offset in the closed-loop response (i.e., the state can only be bounded in a small neighborhood around the origin instead of converging to it). Therefore, ε should be chosen carefully to improve the smoothness of the control action while maintaining a sufficiently small offset.

In the subsequent chapters, we will often use a stabilizing feedback controller, for example, the control law of Eqs. 2.14–2.15, in the design of model predictive control schemes based on the following assumptions.

Assumption 2.1 There exists a feedback controller $\Phi(x) \in U$ with $\Phi(0) = 0$ that renders the origin of the nominal closed-loop system of Eq. 2.9 with $u = \Phi(x)$ and $w \equiv 0$ **asymptotically** stable for all $x \in D \subset \mathbf{R}^n$ where D is an open neighborhood of the origin.

According to converse Lyapunov theorems [90, 123], Assumption 2.1 implies that a $\mathscr{C}^1$ Lyapunov function $V : D \rightarrow \mathbf{R}_+$ exists for the closed-loop system of Eq. 2.9 with $w \equiv 0$ and $u = \Phi(x) \in U$ that satisfies the following inequalities for all x in D:

$$\alpha_1(|x|) \le V(x) \le \alpha_2(|x|), \quad (2.16a)$$

$$\frac{\partial V(x)}{\partial x} f(x, \Phi(x), 0) \le -\alpha_3(|x|), \quad (2.16b)$$

$$\left| \frac{\partial V(x)}{\partial x} \right| \le \alpha_4(|x|), \quad (2.16c)$$

where D is an open neighborhood of the origin, and α_i, $i = 1, 2, 3, 4$ are class $\mathscr{K}$ functions. We define a level set of $V(x)$ inside D (ideally, the largest subset contained in D) as $\Omega_\rho = \{x \in D \mid V(x) \le \rho, \ \rho > 0\}$, which is taken to be the stability region of the closed-loop system under the controller $\Phi(x)$. Since Ω_ρ is a forward invariant set in D where Eq. 2.16 is satisfied, given any initial states $x_0 \in \Omega_\rho$, it is guaranteed that for all $t \geq t_0$, $x(t)$ of the system of Eq. 2.9 with $w(t) \equiv 0$ remains in Ω_ρ, and the origin can be rendered asymptotically stable under the control law $u = \Phi(x) \in U$.

While in general there is no systematic method for constructing Lyapunov functions for broad classes of nonlinear systems with constraints, the sum of squares decomposition [148] and Zubov's method [60] are generally used to construct Lyapunov functions for certain classes of systems. Additionally, there are natural Lyapunov function candidates such as energy functions in physical systems. Within the context of chemical process control, quadratic Lyapunov functions have been demonstrated to yield very good estimates of the closed-loop stability region for nonlinear systems (see, for example, examples in [50] and the example in the last section of this chapter).

Additionally, we can design a feedback controller $\Phi(x)$ based on a stronger stabilizability assumption as follows.

Assumption 2.2 There exists a feedback controller $\Phi(x) \in U$ with $\Phi(0) = 0$ that renders the origin of the nominal closed-loop system of Eq. 2.9 with $u = \Phi(x)$ and $w \equiv 0$ **exponentially** stable for all $x \in D$ where D is an open neighborhood of the origin.

Assumption 2.2 implies that a $\mathscr{C}^1$ Lyapunov function $V : D \to \mathbf{R}_+$ exists for the closed-loop system of Eq. 2.9 with $w \equiv 0$ and $u = \Phi(x) \in U$ such that the following inequalities hold for all x in D:

$$c_1|x|^2 \leq V(x) \leq c_2|x|^2, \tag{2.17a}$$

$$\frac{\partial V(x)}{\partial x} f(x, \Phi(x), 0) \leq -c_3|x|^2, \tag{2.17b}$$

$$\left|\frac{\partial V(x)}{\partial x}\right| \leq c_4|x|, \tag{2.17c}$$

where $c_i, i = 1, 2, 3, 4$ are positive real numbers. The stability region Ω_ρ is designed following the same approach as discussed above, and it is demonstrated that given any initial states $x_0 \in \Omega_\rho$, it is guaranteed that for all $t \geq t_0$, $x(t)$ of the system of Eq. 2.9 with $w(t) \equiv 0$ remains in Ω_ρ, and the origin can be rendered exponentially stable under the control law $u = \Phi(x) \in U$. Moreover, the smoothness property of the vector field f in the nonlinear system of Eq. 2.9 combined with the bounds on u and w implies that there exists a positive constant M such that the following inequality holds for all $x \in \Omega_\rho$, $u \in U$ and $w \in W$:

$$|f(x, u, w)| \leq M. \tag{2.18}$$

In addition, since the Lyapunov function $V(x)$ is a continuous differentiable function and f is a sufficiently smooth function, there exist positive constants L_x, L_w, $L_x^{'}$, $L_w^{'}$ such that the following inequalities hold:

$$|f(x, u, w) - f(x', u, 0)| \leq L_x|x - x'| + L_w|w| \tag{2.19a}$$

$$\left|\frac{\partial V(x)}{\partial x} f(x, u, w) - \frac{\partial V(x')}{\partial x} f(x', u, 0)\right| \leq L'_x|x - x'| + L'_w|w|, \tag{2.19b}$$

for all $x, x' \in \Omega_\rho$, $u \in U$, and $w \in W$.

2.3.2 *Model Predictive Control*

While an explicit feedback controller such as the Sontag control law of Eq. 2.14 that satisfies Assumption 2.1 (Assumption 2.2) can asymptotically (exponentially) stabilize the system of Eq. 2.9 at the origin, it may not be the optimal controller in general since process performance and system constraints are not explicitly taken into account. To overcome the shortcomings of explicit feedback controllers, model predictive control (MPC), also referred to as receding horizon control, has been proposed to control nonlinear processes and take process performance and constraints into considerations [42, 66, 124, 133, 160, 165]. MPC is essentially an optimization-based control method that minimizes/maximizes a cost function or a performance index subject to system/process constraints over a prediction horizon based on a prediction model, i.e., the process model of Eq. 2.9 with $w \equiv 0$. A brief overview of MPC is presented below.

2.3.2.1 Main Components of MPC

As shown in Fig. 2.1, MPC typically optimizes the input trajectory (i.e., control actions) over the prediction horizon to track a set-point or a reference trajectory. The main components of MPC are listed as follows [62]:

1. A process/system model that can predict the evolution of future state trajectories over a time horizon termed the prediction horizon.
2. A cost functional or performance index that measures process performance as a real number based on process/system (output, input, and state) trajectories over the prediction horizon. This is the objective function of the optimization problem.
3. Constraints on the process/system, e.g., physical constraints on control actuators and system states/outputs, e.g., stability and safety constraints.
4. A receding horizon control approach to sampled-data implementation of controllers for continuous-time systems.

Compared to a proportional–integral–derivative (PID) controller, MPC has a number of advantages, which are summarized as follows. (1) As MPC allows the current timeslot (i.e., from t_k to t_{k+1} in Fig. 2.1) to be optimized, while taking future timeslots (i.e., the remaining part of the prediction horizon) into account, the performance of the closed-loop system such as energy consumption and speed of convergence to

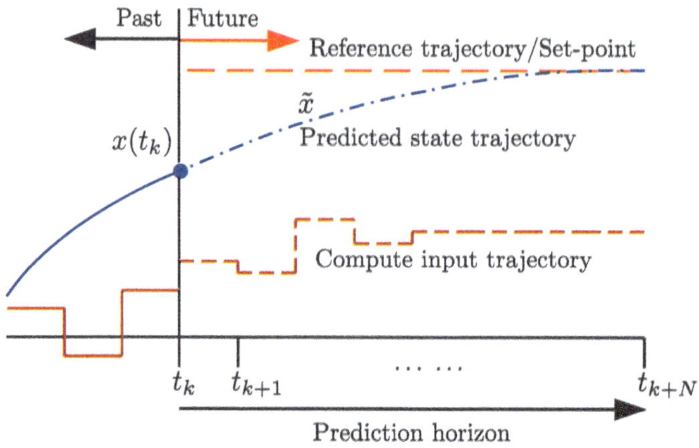

Fig. 2.1 General concept for model predictive control (MPC)

the set-point in chemical processes is improved under MPC compared to the PID controller that does not have predictive ability. (2) MPC has superior performance for processes with a large number of process variables (e.g., manipulated and controlled variables). (3) MPC allows constraints to be imposed on both manipulated and controlled variables while an integral windup often occurs in PID controllers as a limitation of physical systems. (4) Moreover, as the model accounts for inherent process characteristics (e.g., nonlinear behavior and multivariable interactions), MPC can accommodate a variety of process dynamics such as time delays, inverse response, and inherent nonlinearities.

2.3.2.2 Process Model

The mathematical model of the process/system is an essential element of an MPC controller as the prediction of process/system behavior is used in both MPC constraints and objective function. Traditionally, industrial MPCs utilized linear (empirical) process models, e.g., input–output model, state-space model, step, and impulse response models [165], because linear models can be considered good representations for many real processes over a small operating range and are computationally cheap to solve. However, considering that chemical processes are inherently nonlinear where nonlinearity comes from, for example, conservation of mass, momentum, and energy, nonlinear process models are preferred in MPCs to improve closed-loop control performance when the processes are operated over a wide operating region. Additionally, MPC takes feedback information from every sampling step to improve the poor performance due to linear models or imperfect nonlinear models used in the prediction. Throughout this book, nonlinear dynamical models will be used in MPC to predict the future evolution of the process/system. The assumption that a nonlinear first-principles-based process model is available to MPC holds for all the MPC schemes contained in this book except the work in Chap. 6 that introduces a system identification approach using machine learning techniques to develop data-driven process models when first-principles models are not available.

Additionally, as nonlinear MPC (NMPC), i.e., the MPC scheme using nonlinear system models in the prediction, are not necessarily convex optimization problems anymore, solving the dynamic optimization problem of NMPC becomes more challenging than that to linear MPC. Since developing optimization techniques NMPC problems is beyond the scope of this book, the interested reader may refer to [62] for a brief review of nonlinear and dynamic optimization, and [31, 32, 36, 115, 141] for a comprehensive and detailed presentation on optimization methods.

2.3.2.3 Receding Horizon Implementation

MPC is implemented in a receding horizon manner in a way that the optimization problem is repeatedly solved to compute the control actions to be applied. Specifically, the continuous-time process is sampled at discrete time steps (i.e., sampling times), where the time interval between two consecutive sampling times, e.g., between t_k and t_{k+1} in Fig. 2.1, is called the sampling period. Process states/outputs are measured or estimated for every sampling time to provide feedback information to MPC. The MPC optimization problem is solved with the state measurement or estimate at each sampling time to compute the optimal control action(s) over the prediction horizon, from which the first control action (i.e., the one optimized for the first sampling period in the prediction horizon) will be applied to the system. The horizon is moving one sampling period forward, and the MPC problem will be resolved at the next sampling time with a new state measurement/estimate. The aforementioned steps are repeated until the end of the operation. By adopting receding horizon implementation, the MPC is able to use process feedback information to improve closed-loop performance and compensate for process disturbances, modeling errors, and other forms of uncertainty. Additionally, it is noted that if the solution to the infinite-horizon MPC (i.e., the MPC formulated with an infinite prediction horizon) exists, it arguably provides the best control actions since chemical process operations are generally operated over long periods of time without a natural termination or shutdown time. However, to improve the computational efficiency of the MPC optimization problem, MPC is generally designed to be a finite-dimensional optimization problem with a finite prediction horizon and optimized variables (i.e., control actions). Therefore, the receding horizon implementation also allows for a better approximation of the solution to the corresponding infinite-horizon MPC optimization problem.

2.3.2.4 Sample-and-Hold Implementation of Controllers

Sample-and-hold implementation has been widely used in analog-to-digital converters that sample the voltage of a continuously varying analog signal and hold its value at a constant level for a period of time. As digital computers are commonly used in industrial control systems, sample-and-hold has also been utilized to integrate continuous-time physical systems with digital controllers. Specifically, given the continuous-time nonlinear system of Eq. 2.9, the following sampled time system is obtained:

$$x(t_{k+1}) \approx x(t_k) + \Delta \cdot f(x(t_k), u(t_k), w(t_k)) \tag{2.20}$$

where $t_k = k\Delta$, $k = 0,\ 1,\ \ldots$, and $\Delta > 0$ is the sampling period. u is a piecewise function of Δ, which means u holds constant within every sampling period, i.e., $u(t) = u(t_k)$, $\forall t \in [t_k,\ t_{k+1})$. Additionally, the explicit Euler method is utilized to integrate the continuous-time system of Eq. 2.9 with a sufficiently small integration time step h_c ($0 < h_c \ll \Delta$) to provide a better approximation for the sampled time system of Eq. 2.20 by iteratively performing the following calculation within one sampling period:

$$x(t_k + h_c) \approx x(t_k) + h_c \cdot f(x(t_k), u(t_k), w(t_k)) \tag{2.21}$$

where Δ is an integer multiple of h_c. The state x, input u, and disturbance w vectors in the function $f(x, u, w)$ are updated for every h_c step, and thus, it takes $\frac{\Delta}{h_c}$ iterations of Eq. 2.21 to derive $x(t_{k+1}) := x(t_k + \Delta)$. Moreover, it should be noted that for the explicit Euler method, there exists an upper bound for the integration time step h_c to ensure numerical stability. Therefore, h_c is chosen to be a sufficiently small positive number in all the simulation studies throughout this book.

2.3.2.5 MPC Formulation

The MPC problem can be formulated as the dynamic optimization problem as follows:

$$\min_{u \in S(\Delta)} \int_{t_k}^{t_{k+N}} l(\tilde{x}(\tau), u(\tau))\, \mathrm{d}\tau \tag{2.22a}$$

$$\text{s.t.} \quad \dot{\tilde{x}}(t) = f(\tilde{x}(t), u(t), 0) \tag{2.22b}$$

$$\tilde{x}(t_k) = x(t_k) \tag{2.22c}$$

$$(x(t), u(t)) \in Z,\ \forall\, t \in [t_k,\ t_{k+N}) \tag{2.22d}$$

where Δ, $S(\Delta)$, N, and $\tilde{x}(t)$ represent the sampling period, the set of piecewise constant functions with period Δ, the number of sampling periods in the prediction horizon, and the predicted state trajectory, respectively. $l(x, u)$ is the cost function of MPC that represents process performance index; in tracking MPC that steers the system to the optimal steady-state or the optimal trajectory, the cost function is typically designed with a quadratic form to minimize the deviations of the process inputs and states from the steady-state value or reference trajectory value over the prediction horizon, i.e., $l(x, u) = |x|^2_{Q_1} + |u|^2_{Q_2}$, where Q_1, Q_2 are positive definite matrices that manage the trade-off between the speed of state convergence to the steady-state and the cost of control action. By designing the cost function in a quadratic form, the minimum value of the cost function is attained at the steady-state. The predicted state trajectory $\tilde{x}$ of Eq. 2.22b is obtained using the nominal process model of Eq. 2.9 (i.e., $w \equiv 0$) under sample-and-hold input profile optimized by MPC. Equation 2.22c defines the initial conditions for the nominal process system of Eq. 2.22b, which are state/output measurements obtained at each sampling period. Equation 2.22d defines the state, input, and other process constraints, where Z is a compact set. Throughout this book, the term *MPC* will refer to tracking MPC that stabilizes a system at steady-state, unless stated otherwise.

MPC is implemented in a receding horizon fashion to compute optimal control actions by solving the optimization problem of Eq. 2.22. Let $u^*(t)$ be the optimal solution of the optimization problem of Eq. 2.22 over the prediction horizon $t \in [t_k, t_{k+N})$. It is assumed that the measurement of the closed-loop states are available at each sampling time. The problem of Eq. 2.22 is solved with a feedback measurement of state $x(t_k)$ at the sampling time t_k to compute optimal control actions. After $u^*(t)$, where $t \in [t_k, t_{k+N})$ is obtained from the MPC optimization problem, only the first control action of $u^*(t)$, i.e., $u^*(t|t_k)$ defined for $t \in [t_k, t_{k+1})$ is sent to the control actuators to be applied over the next sampling period. Then, the MPC optimization problem is re-initialized at the next instance of time $t_{k+1} := t_k + \Delta$ with an updated state measurement, and the optimization problem is solved again by rolling the horizon one sampling period forward.

However, since the MPC scheme of Eq. 2.22 is developed with a finite prediction horizon, i.e., $N \neq \infty$, it is possible that the MPC scheme of Eq. 2.22 is not stabilizing, e.g., [124]. Therefore, to ensure stabilization of the closed-loop system with a finite N prediction horizon, additional constraints or variations to the cost function should be employed. For example, we can design an MPC by incorporating terminal constraints, with a sufficiently long prediction horizon, or using contractive constraints that will be discussed in more detail in the next section.

2.3.3 Lyapunov-Based MPC

In Sect. 2.3.1, we have introduced the design of stabilizing control law using Lyapunov techniques that provides an explicitly characterized set of initial conditions from which closed-loop stability is guaranteed. Despite the well-characterized stability properties, the Lyapunov-based controllers are not guaranteed to be optimal as performance considerations are not accounted for in the calculation of control actions. Therefore, to ensure stability of the closed-loop system in MPC, Lyapunov-based controller meeting asymptotic (exponential) stabilizability assumption in Assumption 2.1 (Assumption 2.2) is utilized to design a contractive constraint in the formulation of MPC [51, 129, 130, 135]. The resulting tracking MPC is termed Lyapunov-based MPC (LMPC) and is represented by the following optimization problem:

$$\min_{u \in S(\Delta)} \int_{t_k}^{t_{k+N}} l_t(\tilde{x}(\tau), u(\tau))\, d\tau \tag{2.23a}$$

$$\text{s.t.} \quad \dot{\tilde{x}}(t) = f(\tilde{x}(t), u(t), 0) \tag{2.23b}$$

$$\tilde{x}(t_k) = x(t_k) \tag{2.23c}$$

$$u(t) \in U, \ \forall\, t \in [t_k, t_{k+N}) \tag{2.23d}$$

$$\frac{\partial V(x(t_k))}{\partial x} f(x(t_k), u(t_k), 0) \leq \frac{\partial V(x(t_k))}{\partial x} f(x(t_k), \Phi(x(t_k)), 0),$$

$$\text{if } x(t_k) \in \Omega_\rho \backslash \Omega_{\rho_{\min}} \tag{2.23e}$$

$$V(\tilde{x}(t)) \leq \rho_{\min}, \ \forall\, t \in [t_k, t_{k+N}), \ \text{if } x(t_k) \in \Omega_{\rho_{\min}} \tag{2.23f}$$

where the notations follow those in Eq. 2.22. The objective function of Eq. 2.23 is the time integral of $l_t(\tilde{x}(t), u(t))$ over the prediction horizon. Equation 2.23b is the nominal process model of Eq. 2.9 with $w(t) \equiv 0$ for predicting future states of the closed-loop system. Equation 2.23c defines the input constraints applied over the entire prediction horizon. The constraint of Eq. 2.23e forces the closed-loop state to move toward the origin by decreasing the Lyapunov function value at least at the rate under the Lyapunov-based controller $\Phi(x(t_k))$ at $t = t_k$, if $x(t_k) \in \Omega_\rho \backslash \Omega_{\rho_{\min}}$. However, if $x(t_k)$ enters $\Omega_{\rho_{\min}}$, which is a small neighborhood around the origin, the states predicted by the nonlinear model of Eq. 2.23b will be maintained in $\Omega_{\rho_{\min}}$ over the prediction horizon under the constraint of Eq. 2.23f.

Figure 2.2 shows an illustration of state trajectory for the closed-loop system under LMPC, where x_s is the steady-state, Ω_ρ is the closed-loop stability region, Ω_{ρ_s} is a small level set close to the origin in which the decay of the Lyapunov function value is not guaranteed due to the sample-and-hold implementation of control actions and the impact of sufficiently small disturbances, and $\Omega_{\rho_{\min}}$ is a small forward invariant set around the origin that ensures ultimate boundedness of the closed-loop state under LMPC.

2.3.3.1 Closed-Loop Stability Under LMPC

Closed-loop stability of the nonlinear system of Eq. 2.9 is guaranteed under the LMPC of Eq. 2.23 in the sense that for any initial condition $x_0 \in \Omega_\rho$, the closed-loop state is guaranteed to be bounded in Ω_ρ for all times and converges to a small neighborhood $\Omega_{\rho_{\min}}$ of the origin and remains in it afterwards. In this section, we provide sufficient conditions for closed-loop stability under LMPC. To begin with, we present the following proposition that defines an upper bound on the deviation between the state trajectory from the actual system and from the nominal model of Eq. 2.9 with $w(t) \equiv 0$ when the same control input trajectories are applied.

Proposition 2.1 (c.f. [76, 131]) *Consider the following systems with initial states* $x_a(t_0) = x_b(t_0) \in \Omega_\rho$.

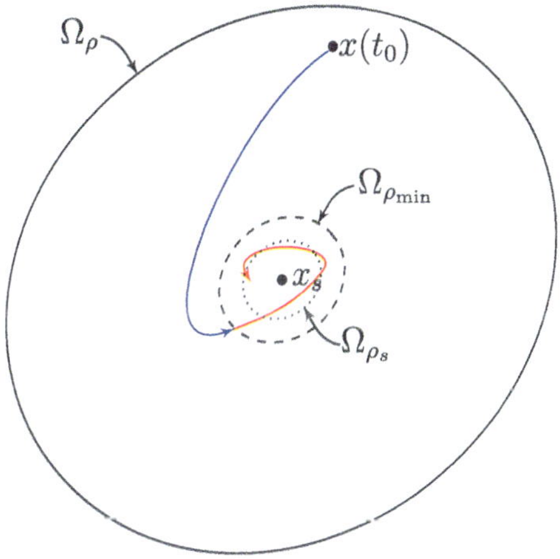

Fig. 2.2 An exemplar closed-loop state trajectory with initial state $x(t_0)$ under LMPC

$$\begin{aligned}\dot{x}_a(t) &= f(x_a(t), u(t), w(t)) \\ \dot{x}_b(t) &= f(x_b(t), u(t), 0).\end{aligned} \tag{2.24}$$

There exists a class $\mathcal{K}$ function $f_W(\cdot)$ such that

$$|x_a(t) - x_b(t)| \leq f_W(t - t_0), \tag{2.25}$$

for all $x_a(t), x_b(t) \in \Omega_\rho$ and all $w(t) \in W$ with

$$f_W(\tau) = \frac{L_w\theta}{L_x}(e^{L_x\tau} - 1). \tag{2.26}$$

The following proposition provides the bounds for the difference between the Lyapunov function values of two different states in Ω_ρ.

Proposition 2.2 (c.f. [76, 131]) *Consider the nonlinear system of Eq. 2.9 with a Lyapunov function $V(\cdot)$. There exists a positive constant M_v and a quadratic function $f_V(\cdot)$ such that the following inequality holds:*

$$V(x) \leq V(\hat{x}) + f_V(|x - \hat{x}|) \tag{2.27}$$

for all $x, \hat{x} \in \Omega_\rho$ with

$$f_V(s) = \alpha_4(\alpha_1^{-1}(\rho))s + M_v s^2 \tag{2.28}$$

where α_1, α_4 are class $\mathcal{K}$ functions that satisfy Eq. 2.16.

The following theorem is derived to show the closed-loop stability of the system under the sample-and-hold implementation of LMPC.

Theorem 2.3 (c.f. [76, 151]) *Consider the closed-loop system of Eq. 2.9 under the LMPC of Eq. 2.23 with the Lyapunov function V that satisfies Eq. 2.16. Let*

Assumption 2.1 hold and Ω_ρ *be the stability region. Then if* $\rho_s < \rho$, θ, *and* Δ *satisfy*

$$-\alpha_3(\alpha_2^{-1}(\rho_s)) + L'_x M\Delta + L'_w\theta \leq -\varepsilon_w/\Delta \tag{2.29}$$

for $\varepsilon_w > 0$, *where* α_2, α_3 *are class* $\mathcal{K}$ *functions that satisfy Eq. 2.16, then for any* $x_0 \in \Omega_\rho$,

$$\dot{V}(x(t)) \leq -\varepsilon_w/\Delta, \quad V(x(t)) \leq V(x(t_k)), \ \forall\, t \in [t_k, t_{k+1}) \tag{2.30}$$

and

$$V(x(t_{k+1})) < V(x(t_k)) \tag{2.31}$$

hold for the closed-loop state trajectory of the sampled-data system

$$\dot{x}(t) = f(x(t), \Phi(x(t_k)), w(t)), \ \forall\, t \in [t_k, t_{k+1}), \ k = 0, 1, \ldots \tag{2.32}$$

when $x(t_k) \in \Omega_\rho \backslash \Omega_{\rho_s}$. *If* $\rho_{\min} < \rho$ *where*

$$\rho_{\min} = \max\{V(x(t+\Delta)) : V(x(t)) \leq \rho_s\}, \tag{2.33}$$

then the closed-loop state $x(t)$ *is bounded in* Ω_ρ *for all times and is (uniformly) ultimately bounded in* $\Omega_{\rho_{\min}}$ *as follows:*

$$\limsup_{t\to\infty} V(x(t)) \leq \rho_{\min} \tag{2.34}$$

where $\rho_{\min}$ is defined to be a superset of ρ_s such that the closed-loop state under any sample-and-hold control action (not necessarily $\Phi(x(t_k))$) satisfying the input constraints does not leave $\Omega_{\rho_{\min}}$ (even in the presence of bounded disturbances) by the end of a sampling time when $x(t_k) \in \Omega_{\rho_s}$. The proofs are omitted here, and the reader is referred to [76, 130, 131] for detailed development of the propositions and theorem above.

Remark 2.2 It is noted that for nonlinear systems with sufficiently small bounded disturbances, the origin of Eq. 2.9 typically cannot be rendered asymptotically stable unless additional conditions hold under the sample-and-hold implementation of the controllers. Therefore, the origin is considered practically stable if the state trajectory of the closed-loop system starting from Ω_ρ remains bounded in Ω_ρ and converges to a small compact set around the origin in which it will be bounded thereafter.

Remark 2.3 As it is computationally impossible to impose the constraint of Eq. 2.23f for all $t \in [t_k, t_{k+N})$, we practically impose the constraint of Eq. 2.23f only at each sampling time of the prediction horizon (i.e., at $t = t_i, i = k, k+1, \ldots, k+N-1$) such that the LMPC of Eq. 2.23 in a sample-and-hold fashion is a finite-dimensional optimization problem. This methodology has been demonstrated to yield acceptable results for the examples contained in this book.

Remark 2.4 A numerical integration method (e.g., Explicit Euler method) will be utilized to solve the dynamic model of Eq. 2.22b in MPC. As a result, discretization and numerical error will occur. To reduce the discretization error, a sufficiently small integration time step (i.e., much smaller than the sampling period Δ for executing MPC) is required such that the resulting numerical error could be bounded by a small bound. Since the numerical error can be considered as a source of (bounded) process disturbances, closed-loop stability results are still guaranteed for the system under MPC as long as the numerical error is sufficiently small.

2.3.3.2 Feasibility Analysis

In addition to closed-loop stability, feasibility is another issue that may arise when there is no solution to the MPC optimization problems. The feasibility issue often comes from the contradiction of MPC constraints, i.e., there does not exist a feasible control action that satisfies all the constraints. Considering that the input constraints that typically represent physical limitation on control actuators, e.g., valve saturation, cannot be violated, one solution is to relax the state/output constraints by introducing slack variables. Many researchers have studied formulating soft constraints within MPC to avoid potential infeasibility issues, e.g., [67, 144, 174]. However, for the optimization problem of Eq. 2.23, it is demonstrated that feasibility is guaranteed for all times given that the closed-loop stability region Ω_ρ is characterized using the Lyapunov-based controller $u = \Phi(x) \in U$. In fact, $u = \Phi(x) \in U$ implemented in a sample-and-hold fashion is always a feasible solution to the LMPC optimization problem of Eq. 2.23 for any $x \in \Omega_\rho$. Specifically, it is readily shown that $u = \Phi(x) \in U$ meets the input constraint of Eq. 2.23d. By letting $u(t_k) = \Phi(x(t_k))$ for any $x(t_k) \in \Omega_\rho \backslash \Omega_{\rho_{\min}}$, the constraint of Eq. 2.23e is satisfied as an equality (i.e., the inequality constraint becomes active). Additionally, for any $x(t_k) \in \Omega_{\rho_{\min}}$, we will show that $u = \Phi(x(t)) \in U, t \in [t_k,\ t_{k+N})$ provides a feasible input trajectory to the constraint of Eq. 2.23f that is applied over $t \in [t_k, t_{k+N})$. First, assuming that $x(t_k) \in \Omega_{\rho_{\min}} \backslash \Omega_{\rho_s}$, it is shown in Eq. 2.31 that $V(x(t_{k+1})) < V(x(t_k))$ holds for any $x(t_k) \in \Omega_\rho \backslash \Omega_{\rho_s}$, which implies that $x(t_{k+1})$ is maintained within $\Omega_{\rho_{\min}}$; however, if $x(t_k) \in \Omega_{\rho_s}$, it is derived from Eq. 2.33 that the state over the next sampling period, i.e., $x(t_{k+1})$, will remain inside $\Omega_{\rho_{\min}}$ under any $u \in U$ (which includes $u = \Phi(x) \in U$). Therefore, in either case, $u = \Phi(x) \in U$ is a feasible solution that maintains $x(t_{k+1})$ within $\Omega_{\rho_{\min}}$. Following the above analysis, it is readily shown that by applying $u = \Phi(x(t)) \in U$ for every sampling period within the prediction horizon, the constraint of Eq. 2.23f is satisfied for all $t \in [t_k,\ t_{k+N})$. Therefore, as the closed-loop state is guaranteed to be bounded in Ω_ρ for all times (Theorem 2.3), given an initial condition $x_0 \in \Omega_\rho$, the LMPC of Eq. 2.23 is both initially feasible, and also recursively feasible at each subsequent sampling period until the end of operation.

2.3.4 Lyapunov-Based Economic MPC

The economic success of the chemical and petrochemical industries relies on optimal process operation which has led to the emergence of an overall process control goal of incorporating process/system economic considerations into feedback control objectives. Figure 2.3 shows a traditional paradigm for optimizing process economics via a two-layer control architecture, where in the upper layer, a real-time optimization (RTO) is solved to obtain economically optimal steady-states, while in the lower layer, tracking MPC or traditional proportional–integral–derivative (PID) control is utilized to drive the process state to the optimal steady-state by computing optimal control actions u^*.

Another approach to addressing integrated process control and dynamic economic optimization problems is to use economic model predictive control (EMPC). EMPC is a model-based advanced control technique that dynamically optimizes process economic performance by operating processes in a time-varying fashion (off steady-state operation). Stability constraints are incorporated in EMPC to guarantee feasibility and closed-loop stability within an explicitly defined estimate of the closed-loop stability region under an appropriate control law (e.g., a Lyapunov-based stabilizing control law $\Phi(x)$ that satisfies Assumption 2.1 or Assumption 2.2) (see, also, Ref. [62] for an overview of recent results on EMPC). The EMPC that incorporates Lyapunov-based constraints in the design is termed Lyapunov-based economic MPC (LEMPC) and is represented by the following optimization problem:

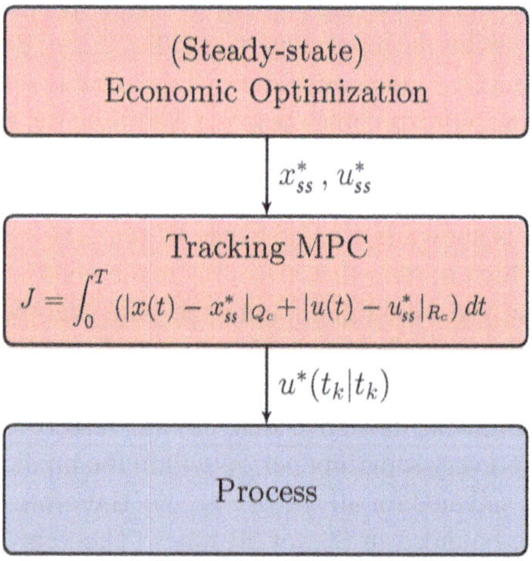

Fig. 2.3 A two-layer paradigm for optimizing process economics within process control

$$\max_{u\in S(\Delta)} \int_{t_k}^{t_{k+N}} l_e(\tilde{x}(\tau), u(\tau))\, d\tau \tag{2.35a}$$

$$\text{s.t.} \quad \dot{\tilde{x}}(t) = f(\tilde{x}(t), u(t), 0) \tag{2.35b}$$

$$\tilde{x}(t_k) = x(t_k) \tag{2.35c}$$

$$u(t) \in U, \ \forall\, t \in [t_k, t_{k+N}) \tag{2.35d}$$

$$V(\tilde{x}(t)) \le \rho_e, \ \forall\, t \in [t_k, t_{k+N}), \ \text{if } x(t_k) \in \Omega_{\rho_e} \tag{2.35e}$$

$$\frac{\partial V(x(t_k))}{\partial x} f(x(t_k), u(t_k), 0) \le \frac{\partial V(x(t_k))}{\partial x} f(x(t_k), \Phi(x(t_k)), 0),$$
$$\text{if } x(t_k) \in \Omega_\rho \backslash \Omega_{\rho_e}, \tag{2.35f}$$

where the notations follow those in Eqs. 2.22 and 2.23. The LEMPC optimization problem maximizes the time integral of the objective function $l_e(x, u)$ that represents the process economic performance over the prediction horizon subject to the constraints of Eqs. 2.35b–2.35f. Specifically, the constraints of Eqs. 2.35b–2.35d are the same as Eqs. 2.23b–2.23d for LMPC. The constraint of Eq. 2.35e (Mode 1 constraint) maintains the predicted states $\tilde{x}$ within Ω_{ρ_e} that is designed to ensure forward invariance of the closed-loop stability region Ω_ρ accounting for the sample-and-hold implementation of control actions and the impact of sufficiently small disturbances w, if the current state $x(t_k)$ at time $t = t_k$ is within Ω_{ρ_e}. However, if the current state leaves Ω_{ρ_e} due to disturbances, the constraint of Eq. 2.35f (Mode 2 constraint) is activated to drive the state toward the origin at least at the speed under the Lyapunov-based controller $\Phi(x(t_k))$ at $t = t_k$ such that it can enter Ω_{ρ_e} within finite sampling steps. An example of the state trajectory of the closed-loop system under LEMPC is shown in Fig. 2.4.

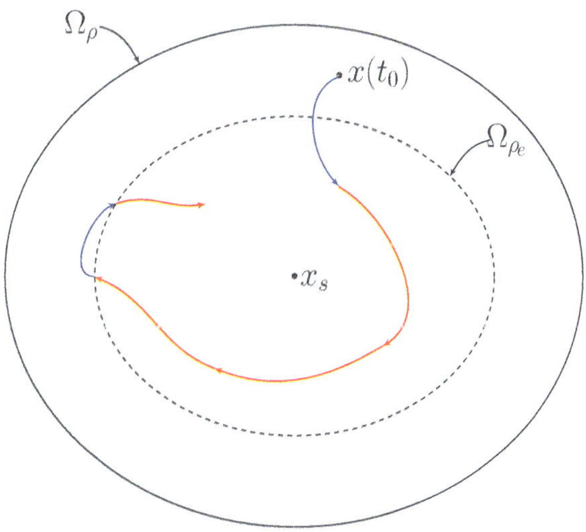

Fig. 2.4 An exemplar closed-loop state trajectory under LEMPC, where the red and the blue trajectories are under Mode 1 and Mode 2 constraints, respectively

2.3.4.1 Closed-Loop Stability Under LEMPC

Closed-loop stability is guaranteed for the nonlinear system of Eq. 2.9 under LEMPC in the sense that for any initial condition $x_0 \in \Omega_\rho$, the closed-loop state is guaranteed to be bounded in Ω_ρ for all times. Additionally, it is demonstrated that the LEMPC of Eq. 2.35 is robust to sufficiently small bounded disturbances due to the design of Ω_{ρ_e} that accounts for its impact within one sampling period [76].

Based on Propositions 2.1 and 2.2, the following theorem presents sufficient conditions for closed-loop stability of the system under LEMPC.

Theorem 2.4 (c.f. [76]) *Consider the closed-loop system of Eq. 2.9 with the Lyapunov function V that satisfies Eq. 2.16 under the LEMPC of Eq. 2.35. Let Assumption 2.1 hold, and Ω_ρ be the closed-loop stability region. If $\varepsilon_w > 0$, $\rho_s < \rho_e < \rho$, θ, and Δ satisfy*

$$\rho_e \leq \rho - f_V(f_W(\Delta)) \tag{2.36}$$

and

$$-\alpha_3(\alpha_2^{-1}(\rho_s)) + L'_x M\Delta + L'_w\theta \leq -\varepsilon_w/\Delta \tag{2.37}$$

where $f_V(\cdot)$ and $f_W(\cdot)$ are given in Eqs. 2.28 and 2.26, then for any $x_0 \in \Omega_\rho$, the closed-loop state $x(t)$ is bounded in Ω_ρ, $\forall t \geq 0$.

Additionally, recursive feasibility is also guaranteed for the LEMPC optimization problem as the stability region Ω_ρ is characterized following the same approach to LMPC. The analysis is similar to that in Sect. 2.3.3.2 and is omitted here. As recursive feasibility and closed-loop stability are two of the most common issues in MPC designs, they will be discussed with rigorous theoretical analysis for much of the work contained in this book.

To conclude, the main difference between LEMPC and LMPC in terms of operation is that the process is operated in an optimal time-varying (off steady-state) fashion under LEMPC while it is forced to and operated at the optimal steady-state under LMPC. With regards to closed-loop stability, LEMPC only requires boundedness of the closed-loop state x in Ω_ρ for any $x_0 \in \Omega_\rho$ while LMPC additionally requires asymptotic/exponential convergence of the process state to the steady-state (boundedness of the state is therefore also guaranteed) for any $x_0 \in \Omega_\rho$.

Chapter 3
Safeness Index-Based MPC and EMPC

3.1 Introduction

In Chap. 2, we have introduced the formulation of MPC and demonstrated its capability of optimizing process performance accounting for system constraints [167]. Among different MPC formulations, LMPC and LEMPC systems have been developed based on an explicitly defined estimate of the closed-loop stability region via a Lyapunov-based control law, for which stabilizability and feasibility have been extensively studied, e.g., [51, 62, 76, 129, 130, 135]. However, in addition to stability, which is an important property of control systems, the problem of incorporating safety considerations in LMPC and LEMPC remains an important research topic as well since safety is another important issue in many chemical process industries.

In this chapter, we first introduce the concept of operational safety in process control, followed by a notion termed Safeness Index that characterizes the "safeness" of a process operating region. Subsequently, LMPC and LEMPC schemes that incorporate Safeness Index-based constraints as hard constraints to define a safe region of operation are developed to maintain the process state in the safe operating region while optimizing process performance at the same time. Stability and operational safety of the closed-loop systems under Safeness Index-based MPC and EMPC schemes are rigorously analyzed and demonstrated through simulations using a chemical process example.

3.1.1 Class of Nonlinear Systems

The following system of nonlinear first-order ordinary differential equations is used to represent the class of nonlinear systems that we consider:

$$\dot{x} = f(x, u, w) \tag{3.1}$$

Z. Wu and P. D. Christofides, *Process Operational Safety and Cybersecurity*, Advances in Industrial Control, https://doi.org/10.1007/978-3-030-71183-2_3

where $x \in D \subset \mathbf{R}^n$ is the state vector, $u \in U \subset \mathbf{R}^m$ is the manipulated input vector bounded by the input constraint $u \in U := \{u_{\min} \le u \le u_{\max}\} \subset \mathbf{R}^m$, where $u_{\min}$ and $u_{\max}$ are the lower and upper bounds for the input vector, respectively. $w \in W$ is a bounded disturbance vector, where $W := \{w \in \mathbf{R}^l \mid |w| \le \theta,\ \theta \ge 0\}$. Without loss of generality, the initial time t_0 is set to zero, and the state x of the system is sampled at time instances $t_k := t_0 + k\Delta$ with $k = 0, 1, 2, \ldots$ and Δ is the sampling period. It is assumed that $f : \mathbf{R}^n \times \mathbf{R}^m \times \mathbf{R}^l \to \mathbf{R}^n$ is a smooth vector function of its arguments with $f(0, 0, 0) = 0$, and thus, the origin is a steady-state of the nominal system of Eq. 3.1 with $w(t) \equiv 0$.

To proceed, we will make the following stabilizability assumption.

Assumption 3.1 We assume that there exists a stabilizing feedback controller $u = \Phi(x) \in U$ for the nominal system of Eq. 3.1 with $w(t) \equiv 0$ that renders the origin of the closed-loop system under continuous implementation of the controller asymptotically stable in the sense that there exist class $\mathcal{K}$ functions $\alpha_i, i = 1, 2, 3, 4$ and a $\mathcal{C}^1$ Lyapunov function $V : D \to \mathbf{R}_+$ such that the following inequalities hold for all x in a neighborhood D around the origin:

$$\alpha_1(|x|) \le V(x) \le \alpha_2(|x|), \tag{3.2a}$$

$$\frac{\partial V(x)}{\partial x} f(x, \Phi(x), 0) \le -\alpha_3(|x|), \tag{3.2b}$$

$$\left| \frac{\partial V(x)}{\partial x} \right| \le \alpha_4(|x|). \tag{3.2c}$$

Under the feedback control law that satisfies Assumption 3.1, the closed-loop stability region Ω_ρ is characterized as a level set of V within the set D where Eq. 3.2 holds, $\Omega_\rho = \{x \in D \mid V(x) \le \rho,\ \rho > 0\}$. One of the candidate controllers that can render the origin of the nominal system of Eq. 3.1 asymptotically stable is given by the saturated Sontag control law of Eqs. 2.14–2.15. When x is maintained within the stability region Ω_ρ, we have from the continuity of x, the smoothness property of f, and the continuous differentiability of $V(x)$ that there exist positive constants M, L_x, L_w, L_x', and L_w' such that the following inequalities hold:

$$|f(x, u, w)| \le M \tag{3.3}$$

$$\left| f(x, u, w) - f(x^*, u, 0) \right| \le L_x \left| x - x^* \right| + L_w |w| \tag{3.4}$$

$$\left| \frac{\partial V(x)}{\partial x} f(x, u, w) - \frac{\partial V(x^*)}{\partial x} f(x^*, u, 0) \right| \le L_x' \left| x - x^* \right| + L_w' |w| \tag{3.5}$$

for all $x, x^* \in \Omega_\rho$, $u \in U$, and $w \in W$.

3.2 Process Operational Safety

In chemical process industries, operational safety is a critical issue and the loss of safety could lead to severe consequences for both property and lives [181]. Despite the safe process/plant design and operation procedures that have been developed, characterized, and standardized over the years, unsafe operations continue to occur, which lead to accidents that cause significant human and capital losses [3, 4].

These consistent accidents throughout chemical process plant history [96, 98] imply that the current hierarchical design of the safety system layers employed in industries (i.e., the safety system comprised of the basic process control, alarms, emergency shutdown, and safety relief systems as shown in Fig. 1.1) is quite limited in preventing unsafe scenarios. Particularly, as process economic considerations motivate integrated system designs that account for both process control and optimization, a systems approach to analyzing process safety should also be used (e.g., [6, 109, 116, 195]) in which unsafe operations are regarded as the result of the process state moving to unsafe operating conditions in state-space over time. While the standard industrial thinking improves the "safeness" of a chemical process by employing individual alarms, or safety relief devices in process design to prevent each possible accident due to disturbances or equipment fault [55, 119], the systems approach of addressing process operational safety accounts for important aspects of the process system, such as physical constraints of process control actuators, multivariable interactions among process variables, and unmonitored process state variables that contain valuable process safety information [109].

As model predictive control (MPC) is able to optimize process operation while accounting for limitations on the capacity of process control actuators and multivariable interactions [124, 165], it provides a potential solution to addressing some of the issues above. However, it still remains challenging to develop a systematic methodology that coordinates control and safety systems together. For example, metrics that indicate the safeness of system operation need to be developed and shared by the safety and control systems. Based on the developed safety metrics, constraints also need to be developed and incorporated in MPC to prevent the process state from migrating to unsafe regions, while closed-loop stability and feasibility must be maintained as well. Overall, the designs of both of these systems rely on a metric that can unify safety and control system considerations. In this work, a metric termed the Safeness Index, which is essentially a function of the closed-loop process state to indicate the relative safeness of the process state in state-space is developed. This index is then incorporated in the control system as well as the safety system by setting thresholds on the value of the Safeness Index upon which the actions of the control and safety systems are based.

To begin with, we assume that an open set $\mathscr{D}$ within which the system is unsafe to operate (e.g., the temperature or pressure is extremely high) exists in the state-space. As a result, $\mathbb{X}_0 := \{x \in \mathbf{R}^n \backslash \mathscr{D}\}$ where $\{0\} \in \mathbb{X}_0$ represents a set from which the set of initial conditions to be considered will be developed. It is noted that the set $\mathbb{X}_0$ has no intersection with the unsafe operating region $\mathscr{D}$, i.e., $\mathbb{X}_0 \cap \mathscr{D} = \emptyset$, and contains the origin $x = 0$ such that the steady-state operation is considered safe. Additionally, considering that the input vector is constrained by physical bounds, i.e., $u \in U$, the safe operating region $\mathscr{U}$ is characterized as a subset of $\mathbb{X}_0$, from which a feasible control action exists to maintain process operational safety for the closed-loop system of Eq. 3.1 in the following sense.

Definition 3.1 Consider the nominal system of Eq. 3.1 (i.e., $w(t) \equiv 0$) with input constraints $u \in U$. For any initial state $x(t_0) = x_0 \in \mathscr{U}$, if there exist control actions $u \in U$ that can render the process states within $\mathscr{U}$ for all times, i.e., $x(t) \in \mathscr{U}$, $\forall\, t \geq t_0$, then the process state is maintained within the safe operating region $\mathscr{U}$ at all times, and we say that process operational safety is achieved for the nominal system of Eq. 3.1.

Following the above definition of process operational safety, the definition of simultaneous closed-loop stability and operational safety for the nonlinear system of Eq. 3.1 under a controller that is designed to steer the system to the steady-state (e.g., tracking MPC) is presented as follows.

Definition 3.2 Consider the nominal system of Eq. 3.1 with $w(t) \equiv 0$ and input constraints $u \in U$. If for any initial state $x(t_0) = x_0 \in \mathscr{U}$, there exists a control action $u \in U$ such that the state trajectories of the closed-loop system satisfy $x(t) \in \mathscr{U}$, $\forall\, t \geq t_0$, and $\lim_{t\to\infty} |x(t)| \leq d$, where $\mathscr{B}_d(0)$ is a small neighborhood around the origin, then we say that closed-loop stability and operational safety are achieved simultaneously in the sense that the process state is maintained within a safe operating region at all times, and can be ultimately driven to the origin.

Unlike tracking MPC, closed-loop stability under EMPC represents boundedness of the state in the stability region Ω_ρ since EMPC may operate processes in an off steady-state manner. Therefore, the definition of simultaneous closed-loop stability and operational safety under EMPC follows Definition 3.1 when the system is required to operate in a safe operating region $\mathscr{U}$ that is a subset of Ω_ρ.

3.2.1 *Safeness Index*

We develop the metric of process Safeness Index in this section that will be later used in the control and safety systems. Safeness Index is a function that indicates the safeness of a plant as a whole. It can be developed as a function of the process (closed-loop) state accounting for interactions between units and multivariable interactions, which cannot be assessed by a typical component-by-component safety analysis that is commonly used in industry. Additionally, the development of Safeness Index functions should account for the fact that has been pointed out by many researchers that a process does not become unsafe automatically, but takes a gradual trajectory in that direction (e.g., [109]). Traditional safety thinking in the process industries pays more attention to the reasons that a state became unsafe based on a cause-and-effect-type relationship rather than the fact that the state is unsafe. In this work, we develop the Safeness Index as a function of the current state only, which allows engineers to characterize where the process is on the safeness spectrum based on the present condition instead of considering every possible failure mechanism of the given system. Additionally, by developing the Safeness Index as a state-based function, it is able to capture safety information even for unmeasured states provided that appropriate state estimation techniques are utilized, which is unachievable by traditional safety system designs that use process measurements only.

Although the development of a Safeness Index shows great potential for improving process safety, it is important to develop a methodology that determines the value of the Safeness Index for a given process. A possible methodology would be to define the Safeness Index as a binary function $S(x)$ that takes 1 for unsafe states and 0 for safe operating states. However, such a binary form may become ineffective for improving the performance of safety systems when it is used as the constraint in optimization-based control systems because a binary Safeness Index does not account for the case that the system is approaching an unsafe state but has not reached it yet, which should also trigger the safety system to prevent an unsafe operation. To address the issues, we develop a systematic methodology for formulating a Safeness Index function in this section based on the following two factors: (1) $S(x)$ is developed as a function of the (closed-loop) process state only to allow engineers to determine whether the system is safe or not based on the current operating condition, which enables a departure from the traditional safety thinking in chemical accident analysis and process safety system design [108]. Additionally, the proposed functional form of $S(x)$ allows the analysis of process operational safety accounting for the controller's limitations and effects. (2) $S(x)$ should be developed to indicate the safeness of a plant as a whole and account for multivariable interactions that cannot be achieved by the component-by-component safety analyses that are commonly used in chemical plants.

Given a chemical process, the proposed methodology requires the results of industrial safety studies, analysis of information on past accidents, process first-principles models, and past operating data (Fig. 3.1) to determine the functional form of $S(x)$ and the states that should be accounted for in $S(x)$. An extensive literature review of accidents and their causes (e.g., [24, 55, 91, 98, 168, 187]) can be initially performed to determine which states should be incorporated in $S(x)$ based on the investigation reports showing which states (e.g., pressure and temperature) took abnormal values in accidents. The literature study and the standard industrial safety analysis methods such as HAZOP studies and what-if analyses will help engineers analyze the types of accidents that may occur at the chemical plant under consideration. Any states that are related to the abnormal situations from the safety analyses and the literature review should be included in $S(x)$. Additionally, a first-principles model may help reveal other safety-critical process states that were neglected in the qualitative analyses at early stages due to complexities in the system. For example, $S(x)$ can be designed following the rules: (1) Incorporates states that may lead to unsafe operation based on the first-principles knowledge of the reactor material limitations (e.g., high pressure or high temperature that can lead to reactor rupture) or the chemistry of the reactions involved (e.g., reactions associated with ignition at certain temperatures [52]); (2) Incorporates states that have a great impact on other safety-critical

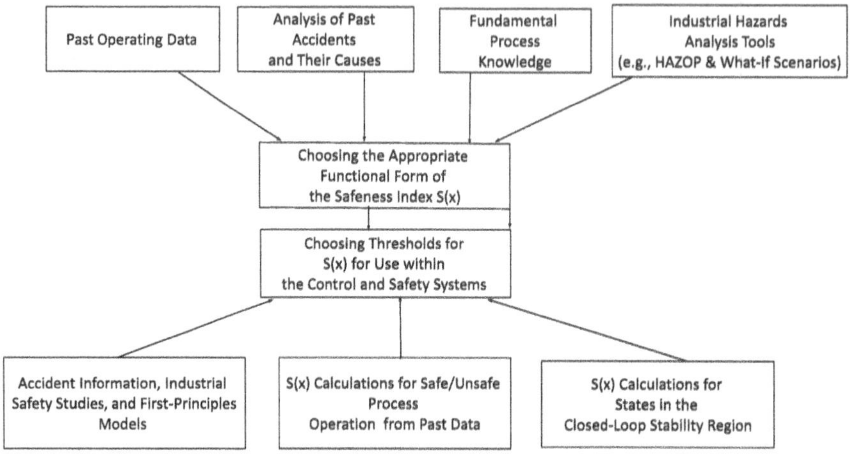

Fig. 3.1 Systematic methodology to construct Safeness Index function $S(x)$ and its thresholds

states in the reactor; (3) Incorporates all states that contribute to the safeness of the process, including the states that are unmeasurable and the states that might only affect the safeness when their values are far beyond the range of normal operating conditions.

To perform the above analyses, closed-loop simulations of the process with various operating conditions can be carried out to generate process operating data that help determine the states in $S(x)$. For example, time-series process data under the following scenarios: normal operation, near-miss (e.g., situations in which the safety system is triggered [149]), and unsafe operating conditions causing accidents can be analyzed to find which states are significantly different from their values under normal operation. Therefore, the states that play a dominant role in near-miss and accident conditions need to be included in the designs of $S(x)$.

After we identify the states that will be included in $S(x)$, the next step is to determine the functional form of $S(x)$ as well as the threshold value that will be used to distinguish between safe and unsafe operating regions in the state-space. Based on the thresholds on $S(x)$, the control and safety systems can be designed to take specific actions based on different regions to ensure safe operation. We have the two primary principles in designing the functional form of $S(x)$: (1) $S(x)$ is designed to have a significantly large value when the closed-loop state is in an unsafe operating region; (2) Controller limitations should also be accounted for in the way that $S(x)$ increase rapidly as the state approaches the boundary of the stability region in which closed-loop stability is guaranteed with a feasible control action. Specifically, in Principle 1, the development of $S(x)$ needs to consider the potential differences in the magnitude of the various states of the process. For example, given a chemical process in which the concentration of corrosive reactant and the temperature play an important role in safe operation, scaling of process states and nonlinear dependence on certain process states are required when designing the functional form of $S(x)$ since the order of magnitude of the concentration is much less than that of the temperature. In other words, $S(x)$ should be developed to account for the unsafe scenario in which the temperature drops while the concentration increases to unsafe conditions, by having a significantly large value in the abnormal range of concentration. Additionally, the development of $S(x)$ may benefit from scaling and nonlinearities when a process state leads to an unsafe operation only when it takes an extreme value, or when the process state can quickly move to the states that pose safety concerns under certain process dynamics. For example, if there exists a certain pressure P_1 for a reactor according to the process dynamics, from which the reactor pressure can quickly elevate to a high level that ruptures the reactor, $S(x)$ should be developed to have a large value when the pressure P_1 is reached.

In Principle 2, closed-loop stability also guides the design of the functional form of the Safeness Index. This allows the safety and control systems to be triggered when the threshold on $S(x)$ is reached, indicating that the closed-loop system may lose controllability. For example, as we characterize the stability region Ω_ρ as a level set of Lyapunov function V (i.e., $\Omega_\rho = \{x \in \mathbf{R}^n \mid V(x) \leq \rho, \ \rho > 0\}$), $S(x)$ can be designed in a quadratic form (e.g., $S(x) = x^T x/\rho$ where ρ is the size of the stability region Ω_ρ) such that $S(x)$ increases as the state approaches the boundary of the stability region, and the value of $S(x)$ lies between 0 and 1. Additionally, when a process state is initiated near an open-loop unstable steady-state and evolves toward an open-loop stable steady-state with a higher temperature, a Safeness Index that assigns a higher value to the states further from the unstable steady-state is preferred since the control system may not be able to prevent the states from reaching unsafe regions under constrained control actions (i.e., actuator constraints) when the state leaves a certain region in state-space. Examples of Safeness Index function construction will be presented in the example at the end of this chapter and also in the large-scale case studies of Chap. 5.

3.2.2 Choosing Thresholds for Safeness Index

After we design the functional form of $S(x)$, the next step is to determine the thresholds on $S(x)$ that can be used to trigger the control and safety systems. The approach for developing the thresholds on $S(x)$ is shown in Fig. 3.1. Specifically, different thresholds on $S(x)$ should be utilized for independent systems (i.e., the control, alarm, emergency shutdown, and relief systems) to be consistent with standard industrial practice in which upper-tier safety systems are activated when the lower-tier safety system cannot maintain the process state within a safe region (for example, the alarms are only activated when the control system does not maintain the process state within a region where all variables instrumented with alarms are within their recommended ranges [119]). Considering that the control system is the first line of defense against unsafe operations, the control system should utilize a lower threshold S_{TH} than that in safety systems to avoid frequently activating alarms and emergency shutdown systems. By computing control actions that maintain the closed-loop state within a safe

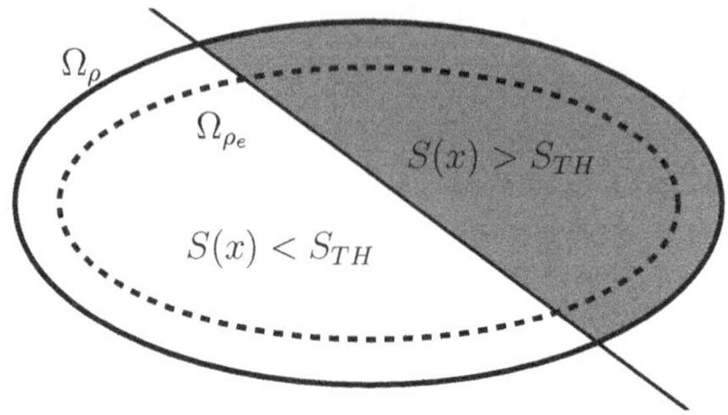

Fig. 3.2 Example of the stability region Ω_ρ partitioned into "safe" ($S(x) < S_{TH}$) and "unsafe" ($S(x) > S_{TH}$) regions, where Ω_{ρ_e} is a subset of Ω_ρ to ensure forward invariance of Ω_ρ in the LEMPC of Eq. 2.35

region where $S(x) < S_{TH}$, false alarms (i.e., activations of the safety system when the controller is able to guarantee closed-loop stability and safety) can be avoided. Motivated by this, this section will present the methods for determining S_{TH}.

To determine the value of S_{TH}, th industrial safety studies, past accident reports, and first-principles models can be utilized to gain insight into which values and magnitudes of the state become large in unsafe operations. Additionally, past operating data can be utilized in determining S_{TH} by (1) labeling the data as "safe" if no alarms were triggered, or only a few (e.g., one or two) alarms were triggered but the closed-loop state quickly re-entered the safe operating region without triggering alarms or emergency shutdown systems, and (2) labeling the data as "unsafe" if a number of alarms sounded during the operating period. Subsequently, we evaluate the value of $S(x)$ for each labeled datasets and choose the threshold S_{TH} with an appropriate value that can help the control system distinguish safe and unsafe operations. For example, we can set the threshold S_{TH} to be the minimum value of $S(x)$ observed in the unsafe datasets that is significantly different from the values of $S(x)$ observed in safe datasets. Additionally, in practical implementation, we should use a more conservative S_{TH} than its theoretical value to allow safety systems to drive the closed-loop state into safe operating regions.

Since the threshold S_{TH} can be later used in an optimization-based control design (e.g., MPC) to ensure process operational safety, another important consideration in setting the threshold S_{TH} is to ensure that there exist states in the closed-loop stability region Ω_ρ that satisfy $S(x) < S_{TH}$. Therefore, before implementing safety constraints in optimization problems, it is important to check the value of $S(x)$ in the stability region Ω_ρ to make sure the targeted region (e.g., a small neighborhood around the targeted steady-state) is characterized as the safe region where $S(x) < S_{TH}$. Additionally, to avoid false alarms of traditional emergency shutdown, or relief systems on individual measured variables, S_{TH} should be developed accounting for the thresholds for those individual measured variables. Figure 3.2 shows an example of "safe" and "unsafe" regions in the state-space determined by a threshold on the Safeness Index, where the $S(x) = S_{TH}$ defines the boundary of the two regions. The proposed method for developing $S(x)$ and S_{TH} will be illustrated using a chemical process example at the end of this chapter as well as the large-scale case studies in Chap. 5.

Remark 3.1 In addition to the traditional triggers that are designed based on individual process variables to prevent unsafe operation, the Safeness Index can be used in the alarm and emergency shutdown systems to allow the safety systems to account for unsafe scenarios associated with multivariable interactions and unmeasurable states. Similarly, the thresholds of the Safeness Index utilized in safety systems can be determined from extensive safety studies and investigation of accident reports. However, unlike the Safeness Index threshold used in optimization-based control systems, to implement the Safeness Index threshold in practical safety systems, it is important to tier the thresholds such that there are various levels of unsafe operations that should trigger tiered thresholds to be consistent with industrial practice.

3.3 Safeness Index-Based MPC and EMPC

Based on the Safeness Index function that characterizes safe and unsafe operating regions of given processes, optimization-based control designs, i.e., MPC and EMPC, have been developed to incorporate hard constraints that maintain the closed-loop state within the region where $S(x) < S_{TH}$ under control actions computed by the controller. Specifically, based on the LMPC of Eq. 2.23 and the LEMPC of Eq. 2.35 discussed in Chap. 2, we present the following formulations of Safeness Index-based MPC, [8, 214, 221]:

$$\min_{u(t)\in S(\Delta)} \int_{t_k}^{t_{k+N}} l_t(\tilde{x}(\tau), u(\tau))\, d\tau \tag{3.6a}$$

$$\text{s.t. } \dot{\tilde{x}}(t) = f(\tilde{x}(t), u(t), 0) \tag{3.6b}$$

$$\tilde{x}(t_k) = x(t_k) \tag{3.6c}$$

$$u(t) \in U,\ \forall\, t \in [t_k, t_{k+N}) \tag{3.6d}$$

$$\frac{\partial V(x(t_k))}{\partial x} f(x(t_k), u(t_k), 0) \le \frac{\partial V(x(t_k))}{\partial x} f(x(t_k), \Phi(x(t_k)), 0),$$
$$\text{if } x(t_k) \in \Omega_\rho \backslash \Omega_{\rho_{\min}} \text{ or } S(x(t_k)) > S_{TH} \tag{3.6e}$$

$$S(\tilde{x}(t)) \le S_{TH},\ \forall\, t \in [t_k, t_{k+N}),\ \text{ if } S(x(t_k)) \le S_{TH} \tag{3.6f}$$

$$V(\tilde{x}(t)) \le \rho_{\min},\ \forall\, t \in [t_k, t_{k+N}),\ \text{ if } x(t_k) \in \Omega_{\rho_{\min}} \tag{3.6g}$$

and of Safeness Index-based EMPC:

$$\max_{u(t)\in S(\Delta)} \int_{t_k}^{t_{k+N}} l_e(\tilde{x}(\tau), u(\tau))\, \mathrm{d}\tau \tag{3.7a}$$

$$\text{s.t. } \dot{\tilde{x}}(t) = f(\tilde{x}(t), u(t), 0) \tag{3.7b}$$

$$\tilde{x}(t_k) = x(t_k) \tag{3.7c}$$

$$u(t) \in U,\ \forall\, t \in [t_k, t_{k+N}) \tag{3.7d}$$

$$V(\tilde{x}(t)) \leq \rho_e,\ \forall\, t \in [t_k, t_{k+N}),\ \text{ if } x(t_k) \in \Omega_{\rho_e} \tag{3.7e}$$

$$S(\tilde{x}(t)) \leq S_{TH},\ \forall\, t \in [t_k, t_{k+N}),\ \text{ if } S(x(t_k)) \leq S_{TH} \tag{3.7f}$$

$$\frac{\partial V(x(t_k))}{\partial x} f(x(t_k), u(t_k), 0) \leq \frac{\partial V(x(t_k))}{\partial x} f(x(t_k), \Phi(x(t_k)), 0),$$
$$\text{if } x(t_k) \in \Omega_\rho \backslash \Omega_{\rho_e} \text{ or } S(x(t_k)) > S_{TH} \tag{3.7g}$$

where the notations follow those in Eqs. 2.23 and 2.35. The control objective of the Safeness Index-based MPC of Eq. 3.6 is to drive the process state to the origin while remaining in the safe operating region where $S(x) \leq S_{TH}$ and $x \in \Omega_\rho$ for all times. Similar to the (tracking) LMPC of Eq. 2.23, the objective function $l_t(x, u)$ of Safeness Index-based MPC is also designed in a quadratic form, i.e., $l_t(x, u) = |x|^2_{Q_1} + |u|^2_{Q_2}$, where Q_1, Q_2 are positive definite matrices, such that the minimum of l_t is achieved at the steady-state, i.e., $x = 0$ and $u = 0$. The constraints of Eqs. 3.6b–3.6d are the same as those of Eqs. 2.23b–2.23d, which represent the nonlinear process model of Eq. 3.1 for prediction, the initial condition for the optimization problem, and the input constraints, respectively. The constraints of Eqs. 3.6e and 3.6g will drive the process state toward the origin when $x(t) \in \Omega_\rho \backslash \Omega_{\rho_{\min}}$ and maintain the state within $\Omega_{\rho_{\min}}$ afterwards. Additionally, in Safeness Index-based MPC, we have the additional constraint of Eq. 3.6f to ensure that the state remains in the safe operating region at all times if starting from an initial condition in the safe region.

The control objective of the Safeness Index-based EMPC of Eq. 3.7 is to optimize process economic performance by maximizing the objective function $l_e(x, u)$ representing process economic benefits over the prediction horizon $t \in [t_k, t_{k+N})$, while ensuring safe operation at all times. All the components of the Safeness Index-based EMPC of Eq. 3.7 are similar to that of the problem of Eq. 2.35 except for the additional constraint of Eq. 3.7f. While EMPC operates processes in a time-varying manner for economic optimality, the constraint of Eq. 3.7f ensures that the state is maintained in the safe operating region. Both Safeness Index-based MPC and EMPC are implemented in a receding horizon fashion, where the first control action of the optimal input sequences will be sent to the actuator to be applied for the next sampling period.

Compared to the LMPC of Eq. 2.23 and the LEMPC of Eq. 2.35, Safeness Index-based MPC and EMPC take process operational safety into account in addition to closed-loop stability, and therefore, can control chemical plants to avoid hazardous operating conditions. However, a shortcoming of Safeness Index-based MPC and EMPC is that closed-loop stability and operational safety may not be achieved simultaneously (definitions of simultaneous stability and operational safety under EMPC and MPC, respectively, can be found in Definitions 3.1 and 3.2). In other words, there might not exist a feasible solution $u \in U$ to the optimization problem of Safeness Index-based MPC (EMPC) that satisfies all the constraints in Eq. 3.6 (Eq. 3.7), respectively.

The infeasibility issue is due to the fact that the safety region defined by $S(x) \leq S_{TH}$ is not necessarily a forward invariant set. Specifically, since the Safeness Index threshold S_{TH} may characterize an irregularly shaped safety region (for example, the $S(x) \leq S_{TH}$ region in Fig. 3.2), it may not be a forward invariant set like the stability region Ω_ρ characterized using Lyapunov functions. As a result, feasibility of the optimization problems of Eq. 3.6 (Eq. 3.7) cannot be guaranteed whenever the constraint of Eq. 3.6f (Eq. 3.7f) is activated (i.e., whenever $S(x(t_k)) \leq S_{TH}$). In other words, it is likely that the closed-loop state leaves the safety region when it moves toward the steady-state under the controller $u = \Phi(x)$. However, it should be noted that to ensure closed-loop stability, the steady-state of the nominal closed-loop system of Eq. 3.1 should be included in the safety region when developing $S(x)$ and S_{TH} (i.e., $S(x) \leq S_{TH}$ when $x = 0$). To address the feasibility issue, we apply the optimal control actions computed by the Safeness Index-based MPC/EMPC when a feasible solution is available, and applies the stabilizing Lyapunov-based controller (i.e., $u = \Phi(x) \in U$) instead when the optimization problems become infeasible. The detailed implementation strategy for the Safeness Index-based MPC/EMPC is summarized as follows:

1. At t_k, the Safeness Index-based MPC/EMPC receives the state measurement $x(t_k)$.Go to Step 2.
2. After the Safeness Index-based MPC/EMPC problems of Eqs. 3.6 and 3.7 are solved using nonlinear optimizers, go to Step 3.
3. If the optimization problem is feasible, then go to Step 3. Else, go to Step 3.
   a. Apply the optimal solution $u^*(t_k|t_k)$ (in a sample-and-hold fashion) from the Safeness Index-based MPC/EMPC to the nonlinear process, and go to Step 4.
   b. Apply the Lyapunov-based controller $u = \Phi(x)$ in a sample-and-hold fashion (i.e., $u(t) = \Phi(x(t_k)) \in U; \forall\, t \in [t_k, t_{k+1})$). Then go to Step 4.
4. Go to Step 1 ($k \leftarrow k + 1$).

Remark 3.2 The threshold S_{TH} could represent a soft threshold for practical systems that should operate below the threshold in general, but short excursions into the unsafe region ($S(x) > S_{TH}$) is acceptable. For example, it is demonstrated in [106] that minor excursions of temperature above the design temperature for a reforming tube of a steam methane reformer, e.g., increasing the temperature by 20 K, may reduce the tube lifetime, but will not result in immediate negative consequences. Therefore, from a process safety perspective, it is perfectly acceptable to allow $S(x)$ to go above S_{TH} for a finite period of time. Additionally, from a process economic perspective, this may also benefit the overall process economic performance by allowing the process state to operate in a larger region of state-space by allowing $S(x)$ above S_{TH}.

3.3.1 Stability, Safety, and Feasibility Analyses

This subsection presents sufficient conditions to show that for any initial condition $x_0 \in \Omega_\rho$, the closed-loop state of Eq. 3.1 under the Safeness Index-based MPC/EMPC implementation strategy is guaranteed to enter the safety region where $S(x) \leq S_{TH}$ in finite time and to remain within the stability region Ω_ρ at all times. Theorem 3.1 below presents sufficient conditions for guaranteed closed-loop stability and operational safety in the sense of boundedness of the state in the safety region for systems under the Safeness Index-based EMPC of Eq. 3.7. The results in Theorems 3.1 and 3.2 further prove the convergence of the state to a small neighborhood around the steady-state under the Safeness Index-based MPC of Eq. 3.6. Before we present the two theorems, we first state the following propositions to define functions and parameters that are needed in Theorems 3.1 and 3.2.

Proposition 3.1 *Consider the systems*

$$\begin{aligned} \dot{x}_a(t) &= f(x_a(t), u(t), w(t)) \\ \dot{x}_b(t) &= f(x_b(t), u(t), 0) \end{aligned} \tag{3.8}$$

with initial states $x_a(t_0) = x_b(t_0) \in \Omega_\rho$. *There exists a class* $\mathcal{K}$ *function* $f_W(\cdot)$ *such that*

$$|x_a(t) - x_b(t)| \leq f_W(t - t_0), \tag{3.9}$$

for all $x_a(t), x_b(t) \in \Omega_\rho$ *and all* $w(t) \in W$ *with*

$$f_W(\tau) = \frac{L_w \theta}{L_x}(e^{L_x \tau} - 1). \tag{3.10}$$

Proposition 3.2 *Consider the Lyapunov function* $V(\cdot)$ *of the system of Eq. 3.1. There exists a quadratic function* $f_V(\cdot)$ *such that*

$$V(x) \leq V(\hat{x}) + f_V(|x - \hat{x}|) \tag{3.11}$$

for all $x, \hat{x} \in \Omega_\rho$ *with*

$$f_V(s) = \alpha_4(\alpha_1^{-1}(\rho))s + M_v s^2 \tag{3.12}$$

where M_v *is a positive constant.*

The following proposition is presented to show guaranteed closed-loop stability of the system of Eq. 3.1 under the stabilizing controller $u = \Phi(x) \in U$ accounting for the impact of the sample-and-hold implementation.

Proposition 3.3 *Let Assumption 3.1 hold, V be the Lyapunov function that satisfies Eq. 3.2, and* Ω_ρ *be the resulting stability region. Then if* $\rho_s < \rho$, θ, *and* Δ *satisfy*

$$-\alpha_3(\alpha_2^{-1}(\rho_s)) + L'_x M\Delta + L'_w\theta \leq -\varepsilon_w/\Delta \tag{3.13}$$

for $\varepsilon_w > 0$*, then for any* $x_0 \in \Omega_\rho$*,*

$$\dot{V}(x(t)) \leq -\varepsilon_w/\Delta, \quad V(x(t)) \leq V(x(t_k)), \; \forall\, t \in [t_k, t_{k+1}) \tag{3.14}$$

and

$$V(x(t_{k+1})) < V(x(t_k)) \tag{3.15}$$

along the closed-loop state trajectory of the sampled-data system

$$\dot{x}(t) = f(x(t), \Phi(x(t_k)), w(t)), \; \forall\, t \in [t_k, t_{k+1}), \; k = 0, 1, \ldots \tag{3.16}$$

when $x(t_k) \in \Omega_\rho \backslash \Omega_{\rho_s}$*. If* $\rho_{\min} < \rho$ *where*

$$\rho_{\min} = \max\{V(x(t+\Delta)) : V(x(t)) \leq \rho_s\}, \tag{3.17}$$

then the closed-loop state is always bounded in Ω_ρ *and is (uniformly) ultimately bounded in* $\Omega_{\rho_{\min}}$ *as follows:*

$$\limsup_{t\to\infty} V(x(t)) \leq \rho_{\min}\,. \tag{3.18}$$

We first consider Safeness Index-based EMPC and demonstrate closed-loop stability and operational safety for the system of Eq. 3.1. The following theorem establishes that under the sample-and-hold implementation of the Safeness Index-based EMPC of Eq. 3.7, the closed-loop state $x(t)$ of Eq. 3.1 is guaranteed to be bounded within Ω_ρ at all times, and to enter the safety region in finite time if $x_0 \in \Omega_\rho$.

Theorem 3.1 *Consider the closed-loop system of Eq. 3.1 under the the Safeness Index-based EMPC of Eq. 3.7 implemented by following Steps 1–4 with a stabilizing controller* $\Phi(x)$ *that satisfies the conditions of Eq. 3.2. Let* $\varepsilon_w > 0$*,* $\Delta > 0$*,* $\rho > \rho_e > \rho_s > 0$ *satisfy*

$$\rho_e \leq \rho - f_V(f_W(\Delta)) \tag{3.19}$$

and

$$-\alpha_3(\alpha_2^{-1}(\rho_s)) + L'_x M\Delta + L'_w\theta < -\varepsilon_w/\Delta. \tag{3.20}$$

If $x_0 \in \Omega_\rho$*,* $\rho_{\min} \leq \rho$ *where* $\rho_{\min}$ *is defined as in Eq. 3.17 and where the compact set* $\Omega_{\rho_{\min}}$ *satisfies*

$$\Omega_{\rho_{\min}} \subseteq \{x \in \Omega_\rho \;:\; S(x) \leq S_{TH}\}, \tag{3.21}$$

then it is guaranteed that the closed-loop state $x(t)$ *of Eq. 3.1 is bounded within* Ω_ρ *at all times, and will enter the safety region in finite time for any* $x_0 \in \Omega_\rho$*.*

Proof The proof consists of two parts. In the first part, we prove that there exists a feasible solution for a nonlinear process operated under the Safeness Index-based EMPC that follows Steps 1–4 when $x_0 \in \Omega_\rho$. In the second part, we prove the results of Theorem 3.1.

Part 1: As shown in the implementation strategy of Steps 1–4 for the Safeness Index-based EMPC, one of the following two cases will occur at each sampling step: 1) a feasible solution is obtained from the Safeness Index-based EMPC optimization problem, from which the first control action $u(t_k|t_k)$ will be applied to the process for the next sampling period $t \in [t_k, t_{k+1})$, and 2) the Safeness Index-based EMPC optimization problem is infeasible, and the stabilizing controller $u = \Phi(x(t_k))$ will be applied for $t \in [t_k, t_{k+1})$. Specifically, it is readily shown that the constraints of Eqs. 3.7b–3.7g are satisfied when a feasible solution is available from the Safeness Index-based EMPC. When the EMPC is infeasible and $\Phi(x)$ is instead applied, it is shown in Proposition 3.3 that the controller $u = \Phi(x)$ is able to stabilize the closed-loop system at the steady-state. Therefore, the conditions met by the control actions for any given sampling period are characterized, and they will be used in the proof of closed-loop stability in Part 2.

Part 2: We now prove the main results of Theorem 3.1. We first prove that the closed-loop state of Eq. 3.1 enters the safety region in finite time for any initial condition in Ω_ρ (including the case that x_0 is outside the safety region, i.e., $S(x_0) > S_{TH}$). Then, we prove that the closed-loop state is bounded in the stability region Ω_ρ at all times for any $x_0 \in \Omega_\rho$.

We discuss the two cases of the implementation strategy of Steps 1–4 (i.e., either a feasible solution from the Safeness Index-based EMPC optimization problem or the stabilizing controller $u = \Phi(x)$) and prove that the state of the closed-loop system of Eq. 3.1 with any initial condition $x_0 \in \Omega_\rho$ will enter the safety region in finite time.

Specifically, when the current state $x(t_k)$ at time $t = t_k$ is outside the safety region, i.e., $S(x(t_k)) > S_{TH}$, and the Safeness Index-based EMPC is solved with a feasible solution that satisfies the contractive constraint of Eq. 3.7g, it is demonstrated in Eq. 3.14 in Proposition 3.3 that the Lyapunov function value is guaranteed to decrease, i.e., $V(x(t)) \leq V(x(t_k)),\ \forall\, t \in [t_k, t_{k+1})$ due to a negative $\dot{V}$ when $x(t_k) \notin \Omega_{\rho_s} \subseteq \Omega_{\rho_{\min}}$. However, if the Safeness Index-based EMPC is infeasible at t_k, the stabilizing controller $\Phi(x(t_k))$ will decrease the Lyapunov function value and drives the closed-loop state into a smaller level set of the Lyapunov function. In both cases, the closed-loop state moves toward the origin within the sampling period until it enters the safety region. Therefore, the contractive constraint of Eq. 3.7g that remains active when $S(x(t_k)) \leq S_{TH}$ will ultimately drive the closed-loop state into the safety region, regardless of its shape, in finite time for any initial condition in Ω_ρ.

Next, we prove that the closed-loop state remains in Ω_ρ at all times (i.e., Ω_ρ is a forward invariant set) under the Safeness Index-based EMPC. Specifically, when the Safeness Index-based EMPC is feasible at all times, closed-loop stability results depend on the Lyapunov-based stability constraints of Eqs. 3.7e and 3.7g. Other constraints such as Eq. 3.7f do not affect closed-loop stability in this case. Specifically, if $x(t_k) \in \Omega_{\rho_e}$, then the constraint of Eq. 3.7e is active, and the predicted state $\tilde{x}(t_{k+1})$ is in Ω_{ρ_e} while the actual state $x(t_{k+1})$ is guaranteed to be bounded in Ω_ρ even in the presence of sufficiently small bounded disturbances (i.e., $|w(t)| \le \theta$) based on the results from Propositions 3.1–3.2, and Eq. 3.19. If $x(t_k) \in \Omega_\rho \backslash \Omega_{\rho_e}$, then Eq. 3.7g is active to decrease the Lyapunov function value within one sampling period such that the closed-loop state enters a lower level set and still remains in Ω_ρ. On the other hand, when there does not exist any feasible solutions to the Safeness Index-based EMPC, the stabilizing controller $\Phi(x(t_k))$ will be applied for the next sampling period to decrease the Lyapunov function value and thus ensure that the closed-loop state does not leave Ω_ρ within that sampling period. If $x_0 \in \Omega_\rho$, then recursive application of the property that $x(t_k) \in \Omega_\rho$ ensures that $x(t_{k+1}) \in \Omega_\rho$. Starting with $k = 0$, it is demonstrated that the Safeness Index-based EMPC implementation strategy of Steps 1–4 maintains the closed-loop state within Ω_ρ at all times. This completes the proof of the boundedness of the closed-loop state in Ω_ρ, for any $x_0 \in \Omega_\rho$ under Safeness Index-based EMPC.

Based on the theoretical developments for Safeness Index-based EMPC, the following theorem provides sufficient conditions for closed-loop stability and operational safety of nonlinear systems under Safeness Index-based MPC.

Theorem 3.2 *Consider the closed-loop system of Eq. 3.1 under the the Safeness Index-based MPC of Eq. 3.6 using the implementation strategy of Steps 1–4 and the controller $\Phi(x)$ that satisfies the conditions of Eq. 3.2. Let $\varepsilon_w > 0$, $\Delta > 0$, $\rho > \rho_{\min} > \rho_s > 0$ satisfy*

$$-\alpha_3(\alpha_2^{-1}(\rho_s)) + L'_x M\Delta + L'_w\theta \le -\varepsilon_w/\Delta. \tag{3.22}$$

If $x_0 \in \Omega_\rho$, $\rho_{\min} \le \rho$ where $\rho_{\min}$ is defined as in Eq. 3.17 and where the compact set $\Omega_{\rho_{\min}}$ satisfies

$$\Omega_{\rho_{\min}} \subseteq \{x \in \Omega_\rho \ : \ S(x) \le S_{TH}\}, \tag{3.23}$$

then it is guaranteed that the closed-loop state $x(t)$ of Eq. 3.1 enters the safety region in finite time when $x_0 \in \Omega_\rho$ remains within Ω_ρ at all times, and ultimately remains inside $\Omega_{\rho_{\min}}$ as $t \to \infty$.

Proof The proof follows closely to that of Theorem 3.1. Specifically, the proof for the existence of an input trajectory with characterizable properties follows exactly the same proof as in *Part* 1 of Theorem 3.1; thus, we will only discuss closed-loop stability and safety properties for the case of Safeness Index-based MPC of Eq. 3.6.

Since the contractive constraint of Eq. 3.6e is activated when $x(t_k) \in \Omega_\rho \backslash \Omega_{\rho_{\min}}$ or the state is out of the safety region, it has been shown in the proof of Theorem 3.1 that for any initial condition in Ω_ρ, the closed-loop state will enter the safety region in finite time, and ultimately enter $\Omega_{\rho_{\min}}$ due to the fact that the Lyapunov function value is decreasing for every sampling period under the constraint of Eq. 3.6e. Therefore, it remains to show that the state will remain inside $\Omega_{\rho_{\min}}$ afterwards under the constraint of Eq. 3.6g. From the definition of $\Omega_{\rho_{\min}}$ of Eq. 3.17, once the closed-loop state enters $\Omega_{\rho_{\min}}$, if $u^*(t_k|t_k)$ that satisfies the EMPC constraint or $\Phi(x(t_k))$ is then applied to the process to drive the closed-loop state toward the origin value until the closed-loop state enters Ω_{ρ_s}, the closed-loop state cannot leave $\Omega_{\rho_{\min}}$ within one sampling period. Additionally, boundedness of the closed-loop state in Ω_ρ can be demonstrated following the same steps as performed in *Part* 2 proof of Theorem 3.1. Furthermore, when the state enters $\Omega_{\rho_{\min}}$, recursive feasibility of the Safeness Index-based MPC is guaranteed at all times since $\Omega_{\rho_{\min}}$ is a subset of the safe operating region, which implies that the constraint of Eq. 3.6f is naturally satisfied if Eq. 3.6g is satisfied. This completes the proof of closed-loop stability and safety for the nonlinear system of Eq. 3.1 under Safeness Index-based MPC.

3.4 Application to a Chemical Process Example

In this section, a chemical process example is provided to illustrate the ability of the Safeness Index-based EMPC of Eq. 3.7 to maintain the closed-loop state within a region where $S(x(t_k)) \leq S_{TH}$ when the standard LEMPC of Eq. 2.35 cannot obtain an input trajectory that achieves this. We revisit the chemical process example discussed in Chap. 1, which is a well-mixed, non-isothermal continuous stirred tank reactor (CSTR) with an irreversible second-order exothermic reaction.

3.4.1 Process Description

The reaction transforms a reactant A to a product B ($A \rightarrow B$) like that depicted in Fig. 1.2. The CSTR first-principles model is given as follows:

$$\frac{\mathrm{d}C_A}{\mathrm{d}t} = \frac{F}{V}(C_{A0} - C_A) - k_0 e^{\frac{-E}{RT}} C_A^2 \tag{3.24a}$$

$$\frac{\mathrm{d}T}{\mathrm{d}t} = \frac{F}{V}(T_0 - T) + \frac{-\Delta H}{\rho_L C_p} k_0 e^{\frac{-E}{RT}} C_A^2 + \frac{Q}{\rho_L C_p V} \tag{3.24b}$$

where the notations and the parameter values can be found in Sect. 1.3.1, and are omitted here. The inlet concentration C_{A0} of the reactant species A and the heat input/removal rate Q are the manipulated inputs. There are three steady-states for the CSTR of Eq. 3.24 associated with the steady-state input values $[C_{A0s}\ Q_s] = [4\ \frac{\text{kmol}}{\text{m}^3}\ 0\ \frac{\text{kJ}}{\text{h}}]$. In this study, the CSTR is operated around a stable steady-state at $[C_{As}\ T_s] = [1.22\ \frac{\text{kmol}}{\text{m}^3}\ 438\ \text{K}]$. The CSTR system of Eq. 3.24 can be further represented by the following nonlinear ODEs:

$$\frac{dx_1}{dt} = -\frac{F}{V}x_1 - k_0 e^{\frac{-E}{R(x_2+T_s)}}(x_1 + C_{As})^2 + \frac{F}{V}(C_{A0s} - C_{As}) + \frac{F}{V}u_1 \quad (3.25a)$$

$$\frac{dx_2}{dt} = -\frac{F}{V}x_2 + \frac{-\Delta H}{\rho_L C_p}k_0 e^{\frac{-E}{R(x_2+T_s)}}(x_1 + C_{As})^2 + \frac{F}{V}(T_0 - T_s) + \frac{Q_s + u_2}{\rho_L C_p V} \quad (3.25b)$$

where $x(t)$ and $u(t)$ denote the state and the manipulated inputs of the CSTR in deviation variable form (i.e., $x^T = [C_A - C_{As} \;\; T - T_s]$ is the state vector and $u^T = [C_{A0} - C_{A0s} \;\; Q - Q_s]$ is the manipulated input vector). It is observed that the dynamic model of Eq. 3.25 is in the general class of nonlinear systems:

$$\dot{x}(t) = \tilde{f}(x(t)) + g_1(x(t))u_1(t) + g_2(x(t))u_2(t) \quad (3.26)$$

where $\tilde{f}^T = [\tilde{f}_1 \; \tilde{f}_2]$ is a vector containing the terms in the CSTR model of Eq. 3.25 that do not include u_1 or u_2, and $g_i^T = [g_{i1} \; g_{i2}]$ $(i = 1, 2)$ is a vector containing the terms in the CSTR model of Eq. 3.25 that multiply u_1 (for $i = 1$) or u_2 (for $i = 2$). The magnitudes of the manipulated inputs are bounded as follows: $|u_1| \le 3.5 \; \frac{\text{kmol}}{\text{m}^3}$ and $|u_2| \le 5 \times 10^5 \; \frac{\text{kJ}}{\text{h}}$. The control objective is to maximize the production rate of B using the following stage cost:

$$l_e(x, u) = k_0 e^{-\frac{E}{RT}} C_A^2 \quad (3.27)$$

The prediction horizon is $N = 10$ and the sampling period is $\Delta = 0.01$ h. In addition, we implement the following material constraint to limit the amount of reactant material available over a certain operating period $t_p = 1.0$ h:

$$\frac{1}{t_p}\int_0^{t_p} u_1(\tau)\, d\tau = 0.0 \text{ kmol/m}^3. \quad (3.28)$$

Equation 3.28 ensures that the averaged reactant material C_{A0} available within one EMPC operating period t_p is equal to its steady-state value, $C_{A0s} = 0$. Additionally, we design the following Safeness Index function $S(x)$ for the CSTR:

$$S(x) = \frac{ax_1 + bx_2}{\max\{ax_1 + bx_2 : V(x) \le \rho\}} \quad (3.29)$$

where a and b are weighting constants. It is shown that the Safeness Index $S(x)$ of Eq. 3.29 varies between -1 and 1, where 1 and -1 indicate the most unsafe state and the safest state, respectively, in the stability region Ω_ρ. In the simulation below, we set the weighting constants a and b to be the same (i.e., both are 1) to put more contributions on temperature than on concentration given that the magnitude of temperature (x_2) is several orders higher than that of concentration (x_1). As a result, the maximum value of $S(x)$, i.e., $\max\{ax_1 + bx_2 : V(x) \le \rho\}$ in the stability region is 74.46. Additionally, we choose the Safeness Index threshold value S_{TH}

to be 0.6 such that the reactor temperatures (in deviation form) above 47 K (i.e., $x_2 \leq 47$ K) are considered unsafe. Subsequently, we construct a Lyapunov-based controller of the form $\Phi(x) = [\Phi_1(x)\ \Phi_2(x)]^T$ to characterize the stability region for the Safeness Index-based EMPC. Specifically, $\Phi_1(x)$ is set to its steady-state value ($\Phi_1(x) = 0.0\ \text{kmol/m}^3$) to meet the material constraint of Eq. 3.28 , and the following feedback law (Sontag control law [183]) is utilized for $\Phi_2(x)$:

$$\Phi_2(x) = \begin{cases} -\dfrac{L_{\tilde{f}}V + \sqrt{L_{\tilde{f}}V^2 + L_{g_2}V^4}}{L_{g_2}V}, & \text{if } L_{g_2}V \neq 0 \\ 0, & \text{if } L_{g_2}V = 0 \end{cases} \tag{3.30}$$

where $L_{\tilde{f}}V$ and $L_{g_2}V$ are the Lie derivatives of the Lyapunov function $V(x)$ with respect to the vector fields $\tilde{f}(x)$ and $g_2(x)$, respectively. Additionally, it is noted that the control law of Eq. 3.30 is saturated to meet the input constraint (i.e., $|\Phi_2(x)| \leq 5 \times 10^5\ \frac{\text{kJ}}{\text{h}}$). In other words, any control action value out of the constraint range will be set to the lower/upper bound value. The stability region of the closed-loop system under the LEMPC of Eq. 2.35 is then estimated using extensive closed-loop simulations under the Lyapunov-based controller $\Phi(x)$. In this study, we use a quadratic Lyapunov function of the form $V(x) = x^T P x$ with the positive definite P matrix given as follows:

$$P = \begin{bmatrix} 1060 & 22 \\ 22 & 0.52 \end{bmatrix}.$$

The stability region is then determined under the Lyapunov function with ρ chosen to be 368 and ρ_e chosen to be 340.

To show that the Safeness Index-based EMPC is superior to the standard LEMPC of Eq. 2.35 without the Safeness Index-based constraint in terms of guaranteed process operational safety, we apply both controllers to the CSTR of Eq. 3.24 for comparison. The optimization problems are solved using the interior-point solver Ipopt [201] at each sampling time. The CSTR was initiated in both cases from the steady-state ($x_0^T = [0\ \frac{\text{kmol}}{\text{m}^3}\ 0\ \text{K}]$) where the Safeness Index $S(x)$ equals zero.

Remark 3.3 In this example, $S(x)$ is designed with an upper bound on the sum of reactant concentration and temperature for the CSTR operated in the stability region (i.e., Ω_ρ) since closed-loop stability is guaranteed for states in Ω_ρ under constrained inputs (i.e., $u \in U$). However, it is noted that in general, $S(x)$ can be designed for a much larger region depending on the operating conditions of practical systems, and $S(x)$ can be developed with both a lower bound and an upper bound. Also, if $S(x)$ will be used in alarm and emergency shutdown systems, the functional form of $S(x)$ and the threshold value should be determined accounting for the traditional trigger value used by the safety systems.

Remark 3.4 To practically apply the material constraint of Eq. 3.28 within the finite-horizon optimization problem of EMPC, we calculate the amount of input energy already used in the operating period and compare it with the total amount of input energy available over the entire operating period. Specifically, under the sample-and-hold implementation of EMPC, the material constraint of Eq. 3.28 is equivalent to the following equation:

$$\frac{1}{t_p}\sum_{i=0}^{M-1} u_1(t_i)\Delta = 0.0\ \text{kmol/m}^3 \tag{3.31}$$

where M is the number of sampling periods in the operating period, i.e., $M = \frac{t_p}{\Delta}$. As the operating period t_p is finite, the horizon of predicted inputs used in the material constraint of Eq. 3.31 will be shrunk when approaching the end of operation. Therefore, a general form of material constraints applied in EMPC at the sampling time t_k is given by the following inequalities:

$$\sum_{i=k}^{min\{k+N,\ M-1\}} u(t_i) + \sum_{i=0}^{k-1} u'(t_i|t_i) \geq -\max\{M-N-k, 0\}u_{\max}, \tag{3.32a}$$

$$\sum_{i=k}^{min\{k+N,\ M-1\}} u(t_i) + \sum_{i=0}^{k-1} u'(t_i|t_i) \leq -\max\{M-N-k, 0\}u_{\min} \tag{3.32b}$$

where $u_{\min}$ and $u_{\max}$ denote the lower and upper bounds for the input value, and N is the prediction horizon. $u(\cdot)$ and $u'(\cdot)$ represent the predicted input and the input energy used, respectively. Equation 3.32 ensures that there is enough input energy for the operating period from t_{k+N} to t_M that is not covered in the prediction horizon by letting the difference between the total available input energy and the total input energy used from the beginning of the operating period through the end of the prediction horizon be bounded by the maximum/minimum allowable inputs that can be applied from t_{k+N} to t_M.

3.4.2 Simulation Results

The closed-loop simulation studies are carried out for the CSTR under the Safeness Index-based EMPC of Eq. 3.7 and the LEMPC of Eq. 2.35. Figure 3.3 shows the input trajectories for the closed-loop CSTR under the two controllers in a one-hour operation. It is observed that the material constraint of Eq. 3.28 is met by both controllers under the dynamic operation that optimizes process economics. Additionally, the heat rate u_2 under both control schemes remained at its steady-state value $u_2 = 0\ \frac{\text{kJ}}{\text{h}}$ for the first 0.8 h operation, and started varying near the end of simulation. Figure 3.4 depicts the trajectories of the reactor temperature and reactant concentration in deviation form from their steady-state values, i.e., $[x_1\ x_2] = [C_A - C_{As}\ T - T_s]$.

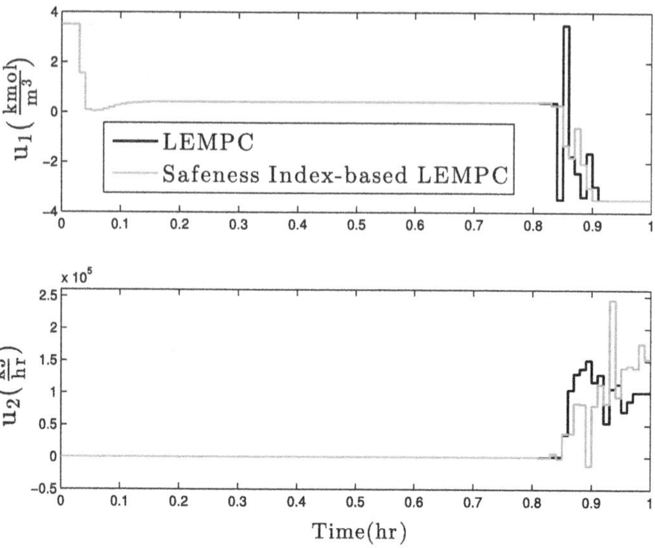

Fig. 3.3 Manipulated input profiles for the closed-loop CSTR under the LEMPC design of Eq. 2.35 and under the Safeness Index-based EMPC design of Eq. 3.7 for the initial condition $x_0^T = [0\ \frac{\text{kmol}}{\text{m}^3}\ 0\ \text{K}]$

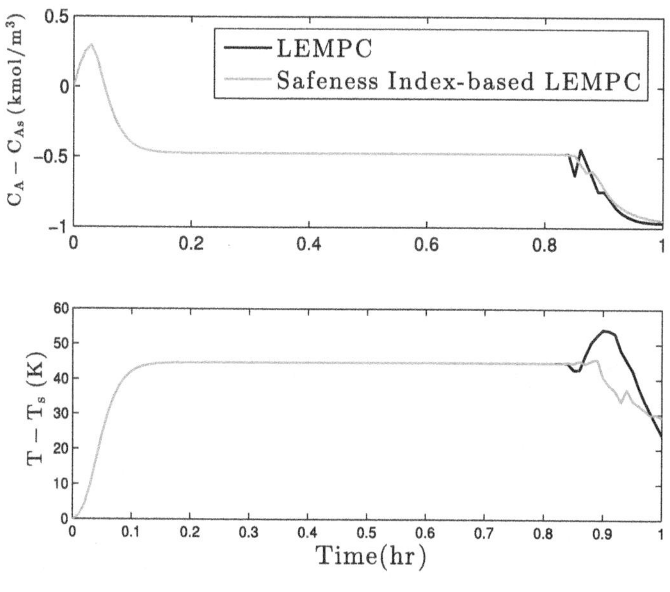

Fig. 3.4 The state profiles for the closed-loop CSTR under the LEMPC design of Eq. 2.35 and under the Safeness Index-based EMPC design of Eq. 3.7 for the initial condition $x_0^T = [0\ \frac{\text{kmol}}{\text{m}^3}\ 0\ \text{K}]$

From Figs. 3.3 and 3.4, it is observed that the closed-loop state and the input trajectories behave similarly under both the Safeness Index-based EMPC scheme and the LEMPC scheme of Eq. 2.35 as they are overlapping before the end of the simulation. This overlap can be explained by the fact that both EMPCs are designed to maximize the production rate of B while maintaining the state in the stability region over the prediction horizon. This is achieved under both EMPCs by maintaining the closed-loop state at $[x_1 \ x_2] = [-0.477 \ \frac{\text{kmol}}{\text{m}^3} \ 44.6 \ \text{K}]$, which is within the safety region (i.e., $S(x) = 0.59 \le 0.6$ where $x = [-0.477 \ \frac{\text{kmol}}{\text{m}^3} \ 44.6 \ \text{K}]$) for much of the period of operation. At the end of the simulation, EMPC ensures that the material constraint of Eq. 3.28 is met before the end of the operating period. When the safety constraint on $S(x)$ is not utilized, the control actions computed by the LEMPC of Eq. 2.35 maximize the process economics but drive the closed-loop state out of the safety region. However, under the Safeness Index-based EMPC, a different trajectory was obtained at the end of the prediction horizon to meet the material constraint and also maximize the process economics while satisfying the constraint that the closed-loop state remains inside the safety region. Specifically, it is shown in Fig. 3.4 that the reactor temperature profile for the CSTR under the LEMPC scheme exceeds the maximum allowable temperature (i.e., the threshold of $S(x)$), while the Safeness Index-based EMPC decreases the temperature at the end to meet the Safeness Index-based constraints.

Additionally, Fig. 3.5 shows the Safeness Index value $S(x)$ for the Safeness Index-based EMPC and the LEMPC of Eq. 2.35 over the operating window, from which it is clearly seen that the $S(x)$ under LEMPC causes S_{TH} to exceed its threshold near the end of the operating window. Figure 3.6 shows the state-space trajectories of the reactant concentration and reactor temperature (i.e., $[x_1 \ x_2] = [C_A - C_{As} \ T - T_s]$), from which it is demonstrated that the closed-loop trajectory under the LEMPC of Eq. 2.35 leaves the safety region (shaded gray), whereas the closed-loop state under the Safeness Index-based EMPC remains in the safety region at all times.

To demonstrate the robustness of the Safeness Index-based EMPC of Eq. 3.7, a bounded disturbance vector $w^T = [w_1 \ w_2]$ was added to the right-hand side of Eq. 3.24. The bounded disturbance vector $w^T = [w_1 \ w_2]$ corresponds to Gaussian white noise with variances $\sigma_1 = 1 \ \frac{\text{kmol}}{\text{m}^3}$ and $\sigma_2 = 40 \ K$ with $|w_1| \le 1 \ \frac{\text{kmol}}{\text{m}^3}$ and $|w_2| \le 40$ K. The closed-loop simulation results are shown in Figs. 3.7 and 3.8, from which it is concluded that the Safeness Index-based EMPC guarantees process operational safety even in the presence of uncertainty.

Remark 3.5 As shown in Fig. 3.6, the closed-loop state under standard LEMPC leaves the safety region for a short time during the LEMPC dynamic operation. While this is considered unsafe in this example, in general it is acceptable to allow a short excursion of the state into the unsafe region ($S(x) > S_{TH}$) in order to improve economic benefits by designing S_{TH} to be a soft threshold (see Remark 3.2). Since the closed-loop state does not leave the stability region Ω_ρ at all times, the short excursion into the unsafe region does not jeopardize closed-loop stability, which implies there always exists a feasible control action that can re-stabilize the system at the steady-state.

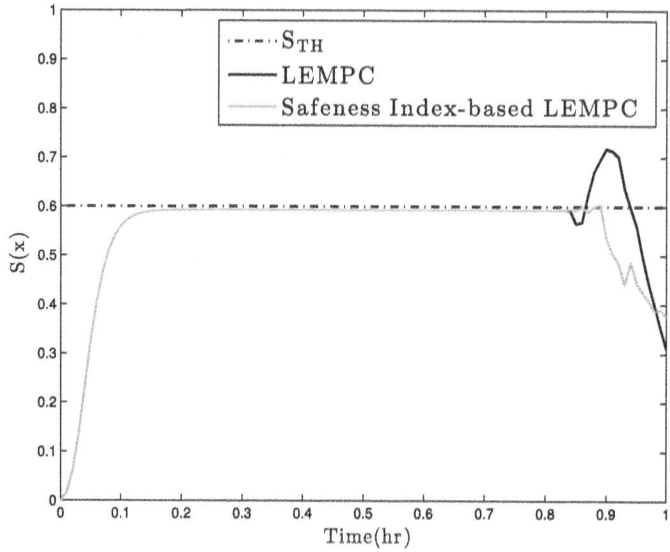

Fig. 3.5 The Safeness Index function $S(x)$ for the closed-loop CSTR under the LEMPC design of Eq. 2.35 and under the Safeness Index-based EMPC design of Eq. 3.7 for the initial condition $x_0^T = [0 \ \frac{\text{kmol}}{\text{m}^3} \ 0 \text{ K}]$

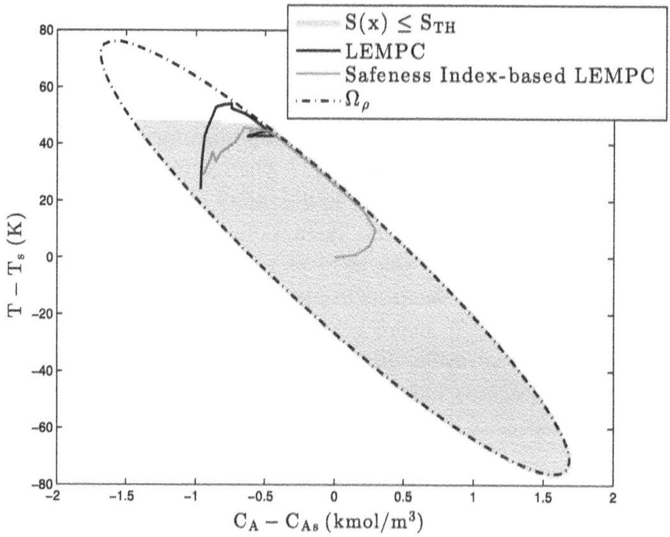

Fig. 3.6 The state-space profile for the closed-loop CSTR under the LEMPC design of Eq. 2.35 (black trajectory) and under the Safeness Index-based EMPC design of Eq. 3.7 (dark gray trajectory) for the initial condition $x_0^T = [0 \ \frac{\text{kmol}}{\text{m}^3} \ 0 \text{ K}]$

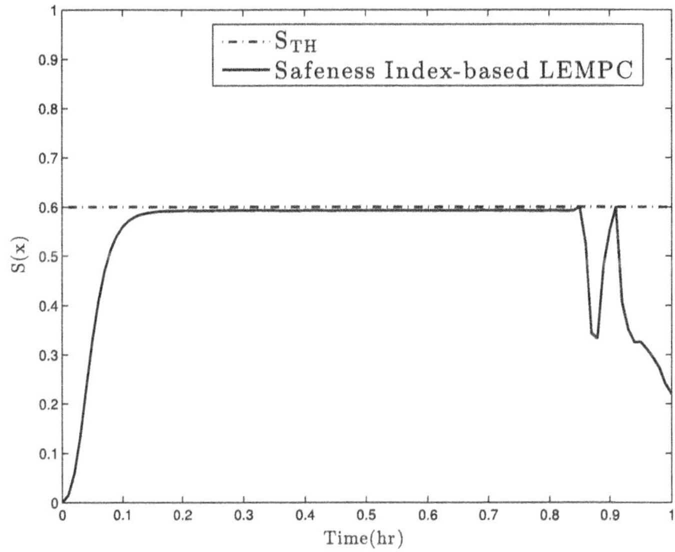

Fig. 3.7 The Safeness Index function $S(x)$ for the closed-loop CSTR under the Safeness Index-based EMPC design of Eq. 3.7 for the initial condition $x_0^T = [0 \ \frac{\text{kmol}}{\text{m}^3} \ 0 \ \text{K}]$ with bounded process disturbances

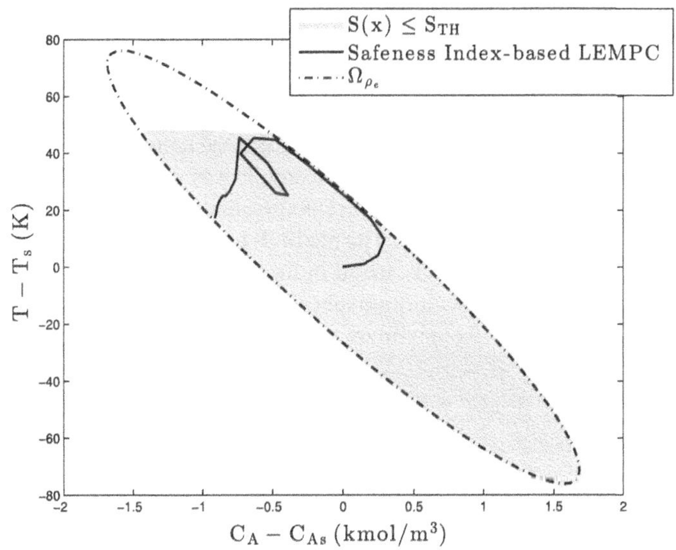

Fig. 3.8 The state-space profile for the closed-loop CSTR under the Safeness Index-based EMPC design of Eq. 3.7 for the initial condition $x_0^T = [0 \ \frac{\text{kmol}}{\text{m}^3} \ 0 \ \text{K}]$ with bounded process disturbances

Remark 3.6 It is noted that the Safeness Index-based EMPC of Eq. 3.7 is implemented following the strategy in Steps 1–4 to address feasibility issues resulting from the lack of the property of forward invariance of the safety region $S(x) \leq S_{TH}$. Another approach that ensures operational safety and feasibility simultaneously is to operate the system in a safety level set (i.e., a smaller level set of the Lyapunov function, $\Omega_{\bar{\rho}} = \{S(x) \leq S_{TH} \mid V(x) \leq \bar{\rho},\ 0 < \bar{\rho} < \rho\}$) within the safety region (the gray region in Fig. 3.6). In this way, closed-loop stability and operational safety are naturally guaranteed for any initial condition $x_0 \in \Omega_{\bar{\rho}}$, and the optimization problem of Safeness Index-based EMPC is guaranteed to be feasible for all times since $\Omega_{\bar{\rho}}$ is a forward invariant set.

3.5 Conclusions

In this chapter, a Safeness Index was developed to coordinate the safety and control systems to ensure process operational safety in nonlinear chemical processes. Specifically, we presented a general approach to designing the functional form of the Safeness Index $S(x)$, and discussed the method for choosing the threshold S_{TH}. Subsequently, we incorporated the Safeness Index-based constraints within MPC and EMPC schemes to integrate process safety, feedback control, and process economics (for EMPC) within a unified framework. To address the feasibility issue of Safeness Index-based MPC, we presented an implementation strategy that can drive the closed-loop state into the safe operating region characterized by the Safeness Index function, along with rigorous analyses for closed-loop stability, process operational safety, and recursive feasibility of nonlinear systems under both MPC and EMPC schemes. Finally, a chemical process example was utilized to demonstrate the guaranteed operational safety, closed-loop stability, and economic optimality of the Safeness Index-based EMPC scheme. The methods for designing the Safeness Index $S(x)$ and its threshold were also discussed in the chemical reactor example, where the reactor temperature that has a dominant effect on the safeness of the process should be maintained below its maximum allowable value at all times.

Chapter 4
Operational Safety Via Control Lyapunov-Barrier Function-Based MPC

4.1 Introduction

As discussed in the previous chapter, maintaining a safe and stable operation is the highest priority of the control systems in many safety-critical processes in chemical industries. Based on the MPC/EMPC schemes that optimize process/system performance (e.g., process economics and energy consumption) while maintaining the closed-loop state trajectory in a well-defined state-space region, Safeness Index-based MPC/EMPC schemes were proposed in the previous chapter to achieve both a safe and stable operation by forcing the closed-loop state to remain in a safe operating region. However, as the Safeness Index function is used as a hard constraint in MPC, a shortcoming of Safeness Index-based MPC/EMPC is that recursive feasibility of MPC solutions may not be guaranteed as the process state may exit the Safeness Index defined set, which could lead to unsafe operation when a feasible solution does not exist.

To address this issue, in this chapter, novel MPC and EMPC designs that take advantage of barrier functions and Lyapunov functions to ensure simultaneous closed-loop stability and process operational safety as well as recursive feasibility are developed. Specifically, a barrier function that is commonly used to enforce safety properties in the context of optimization-based safety-critical controllers is first introduced. Subsequently, a new function termed *control Lyapunov-barrier function (CLBF)* is designed by combining a control barrier function with a control Lyapunov function through weighted sum, for which a rigorous stability and safety analysis is presented. Based on the CLBF-based controller that guarantees simultaneous stability and safety of nonlinear systems, CLBF-based MPC and EMPC are developed and applied to chemical process examples to demonstrate, evaluate, and analyze the closed-loop stability and safety properties of nonlinear systems.

Z. Wu and P. D. Christofides, *Process Operational Safety and Cybersecurity*,
Advances in Industrial Control, https://doi.org/10.1007/978-3-030-71183-2_4

4.1.1 Class of Nonlinear Systems

The following system of nonlinear first-order ordinary differential equations is used to represent the class of nonlinear systems that we consider:

$$\dot{x} = f(x) + g(x)u + h(x)w, \; x(t_0) = x_0 \tag{4.1}$$

where $x \in D \subset \mathbf{R}^n$ and $u \in U \subset \mathbf{R}^m$ are the state vector and the manipulated input vector, respectively. $w \in W$ is the disturbance vector bounded by $W := \{w \in \mathbf{R}^l \mid |w| \leq \theta, \; \theta \geq 0\}$. The control action constraint is defined by $u \in U := \{u_{\min} \leq u \leq u_{\max}\} \subset \mathbf{R}^m$, where $u_{\min}$ and $u_{\max}$ are the lower and upper bounds for the input vector, respectively. It is assumed that the functions $f(\cdot)$, $g(\cdot)$, and $h(\cdot)$ of dimensions $n \times 1$, $n \times m$, and $n \times l$, respectively, are sufficiently smooth. Additionally, it is assumed that $f(0) = 0$ such that the origin is a steady-state of the nominal system of Eq. 4.1 with $w(t) \equiv 0$. The measurement of $x(t)$ is assumed to be available for feedback at each sampling time $t_k = t_0 + k\Delta$, $k = 0, 1, \ldots$, where Δ is the sampling period. It is noted that we consider the control-affine nonlinear system in the form of Eq. 4.1 to simplify the discussion of explicit stabilizing controller design; however, the control Lyapunov-barrier function design and its incorporation in MPC/EMPC in this chapter are not restricted to systems of Eq. 4.1, and can be generalized to the class of a continuous-time nonlinear system in a general form:

$$\dot{x} = f(x, u, w) \tag{4.2}$$

where $f : \mathbf{R}^n \times \mathbf{R}^m \times \mathbf{R}^l \rightarrow \mathbf{R}^n$ is a smooth vector function of its arguments with $f(0, 0, 0) = 0$.

4.1.2 Characterization of Unsafe Regions

We assume that there exists a set of unsafe states $\mathscr{D}$ in state-space within which it is unsafe to operate the system, and a safe operating region $\mathscr{U}$ that has no intersection with $\mathscr{D}$, i.e., $\mathscr{U} \cap \mathscr{D} = \emptyset$. The definition of process operational safety for the closed-loop system of Eq. 4.1 is restated here for convenience.

Definition 4.1 (*Definition* 3.1) Consider the nominal system of Eq. 3.1 (i.e., $w(t) \equiv 0$) with input constraints $u \in U$. For any initial state $x(t_0) = x_0 \in \mathscr{U}$, if there exist control actions $u \in U$ that can render the process states within $\mathscr{U}$ for all times, i.e., $x(t) \in \mathscr{U}, \forall \, t \geq t_0$, then we say that process operational safety is achieved for the nominal system of Eq. 3.1 under the control law $\Phi(x)$.

To ensure process operational safety, we should first characterize the unsafe region $\mathscr{D}$ through the analysis of process safeness based on past operating data and first-principles models. For example, the Safeness Index $S(x)$ proposed in Chap. 3 provides a solution to indicating the safeness of a process based on current state measurements. $S(x)$ can be developed either from first-principles knowledge of process safety or from an extensive review of past chemical accidents and their causes. As discussed in Chap. 3, $S(x)$ is typically developed as a function of (closed-loop) process states accounting for the interactions among different variables and process units. Then, the safe and unsafe regions are characterized by the threshold S_{TH} (i.e., $S(x) \leq S_{TH}$ is the set of safe states, and $S(x) > S_{TH}$ is the set of unsafe states).

We consider two common types of unsafe regions in this chapter: (1) unbounded sets and (2) bounded sets. Unbounded sets are often encountered in chemical plants, where, for example, there exists an unsafe region consisting of all the operating conditions (e.g., reactor temperature or pressure) above a threshold that is considered unsafe. Bounded sets are often characterized for a multivariable system where the interaction among different variables plays a role in determining whether the system is safe or not. For example, in a chemical reactor example, such a bounded set of unsafe states can be characterized based on the combination of temperature and concentration of reactants that reflect reaction rates. Additionally, bounded unsafe sets often occur in motion planning for self-driving cars and robots that attempt to address obstacle avoiding problems, which can be found, for example, in [126]. Throughout this chapter, we will discuss both unbounded unsafe regions (denoted by $\mathscr{D}_u$) and bounded unsafe regions (denoted by $\mathscr{D}_b$), and prove closed-loop stability and process operational safety for the nonlinear system of Eq. 4.1 under CLBF-based controllers.

4.2 Control Barrier Function

Consider the unforced nonlinear systems described by the following system of first-order nonlinear ordinary differential equations (ODEs):

$$\dot{x} = f(x) \tag{4.3}$$

where $x \in D \subset \mathbf{R}^n$, and $f : D \to \mathbf{R}^n$ is a smooth function of x. Barrier certificates were proposed in [157, 158] to ensure safety for the nonlinear system of Eq. 4.3 in the sense that the system is able to avoid undesirable regions. Since then, they have been successfully applied to solve safety-critical control problems for cyber-physical systems, for example, obstacle avoidance problems for autonomous vehicles and collisions-free multi-robot systems [47, 205]. Specifically, given a safe operating region $\mathscr{U}$ in the state-space, there are two types of barrier certificates/functions $B(x)$ that are commonly used: one is the reciprocal barrier function that satisfies $B(x) \to \infty$ as $x \to \partial\mathscr{U}$, where $\partial\mathscr{U}$ represents the boundary of $\mathscr{U}$, and the other one is termed zeroing barrier function where $B(x) \to 0$ as $x \to \partial\mathscr{U}$. To ensure process safety for the system of Eq. 4.3, we show that the safe operating region $\mathscr{U}$

defined as a superlevel set[1] of a $\mathscr{C}^1$ function $r : D \to \mathbf{R}$ that satisfies the following conditions [16, 17]:

$$\mathscr{U} = \{x \in D \subset \mathbf{R}^n \mid r(x) \geq 0\}, \tag{4.4a}$$
$$\partial\mathscr{U} = \{x \in D \subset \mathbf{R}^n \mid r(x) = 0\}, \tag{4.4b}$$
$$Int(\mathscr{U}) = \{x \in D \subset \mathbf{R}^n \mid r(x) > 0\} \tag{4.4c}$$

is an invariant set, where Int($\mathscr{U}$) represents the interior of the set $\mathscr{U}$. Specifically, we design a reciprocal barrier function as $B(x) = -log(\frac{r(x)}{1+r(x)})$ and impose a condition on the time-derivative of $B(x)$: $\dot{B} \leq \frac{\gamma}{B}$, where $\gamma > 0$. It is readily shown that the following conditions hold for $B(x)$:

$$\inf_{x \in Int(\mathscr{U})} B(x) > 0, \quad \lim_{x \to \partial\mathscr{U}} B(x) = \infty. \tag{4.5}$$

Additionally, we derive the following inequality for $\dot{r}$ by differentiating $B(x)$ and using the constraint $\dot{B} \leq \frac{\gamma}{B}$:

$$\dot{r} \geq \frac{\gamma(h + h^2)}{log(\frac{h}{1+h})}. \tag{4.6}$$

Using the Comparison Lemma [90], it is demonstrated that for any $r(x_0) > 0$, $r(x(t)) > 0$ holds for $t \geq 0$. Therefore, for any initial condition $x_0 \in \mathscr{U}$, the state remains inside $\mathscr{U}$ for all $t \geq 0$ (see [17] for the detailed proof).

Inspired by control Lyapunov functions (CLF) that were proposed for the nonlinear system with control inputs (e.g., Eq. 4.1) based on the Lyapunov function for the unforced system of Eq. 4.3 (see Sect. 2.3.1), barrier function was also extended to control barrier function (CBF) for the nonlinear affine control system of Eq. 4.1 with $w(t) \equiv 0$ in [210]. The definition of a CBF in [210] is presented below. For a comprehensive review on CBFs, the reader is referred to the review [15].

Definition 4.2 Given a set of unsafe states in state-space $\mathscr{D}$, a $\mathscr{C}^1$ function $B(x) : \mathbf{R}^n \to \mathbf{R}$ is a control barrier function if the following properties are satisfied:

$$B(x) > 0, \quad \forall\, x \in \mathscr{D} \tag{4.7a}$$
$$L_f B(x) \leq 0, \ \forall\, x \in \{z \in \mathbf{R}^n \backslash \mathscr{D} \mid L_g B(z) = 0\} \tag{4.7b}$$
$$\mathscr{U} := \{x \in \mathbf{R}^n \mid B(x) \leq 0\} \neq \emptyset. \tag{4.7c}$$

Additionally, a number of recent works, e.g., [170, 210, 228] have demonstrated a control law that guarantees that the safe operation of the process at all times can be found if a CBF can be found for the system. The following theorem provides sufficient conditions under which the existence of a CBF of Eq. 4.7 for the nominal system of Eq. 4.1 (i.e., $w(t) \equiv 0$) under the control law $u = \Phi_b(x)$ of Eq. 4.8 guarantees process operational safety of the closed-loop system for any initial condition $x_0 \in \mathscr{U}$.

[1] $\{x \in \mathbf{R}^n \mid f(x) \geq c\}$ is called a superlevel set of $f : \mathbf{R}^n \to \mathbf{R}$, where c is a constant.

Theorem 4.1 *Assume that the nominal system of Eq. 4.1 (i.e., $w(t) \equiv 0$) with no constraints on the control input u has a $\mathscr{C}^1$ CBF $B(x) : \mathbf{R}^n \to \mathbf{R}$ associated with an unsafe region $\mathscr{D}$ in state-space. The control law of Eq. 4.8 guarantees that the closed-loop state is bounded in the safe region $\mathscr{U}$ for all times if the initial condition x_0 is in $\mathscr{U}$.*

$$\Phi_b(x) = \begin{cases} -\dfrac{p + \sqrt{p^2 + \gamma |q|^4}}{|q|^2} q_i & \text{if } q \neq 0 \\ 0 & \text{if } q = 0 \end{cases} \tag{4.8}$$

where p denotes $L_f B(x)$, q_i denotes $L_{g_i} B(x)$, $q = [q_1 \cdots q_m]^T$, $f = [f_1 \cdots f_n]^T$, $g_i = [g_{i1} \cdots g_{in}]^T$, $(i = 1, 2, \ldots, m)$, and $\gamma > 0$.

Proof By substituting the control law $u = \Phi_b(x)$ into the closed-loop system of Eq. 4.1, we can derive the following equation:

$$\dot{B}(x) = \frac{\partial B}{\partial x}(f(x) + g(x)\Phi_b(x)) = \begin{cases} -\sqrt{p^2 + \gamma |q|^4} & \text{if } q \neq 0 \\ p & \text{if } q = 0. \end{cases} \tag{4.9}$$

Since the CBF $B(x)$ satisfies Eq. 4.7b showing that $p \leq 0$ holds for all $x \in \mathbf{R}^n \backslash \mathscr{D}$ when $q = 0$, $\dot{B}(x)$ in Eq. 4.9 is guaranteed to be nonpositive for all $x \in \mathbf{R}^n \backslash \mathscr{D}$. Therefore, if the state starts from $\mathscr{U} \subset (\mathbf{R}^n \backslash \mathscr{D})$, the value of $B(x)$ is guaranteed to be non-increasing along the trajectory of x. This completes the proof that the safe operating region $\mathscr{U}$ is an invariant set under $u = \Phi_b(x)$.

4.3 Control Lyapunov-Barrier Function

To address simultaneously the tasks of safety, stability, and other considerations such as process economic optimality, control barrier functions (CBF) and control Lyapunov functions (CLF) are utilized to design process control systems. Specifically, the controller developed based on CBFs that satisfy Lyapunov-like conditions can guarantee the boundedness of the state in a safe operating region (i.e., process operational safety) [139, 188, 210]. To further ensure closed-loop stability, CLFs can be naturally unified with CBFs to formulate a quadratic program, which allows for the satisfaction of the control objectives of safety and stability (see, for example, [16, 17, 85]). Additionally, another approach to solving the problem of stabilization of a nonlinear process with guaranteed safety is to use control Lyapunov-barrier functions (CLBF), which are functions that combine CBFs and CLFs via weighted sum.

In this section, we introduce the definition of CLBFs for an input-constrained system of Eq. 4.1, followed by the construction method that separates the control design for achieving the asymptotic stability and safety by designing the CLF and CBF, independently, and then combines them together [170]. The design of the CLBF-based controller for the nonlinear system of Eq. 4.1 will be discussed with a rigorous theoretical analysis on closed-loop stability and operational safety.

4.3.1 Stabilization and Safety via Control Lyapunov-Barrier Function

4.3.1.1 Stabilizability Assumptions

Assumption 4.1 We assume that there exists a positive definite and proper CLF V that satisfies the following condition for the nominal system of Eq. 4.1 with $w(t) \equiv 0$:

$$L_f V(x) < 0, \forall x \in \{z \in \mathbf{R}^n \backslash \{0\} \mid L_g V(z) = 0\}. \tag{4.10}$$

We also assume that V satisfies the small control property, i.e., for every $\varepsilon > 0, \exists\, \delta > 0$, s.t. $\forall\, x \in \mathscr{B}_\delta(0)$, there exists u that satisfies $|u| < \varepsilon$ and $L_f V(x) + L_g V(x)u < 0$.

The assumption that a CLF exists for the nominal system of Eq. 4.1 implies that there also exists a stabilizing feedback control law $\Phi(x)$ that can render the origin of the nominal system of Eq. 4.1 asymptotically stable. The following controller is an example of the feedback control law that renders the origin asymptotically stable and is continuous for all x in a neighborhood of the origin [110]:

$$k_i(x) = \begin{cases} -\dfrac{p + \sqrt{p^2 + \gamma |q|^4}}{|q|^2} q_i & \text{if } q \neq 0 \\ 0 & \text{if } q = 0 \end{cases} \tag{4.11a}$$

$$\Phi_i(x) = \begin{cases} u_{\min} & \text{if } k_i(x) < u_{\min} \\ k_i(x) & \text{if } u_{\min} \leq k_i(x) \leq u_{\max} \\ u_{\max} & \text{if } k_i(x) > u_{\max} \end{cases} \tag{4.11b}$$

where p denotes $L_f V(x)$, q_i, $i = 1, 2, \ldots, m$, denotes $L_{g_i} V(x)$, $q = [q_1 \cdots q_m]^T$, $f = [f_1 \cdots f_n]^T$, $g_i = [g_{i1} \cdots g_{in}]^T$, and $\gamma > 0$. Based on the Sontag control law in [110], the controller $k_i(x)$ of Eq. 4.11a is developed to represent the i_{th} component of the control law $\Phi(x)$ without considering any constraints on the control action. To further account for the input constraint $u \in U$, $\Phi_i(x)$ of Eq. 4.11b is developed to represent the i_{th} component of the saturated control law $\Phi(x)$.

4.3.1.2 Stabilization and Safety via CLBF

Control Lyapunov-barrier function (CLBF) was originally proposed in [170], where the stabilization and safety-related results were established for $u \in \mathbf{R}^m$ only (i.e., no constraints on control actions). Considering the fact that practical nonlinear systems are often subject to input constraints $u \in U$ as assumed in Eq. 4.1, we propose a modified CLBF (termed constrained CLBF or simply CLBF in this chapter) based on the original definition of a CLBF in [170] to account for the input constraints $u \in U$ in the nonlinear system of Eq. 4.1. The definition of a constrained CLBF is given as follows.

Definition 4.3 Consider the nominal system of Eq. 4.1 (i.e., $w(t) \equiv 0$) with a set of unsafe states in state-space (i.e., $\mathscr{D}$); a proper, lower-bounded, and $\mathscr{C}^1$ function $W_c(x) : \mathbf{R}^n \to \mathbf{R}$ is a constrained CLBF if $W_c(x)$ has a minimum at the origin and also satisfies the following properties:

$$
\begin{aligned}
& W_c(x) > \rho_c, \ \ \forall\, x \in \mathscr{D} \subset \phi_{uc} && (4.12a)\\
& L_f W_c(x) < 0, \ \ \forall\, x \in \{z \in \phi_{uc} \backslash (\mathscr{D} \cup \{0\} \cup \mathbb{X}_e) \mid L_g W_c(z) = 0\} && (4.12b)\\
& \mathscr{U}_{\rho_c} := \{x \in \phi_{uc} \mid W_c(x) \le \rho_c\} \neq \emptyset && (4.12c)\\
& \overline{\phi_{uc} \backslash (\mathscr{D} \cup \mathscr{U}_{\rho_c})} \cap \overline{\mathscr{D}} = \emptyset && (4.12d)
\end{aligned}
$$

where $\rho_c \in \mathbf{R}$, ϕ_{uc} is a neighborhood around the origin, and $\mathbb{X}_e := \{x \in \phi_{uc} \backslash (\mathscr{D} \cup \{0\}) \mid \frac{\partial W_c(x)}{\partial x} = 0\}$ is a set of states where $L_f W_c(x) = 0$ (for $x \neq 0$) due to $\partial W_c(x)/\partial x = 0$.

A feedback control law $u = \Phi(x) \in U$ that maintains the closed-loop state of the nominal system of Eq. 4.1 in a level set of $W_c(x)$ in an open neighborhood D_0 around the origin is assumed to exist in the sense that there exists a $\mathscr{C}^1$ constrained CLBF $W_c(x)$ that has a minimum at the origin and satisfies the following inequalities for all $x \in D_0$:

$$\alpha_1(|x|) \le W_c(x) - \rho_0 \le \alpha_2(|x|), \tag{4.13a}$$

$$
\begin{aligned}
& \frac{\partial W_c(x)}{\partial x} F(x, \Phi(x), 0) \le -\alpha_3(|x|), \forall x \in D_0 \backslash \mathscr{B}_\delta(x_e)\\
& \frac{\partial W_c(x)}{\partial x} F(x, \Phi(x), 0) \le 0, \forall x \in \mathscr{B}_\delta(x_e)
\end{aligned}
\tag{4.13b}
$$

$$\left| \frac{\partial W_c(x)}{\partial x} \right| \le \alpha_4(|x|) \tag{4.13c}$$

where $\alpha_j(\cdot)$, $j = 1, 2, 3, 4$ are class $\mathscr{K}$ functions, $W_c(0) = \rho_0$ is the global minimum value of $W_c(x)$ in D_0, and $\mathscr{B}_\delta(x_e)$ is a small neighborhood around the stationary points $x_e \in \mathbb{X}_e$. $F(x, u, w) := f(x) + g(x)u + h(x)w$ is the nonlinear system of Eq. 4.1. It is noted that in Eq. 4.13b, $\frac{\partial W_c(x)}{\partial x} F(x, \Phi(x), 0) \le -\alpha_3(|x|)$ does not hold for all $x \in \mathscr{B}_\delta(x_e)$ since $\frac{\partial W_c(x)}{\partial x}$ is close to zero in the neighborhood around x_e, where $\frac{\partial W_c(x_e)}{\partial x} = 0$ at $x = x_e$. Such a CLBF-based feedback control law can be developed following the universal Sontag control law of Eq. 4.11 with $W_c(x)$ replacing $V(x)$. Additionally, based on the continuity and the smoothness properties assumed for the functions f, g, and h in the nonlinear system of Eq. 4.1, there exist positive constants L_x, L_w, M, $L_x^{'}$, $L_w^{'}$ such that the following inequalities hold for all $x, x' \in \mathscr{U}_{\rho_c} \subset D_0$, $u \in U$, and $w \in W$:

$$|F(x, u, w)| \leq M \tag{4.14a}$$

$$|F(x, u, w) - F(x', u, 0)| \leq L_x|x - x'| + L_w|w| \tag{4.14b}$$

$$\left|\frac{\partial W_c(x)}{\partial x}F(x, u, w) - \frac{\partial W_c(x')}{\partial x}F(x', u, 0)\right| \leq L_x'|x - x'| + L_w'|w|. \tag{4.14c}$$

Based on the stabilizability and safety requirements of Eq. 4.13, a positive real number a can be found such that $\{x \in D_0 \mid \dot{W}_c(x(t)) = L_f W_c + L_g W_c u < -a|W_c(x) - W_c(0)|\}$ is not an empty set under the stabilizing control law $u = \Phi(x) \in U$ (e.g., the Sontag control law of Eq. 4.11 with $W_c(x)$ replacing $V(x)$). Therefore, ϕ_{uc} is defined to be the union of the above set, the origin, and $\mathscr{B}_\delta(x_e)$, i.e., $\phi_{uc} := \{x \in D_0 \mid \dot{W}_c(x(t)) < -a|W_c(x) - W_c(0)|, u = \Phi(x) \in U\} \cup \{0\} \cup \mathscr{B}_\delta(x_e)$. Additionally, we define the set of initial conditions by $\mathbb{X}_0 := \{x \in \phi_{uc} \backslash \mathscr{D}\}$ where $(\{0\} \cup \mathbb{X}_e) \in \mathbb{X}_0$, and thus, it is readily shown that the set $\mathscr{U}_{\rho_c}$ defined by Eq. 4.12c is a subset of $\mathbb{X}_0$. From now on, $\dot{W}_c(x(t))$ will be simply denoted by $\dot{W}_c$ unless stated otherwise.

4.3.1.3 Closed-Loop Stability and Safety Under CLBF-Based Controller

Closed-loop stability and safety are analyzed for the following two cases: a bounded unsafe region $\mathscr{D}_b$ and an unbounded unsafe region $\mathscr{D}_u$ in state-space. The definition of simultaneous operational safety and closed-loop stability for the nonlinear system of Eq. 4.1 is restated here for convenience.

Definition 4.4 (*Definition* 3.2) Consider the nominal system of Eq. 4.1 (i.e., $w(t) \equiv 0$) with input constraints $u \in U$. If for any initial state $x(t_0) = x_0 \in \mathscr{U}$, there exists a control action $u \in U$ such that the state trajectories of the closed-loop system satisfy $x(t) \in \mathscr{U}$, $\forall\, t \geq t_0$, and $\lim_{t\to\infty} |x(t)| \leq d$, where $\mathscr{B}_d(0)$ is a small neighborhood around the origin, then we say that operational safety and closed-loop stability are achieved simultaneously in the sense that the process state is maintained within a safe operating region at all times, and can be ultimately driven to the origin.

Case 1: Consider the nominal system of Eq. 4.1 with an unsafe region characterized as a bounded set $\mathscr{D}_b$ in state-space. It is pointed out in [37] that the continuous control law $u = \Phi(x) \in U$ cannot render the origin asymptotically stable because there exist stationary points other than the origin in state-space (i.e., $x_e \in \mathbb{X}_e$ and $x_e \neq 0$). In other words, for some $x_0 \in \mathbb{X}_0$, the closed-loop state may be trapped in the stationary points x_e (x_e can be either saddle points or local minima of $W_c(x)$) instead of the origin which has the global minimum of $W_c(x)$ under $u = \Phi(x)$. For example, Fig. 4.1 shows that there may exist initial states $x_0 \in \mathscr{U}_{\rho_c} \subset \phi_{uc}$, under which the state will first evolve toward x_e, and a discontinuous control action is needed at x_e to drive the state around the unsafe region $\mathscr{D}_b$ in all possible directions. Additionally, in order for the state to escape from the stationary points x_e and converge to the origin, the shapes and functional forms of $W_c(x)$ need to be carefully

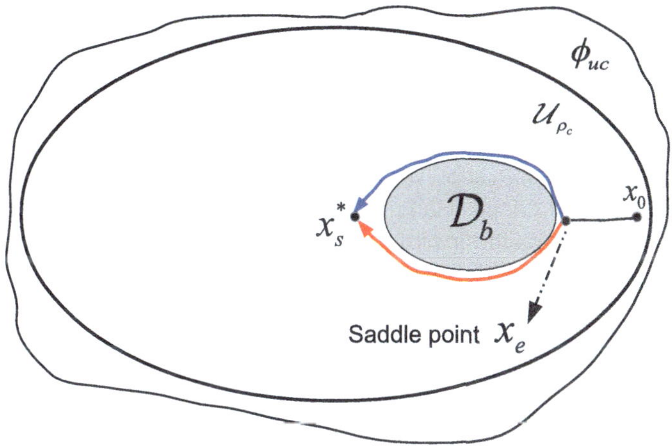

Fig. 4.1 A schematic showing an initial condition x_0 from which the state trajectory converges to x_e and passes around a bounded unsafe set $\mathscr{D}_b$ embedded within the operating region either in the up or down direction using a discontinuous control action

designed such that x_e is a saddle point rather than a local minimum in state-space. After we design the functional form of $W_c(x)$ and find all the stationary points x_e in state-space, we will design a set of control actions $\bar{u}$ that can drive the state away from the saddle point in the direction of decreasing $W_c(x)$ in advance. This set of control actions will then be applied in closed-loop simulation when the state approaches x_e.

Theorem 4.2 below provides sufficient conditions for guaranteeing process operational safety when there exists a constrained CLBF of Eq. 4.12 for the nominal system of Eq. 4.1 (i.e., $w(t) \equiv 0$) under the control law $\Phi(x)$. The proof of the theorem follows from the results in [213, 217].

Theorem 4.2 *Consider the nominal system of Eq. 4.1 (i.e., $w(t) \equiv 0$) with a constrained CLBF $W_c(x) : \mathbf{R}^n \rightarrow \mathbf{R}$ that has a minimum at the origin and is designed with respect to a bounded unsafe region $\mathscr{D}_b$ in state-space. It is guaranteed that the closed-loop state stays in $\mathbb{X}_0$ and does not enter $\mathscr{D}_b$ for all times for $x(0) = x_0 \in \mathbb{X}_0$ under the feedback control law $u = \Phi(x) \in U$.*

Proof First, we prove that if $x_0 \in \mathbb{X}_0$ where $\mathbb{X}_0 := \{x \in \phi_{uc} \backslash \mathscr{D}_b\}$, then the closed-loop state will remain outside of the unsafe region $\mathscr{D}_b$, for all $t \geq 0$. We first consider the initial condition in $\mathscr{U}_{\rho_c}$, i.e., $x_0 \in \mathscr{U}_{\rho_c} \subset \mathbb{X}_0$. By the definition of ϕ_{uc}, $\dot{W}_c$ is guaranteed to be negative everywhere in the set $\mathbb{X}_0 \backslash (\{0\} \cup \mathbb{X}_e)$. Specifically, if $L_g W_c(x) \neq 0$, it follows that $\dot{W}_c = -\sqrt{L_f W_c^2 + \gamma |L_g W_c|^4} < 0$ using the Sontag control law of Eq. 4.11 with $W_c(x)$ replacing $V(x)$; if $L_g W_c(x) = 0$, $\dot{W}_c(x) = L_f W_c(x) < 0$ holds based on the definition of $W_c(x)$. Additionally, if $x \in \mathbb{X}_e$, $\dot{W}_c(x) = 0$ holds. Therefore, we can show that $x(t)$ stays in the set $\mathscr{U}_{\rho_c}$ for all $t \geq 0$ if $x_0 \in \mathscr{U}_{\rho_c}$ since $W_c(x(t)) \leq W_c(x(0))$ holds for all $x(t) \in \mathscr{U}_{\rho_c}$ (i.e., $\dot{W}_c \leq 0$).

Also, owing to the property $\dot{W}_c \leq 0$ and the properness of W_c, $\mathscr{U}_{\rho_c}$ is a compact invariant set. Since $\mathscr{U}_{\rho_c} \cap \mathscr{D}_b = \emptyset$, it follows that for any $x_0 \in \mathscr{U}_{\rho_c}$, the closed-loop state does not enter the set of unsafe states at any time (i.e., it is maintained within the set of safe states at all times). Additionally, since any subset of $\mathcal{U}_{\rho_c}$, $\mathcal{U}_\rho := \{x \in \phi_{uc} \mid W_c(x) \leq \rho\}$ where $\rho \leq \rho_c$, is also a compact invariant set, we can show that if $x_0 \in \mathcal{U}_\rho$, it holds that $x(t) \in \mathcal{U}_\rho$, $\forall t \geq 0$. It remains to be shown that $x(t) \notin \mathcal{D}_b, \forall\, t \geq 0$ holds for all other initial states $x_0 \in \overline{\phi_{uc} \backslash (\mathcal{D}_b \cup \mathcal{U}_{\rho_c})}$. Given an initial state x_0 that is in the set $\overline{\phi_{uc} \backslash (\mathcal{D}_b \cup \mathcal{U}_{\rho_c})}$, $W_c(x_0) > \rho_c$ holds because the state is not within the set $\mathcal{U}_{\rho_c}$ defined in Eq. 4.12c. However, since Eq. 4.12b holds within $\phi_{uc} \backslash (\mathcal{D}_b \cup \{0\})$, it is straightforward to show that $\dot{W}_c(x)$ remains negative along the trajectory of $x(t)$ following the same steps as performed for the initial state $x_0 \in \mathcal{U}_{\rho_c}$. Furthermore, since the set $\overline{\phi_{uc} \backslash (\mathcal{D}_b \cup \mathcal{U}_{\rho_c})}$ does not intersect with $\overline{\mathcal{D}_b}$, any trajectory starting in $\overline{\phi_{uc} \backslash (\mathcal{D}_b \cup \mathcal{U}_{\rho_c})}$ will reach the boundary of $\overline{\phi_{uc} \backslash (\mathcal{D}_b \cup \mathcal{U}_{\rho_c})}$ before reaching the boundary of $\mathcal{D}_b$. Because Eq. 4.12d holds (i.e., $\overline{\phi_{uc} \backslash (\mathcal{D}_b \cup \mathcal{U}_{\rho_c})} \cap \overline{\mathcal{D}_b} = \emptyset$), it must hold that $\overline{\phi_{uc} \backslash (\mathcal{D}_b \cup \mathcal{U}_{\rho_c})} \cap \mathcal{U}_{\rho_c}$ is a nonempty set. Because $W_c(x) > \rho_c$ within $\phi_{uc} \backslash (\mathcal{D}_b \cup \mathcal{U}_{\rho_c})$ but $W_c(x) \leq \rho_c$ within $\mathcal{U}_{\rho_c}$ from Eq. 4.12c, $W_c(x) = \rho_c$, $\forall x \in \partial\phi_{uc} \backslash (\mathcal{D}_b \cup \mathcal{U}_{\rho_c})$ due to the continuity of W_c. This implies that the state trajectory will enter and remain in $\mathcal{U}_{\rho_c}$ after it reaches the boundary of $\overline{\phi_{uc} \backslash (\mathcal{D}_b \cup \mathcal{U}_{\rho_c})}$. This completes the proof that $x(t) \notin \mathcal{D}_b, \forall\, t \geq 0$ for any $x_0 \in \mathbb{X}_0$.

Remark 4.1 Theorem 4.2 proves simultaneous operational safety and closed-loop stability (boundedness of the closed-loop state) for the nominal system of Eq. 4.1 with any initial state $x_0 \in \mathbb{X}_0$ under the control law $u = \Phi(x)$. We discuss the initial condition $x_0 \in \mathbb{X}_0$ in two scenarios: $x_0 \in \mathcal{U}_{\rho_c}$ ($\mathcal{U}_{\rho_c}$ is defined by Eq. 4.12c, in which $W_c(x) \leq \rho_c$ is satisfied), and $x_0 \in \overline{\phi_{uc} \backslash (\mathcal{D}_b \cup \mathcal{U}_{\rho_c})}$. Specifically, since Eq. 4.12d is only needed for showing operational safety when $x_0 \in \overline{\phi_{uc} \backslash (\mathcal{D}_b \cup \mathcal{U}_{\rho_c})}$, the conditions of a constraint CLBF in Eq. 4.12 can be reduced to Eqs. 4.12a–4.12c if we restrict the initial conditions to $\mathcal{U}_{\rho_c}$ or any subset of it. However, if the set $\overline{\phi_{uc} \backslash (\mathcal{D}_b \cup \mathcal{U}_{\rho_c})}$ is considered as a part of initial conditions, then the CLBF W_c should be developed satisfying all the conditions in Eq. 4.12. Additionally, the condition of Eq. 4.12d for the initial condition $x_0 \in \overline{\phi_{uc} \backslash (\mathcal{D}_b \cup \mathcal{U}_{\rho_c})}$ also implies that $W_c(x) = \rho_c$ holds for any $x \in \partial\mathcal{D}_b$. This can be readily shown by contradiction and is omitted here.

Case 2: Consider the nominal system of Eq. 4.1 with an unbounded unsafe region $\mathcal{D}_u$. Since there does not exist any stationary point $x_e \neq 0$ for the case of unbounded unsafe regions according to [37], Eq. 4.12 can be simplified with $\mathbb{X}_e = \emptyset$. As a result, the controller $u = \Phi(x) \in U$ guarantees that $\dot{W}_c < 0$ holds for all $x \in \mathcal{U}_{\rho_c} \backslash \{0\}$. It is shown in Fig. 4.2 that in this case, the trajectories from $x_0 \in \mathcal{U}_{\rho_c}$ converge to the origin while avoiding $\mathcal{D}_u$ in one direction. Additionally, from now on, we will restrict the set of initial conditions to be in $\mathcal{U}_{\rho_c}$ (i.e., $x_0 \in \mathcal{U}_{\rho_c}$) to simplify the discussion. The theorem below demonstrates that process operational safety and closed-loop stability are achieved simultaneously for the nominal system of Eq. 4.1 (i.e., $w(t) \equiv 0$) under $u = \Phi(x) \in U$.

Theorem 4.3 *Consider the nominal system of Eq. 4.1 (i.e., $w(t) \equiv 0$) with a constrained CLBF $W_c(x) : \mathbf{R}^n \rightarrow \mathbf{R}$ that has a minimum at the origin and is designed with respect to an unbounded unsafe region $\mathcal{D}_u$ in state-space. It is guaranteed that*

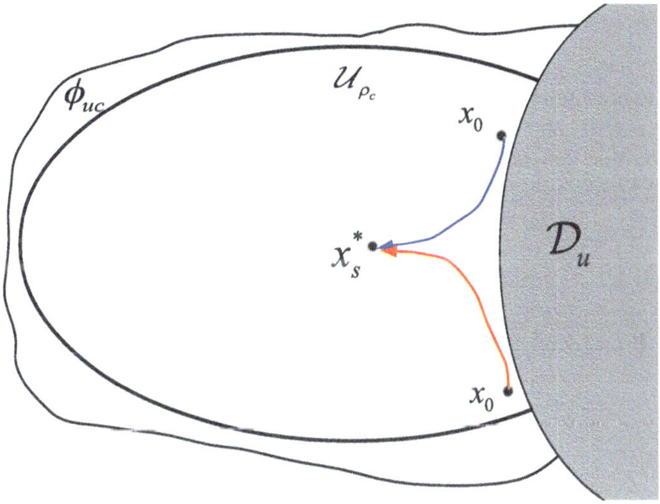

Fig. 4.2 A schematic representing an unbounded unsafe set $\mathscr{D}_u$ in state-space, where the trajectories start from any initial condition x_0 avoid $\mathscr{D}_u$ and converge to the origin x_s^*

the closed-loop state remains in the set $\mathscr{U}_{\rho_c}$ for all times and the origin can be rendered asymptotically stable for any $x_0 \in \mathscr{U}_{\rho_c}$, under the continuous feedback control law $u = \Phi(x) \in U$.

Proof It is straightforward to show that $\dot{W}_c < 0$ holds for all $x \in \phi_{uc}\backslash\{0\}$ under $u = \Phi(x) \in U$ following the first part of the proof for Theorem 4.2. This implies that $\forall x_0 \in \mathscr{U}_{\rho_c} \subset \phi_{uc}$, the state stays in $\mathscr{U}_{\rho_c}$ for all times and can be ultimately stabilized at the origin because $\dot{W}_c < 0$ holds for all $x \in \mathscr{U}_{\rho_c}\backslash\{0\}$.

Remark 4.2 Lyapunov function $V(x)$ and control Lyapunov-barrier function $W_c(x)$ are similar in that the level sets of $V(x)$ and $W_c(x)$ are both invariant sets and both of them have a global minimum at the origin of state-space. However, unlike Lyapunov functions that have a unique minimum at the origin and are positive definite, CLBFs may have multiple stationary points (other than the origin) and can have negative upper bounds for the level set (i.e., $\rho_c < 0$). On the other hand, Lyapunov function $V(x)$ is typically used to design controllers with closed-loop stability properties (e.g., the Sontag control law of Eq. 4.11 guarantees convergence of the state to the origin), while the CLBF is used to design the controllers that guarantee the boundedness of the state and avoidance of the unsafe region in a level set of $W_c(x)$ (e.g., the Sontag control law of Eq. 4.11 in terms of $W_c(x)$). Additionally, if there exists a discontinuous control law that can drive the states away from other stationary points (i.e., saddle points), the CLBF-based control law can further guarantee the convergence of the state to the origin.

4.3.2 Design of Constrained CLBF

This section presents the method for constructing a constrained CLBF. Specifically, we design a CBF and a CLF separately, and construct a CLBF through a linear combination that satisfies the properties in Eq. 4.12. The guidelines for designing the CBF and CLF, and developing the CLBF $W_c(x)$ that has the global minimum at the origin, are given by Proposition 4.1 below.

Proposition 4.1 *Consider the nominal system* $\dot{x} = f(x) + g(x)u + h(x)w$ *with* $w(t) \equiv 0$ *with an open set* $\mathscr{D}$ *of unsafe states. Assume that there exists a* $\mathscr{C}^1$ *CBF* $B : \mathbf{R}^n \rightarrow \mathbf{R}$ *and a* $\mathscr{C}^1$ *CLF* $V : \mathbf{R}^n \rightarrow \mathbf{R}_+$, *such that the following conditions hold:*

$$c_1 |x|^2 \leq V(x) \leq c_2 |x|^2, \ \forall x \in \mathbf{R}^n, c_2 > c_1 > 0 \tag{4.15}$$

$$\mathscr{D} \subset H \subset \phi_{uc}, \ 0 \notin H \tag{4.16}$$

$$B(x) = -\eta < 0, \ \forall x \in \mathbf{R}^n \backslash H; \ B(x) > 0, \ \forall x \in \mathscr{D} \tag{4.17}$$

where H *is a connected and compact set within* ϕ_{uc}. *Let* $W_c(x)$ *have the the form of* $W_c(x) := V(x) + \mu B(x) + \nu$, *where*

$$L_f W_c(x) < 0, \ \ \forall x \in \{z \in \phi_{uc} \backslash (\mathscr{D} \cup \{0\} \cup \mathbb{X}_e) \mid L_g W_c(z) = 0\} \tag{4.18}$$

$$\mu > \frac{c_2 c_3 - c_1 c_4}{\eta}, \tag{4.19a}$$

$$\nu = \rho_c - c_1 c_4, \tag{4.19b}$$

$$c_3 := \max_{x \in \partial H} |x|^2, \tag{4.19c}$$

$$c_4 := \min_{x \in \partial \mathscr{D}} |x|^2. \tag{4.19d}$$

Then for initial states $x_0 \in \phi_{uc} \backslash \mathscr{D}_H$, *where* $\mathscr{D}_H := \{x \in H \mid W_c(x) > \rho_c\}$; *the control law* $\Phi(x)$ *of Eq. 4.11 that replaces* $V(x)$ *by* $W_c(x)$ *guarantees the boundedness of the closed-loop state in* $\phi_{uc} \backslash \mathscr{D}_H$ *such that the state avoids the unsafe region* $\mathscr{D}_H$ *at all times.*

Proof We first define an expanded unsafe region $\mathscr{D}_H$ and a new connected and compact set H that satisfies Eq. 4.16 such that all the states inside the region H that satisfies $W_c(x) > \rho_c$ are included in $\mathscr{D}_H$. Figure 4.3 shows a schematic describing the relationship among the aforementioned sets. Since the unsafe region $\mathscr{D}$ we considered is a subset of $\mathscr{D}_H$, the proof results with respect to $\mathscr{D}_H$ that we will show below also hold for the unsafe set $\mathscr{D}$ since it is guaranteed that the state remains outside of $\mathscr{D}$ under the proposed CLBF if it does not enter $\mathscr{D}_H$. Therefore, we prove that the

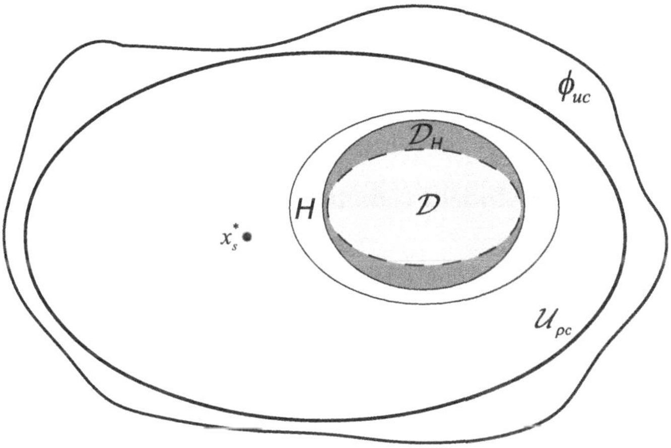

Fig. 4.3 A schematic showing the relationship among the sets ϕ_{uc}, $\mathscr{D}$, $\mathscr{D}_H$, and H, where $\mathscr{U}_{\rho_c}$ is the invariant set shown as an ellipse subtracting $\mathscr{D}_H$

proposed constrained CLBF $W_c(x)$ with $\mathscr{D}_H$ has a global minimum at the origin and satisfies all the conditions of Eq. 4.12. Firstly, from the definition of $\mathscr{D}_H$, it is straightforward to show that Eq. 4.12a holds. Next, we show that $W_c(x) > \rho_c$ holds for all $x \in \mathscr{D}$, using Eqs. 4.15, 4.17, and 4.19:

$$
\begin{aligned}
W_c(x) &= V(x) + \mu B(x) + \nu \\
&> c_1 |x|^2 + \rho_c - c_1 c_4 \\
&> \rho_c.
\end{aligned}
\tag{4.20}
$$

Additionally, Eq. 4.12b is trivially satisfied since it is one of the required properties of CLBF as shown in Eq. 4.18. Finally, the following inequalities are derived to show that Eq. 4.12c holds for all $x \in \partial H$:

$$
\begin{aligned}
W_c(x) &= V(x) + \mu B(x) + \nu \\
&\leq c_2 |x|^2 - \mu\eta + \rho_c - c_1 c_4 \\
&< \rho_c.
\end{aligned}
\tag{4.21}
$$

Hence, Eq. 4.12c holds because $\mathscr{U}_{\rho_c}$ is not an empty set required by Eq. 4.21. This also implies that $\partial H \cap \partial \mathscr{D}_H = \emptyset$. Additionally, we can further derive the following relationship: $\mathscr{D}_H \subset H \subset (\mathscr{D}_H \cup \mathscr{U}_{\rho_c})$, which implies the boundary of $\mathscr{D}_H$ does not intersect with the boundary of $\phi_{uc} \backslash (\mathscr{D}_H \cup \mathscr{U}_{\rho_c})$ (i.e., Eq. 4.12d holds, $\overline{\phi_{uc} \backslash (\mathscr{D}_H \cup \mathscr{U}_{\rho_c})} \cap \overline{\mathscr{D}_H} = \emptyset$). Additionally, it is shown that the global minimum of $W_c(x)$ is achieved at the origin since both the minimums of $B(x)$ and $V(x)$ are attained at the origin. This completes the proof showing that the proposed CLBF construction method satisfies all the CLBF conditions in Eq. 4.12. As a result, the closed-loop state is guaranteed to be bounded in $\phi_{uc}\backslash\mathscr{D}_H$ for any initial states $x_0 \in \phi_{uc}\backslash\mathscr{D}_H$, and the state does not enter $\mathscr{D}_H$ (also $\mathscr{D}$) for all times under the CLBF-based control law $\Phi(x)$.

4.4 CLBF-Based Model Predictive Control

In this section, a CLBF-based model predictive control (CLBF-MPC) scheme that incorporates CLBF-based stability and safety constraints is proposed to regulate the nonlinear system of Eq. 4.1 to the steady-state while avoiding the unsafe operation at the same time. We first discuss the impact of the sample-and-hold implementation of control actions on the stability and safety properties derived by the continuous controller $u = \Phi(x) \in U$. Subsequently, a rigorous theoretical treatment of operational safety and closed-loop stability properties for the system of Eq. 4.1 with the control architecture is provided.

4.4.1 Sample-and-Hold Implementation of CLBF-Based Controller

It was noted in Theorem 4.2 that the CLBF-based controller $\Phi(x)$ implemented in continuous time can ensure process operational safety for the nominal system of Eq. 4.1, i.e., $\dot{x} = f(x) + g(x)u + h(x)w$ with $w(t) \equiv 0$ by maintaining the state in a safe region of operation. Since later in this section, we will use CLBFs to design safety and stability constraints in MPC that implement control actions in a sample-and-hold manner, the sample-and-hold properties of the controller $\Phi(x)$ for the nonlinear system are investigated in the following proposition.

Proposition 4.2 *Consider the nominal system of Eq. 4.1 (i.e., $w(t) \equiv 0$) with a constrained CLBF W_c that has a minimum at the origin. Let $u(t) = \Phi(x(t_k))$, $t_k \leq t < t_{k+1}$, for any $x(t_k) \in \mathscr{U}_{\rho_c}\backslash\mathscr{B}_\delta(x_e)$ where $\delta > 0$, $x_e \in \mathbb{X}_e$ and t_k represents the time instance, i.e., $t_k = k\Delta$, $k = 0, 1, 2, \ldots$. Let $u(t) = \bar{u}(x) \in U$ such that if $x(t_k) \in \mathscr{B}_\delta(x_e)$, $W_c(x(t_{k+1})) < W_c(x(t_k))$ holds for any $\Delta > 0$ under $\bar{u}(x)$. Then, given any positive real number d, we can show that a positive real number Δ^* exists, such that $x(t) \in \mathscr{U}_{\rho_c}$, $\forall t \geq 0$, and $\lim_{t\to\infty} |x(t)| \leq d$, if $x_0 \in \mathscr{U}_{\rho_c}$ and $\Delta \in (0, \Delta^*]$.*

Proof We need to show that any states originating in $\mathscr{U}_{\rho_c}$ can be driven into a small neighborhood around the origin (i.e., the level set of $W_c(x)$: $\mathscr{U}_{\rho_{\min}} = \{x \in \phi_{uc} \mid W_c(x) \leq \rho_{\min}\}$) as $t \to \infty$ where $\rho_{\min} < \rho_c$, under the sample-and-hold implementation of the controller $u = \Phi(x) \in U$. Then, by the continuity of $W_c(x)$, it is trivial to show that $\lim_{t\to\infty} |x(t)| \leq d$ since $x(t) \in \mathscr{U}_{\rho_{\min}}$ as $t \to \infty$. To prove the convergence of state into $\mathscr{U}_{\rho_{\min}}$, we show that $\dot{W}_c(x(t), u(t)) < -\varepsilon$ holds for all $t \in [t_k, t_k + \Delta^*)$ and for all states in the set $\mathbb{Z} := \{\mathscr{U}_{\rho_c}\backslash(\mathscr{U}_{\rho_s} \cup \mathscr{B}_\delta(x_e))\}$ with $u(t) = u(t_k) = \Phi(x(t_k))$, where $\rho_s < \rho_{\min} < \rho_c$ as follows:

$$\begin{aligned}\dot{W}_c(x(t), u(t)) =& \dot{W}_c(x(t_k), u(t_k)) + (\dot{W}_c(x(t), u(t)) - \dot{W}_c(x(t_k), u(t_k)))\\ =& L_f W_c(x(t_k)) + L_g W_c(x(t_k))u(t_k) + (L_f W_c(x(t)) - L_f W_c(x(t_k)))\\ &+ (L_g W_c(x(t)) - L_g W_c(x(t_k)))u(t)\end{aligned} \tag{4.22}$$

where $\dot{W}_c(x, u)$ represents $\frac{\partial W_c(x)}{\partial x}(f(x) + g(x)u)$. Since $f(\cdot)$ and $g(\cdot)$ are smooth functions, and $W_c(x)$ is a $\mathscr{C}^1$ function satisfying Eq. 4.13c, we can show that there exist positive real numbers k_1 and k_2 that satisfy $|(L_f W_c(x(t)) - L_f W_c(x(t_k))| \leq k_1|x(t) - x(t_k)|$, $|(L_g W_c(x(t)) - L_g W_c(x(t_k)))u(t))| \leq k_2|x(t) - x(t_k)|$ for all $x \in \mathscr{U}_{\rho_c}$. Additionally, since $\mathbb{Z}$ is bounded and $f(x)$ and $g(x)$ are continuous functions, there exists a a sampling period Δ' and a positive real number k_4, such that $|x(t) - x(t_k)| \leq k_4\Delta'$ holds for all $t \in [t_k, t_k + \Delta')$. Also, it follows from the the definition of ϕ_{uc} that $\dot{W}_c(x(t_k)) < -a|W_c(x) - W_c(0)| < -a\rho_m$ holds for all $x \in \mathbb{Z}$, where $\rho_m := \min_{x \in \mathbb{Z}} |W_c(x) - W_c(0)|$. Let $\Delta' < \frac{a\rho_m - \varepsilon}{k_4(k_1+k_2)}$ and $0 \leq \varepsilon < a\rho_m$, where $a > 0$ is used to characterize the set ϕ_{uc}. Then, the following inequalities are obtained by substituting the above inequalities from Lipschitz continuity into Eq. 4.22:

$$\begin{aligned}\dot{W}_c(x(t), u(t)) &\leq \dot{W}_c(x(t_k), u(t_k)) + k_4(k_1 + k_2)\Delta'\\ &< -a\rho_m + k_4(k_1 + k_2)\Delta'\\ &< -\varepsilon.\end{aligned} \tag{4.23}$$

Equation 4.23 implies that $W_c(x(t)) < W_c(x(t_k)) \leq \rho_c, \forall\, t > t_k$ and the closed-loop state trajectory $x(t)$ will enter $\mathscr{U}_{\rho_s}$ within finite steps. Hence, it is shown that $x(t)$ is bounded in $\mathscr{U}_{\rho_c}$, for all $t \in [t_k, t_k + \Delta')$.

Additionally, consider $x(t_k) \in \mathscr{B}_\delta(x_e)$ where x_e are designed to be saddle points. Since a set of control actions $\bar{u}(x)$ that decreases $W_c(x)$ are assumed to exist and characterized in advance, the state $x(t_{k+1})$ at the next sampling step is able to enter a smaller level set of $W_c(x)$ and leaves $\mathscr{B}_\delta(x_e)$ within finite sampling steps. Moreover, it is guaranteed that $x(t)$ does not return to $\mathscr{B}_\delta(x_e)$ once it leaves since Eq. 4.23 (i.e., $W_c(x(t)) < W_c(x(t_k)), \forall\, t > t_k$) holds thereafter.

Next, we show that given $x(t_k) \in \mathscr{U}_{\rho_s}$, the trajectory of $x(t)$ will stay in $\mathscr{U}_{\rho_{\min}}, \forall\, t \in [t_k, t_k + \Delta'')$. Consider Δ'' such that

$$\rho_{\min} = \max_{\Delta t \in [0, \Delta'')} \{W_c(x(t_k + \Delta t)) \mid x(t_k) \in \mathscr{U}_{\rho_s},\ u \in U\}. \tag{4.24}$$

Again, a sufficiently small Δ'' exists such that Eq. 4.24 holds. Thus, let $\Delta^* = \min\{\Delta', \Delta''\}$, and we can show that $x(t)$ will move toward $\mathscr{U}_{\rho_{\min}}$ and remain in $\mathscr{U}_{\rho_c}$ during one sampling period $t \in [t_k,\ t_{k+1})$ for any state $x(t_k) \in \mathscr{U}_{\rho_c}$, where $\Delta \in (0, \Delta^*]$ and $t_{k+1} := t_k + \Delta$. Figure 4.4 illustrates the relationship among the sets $\mathscr{U}_{\rho_c}$, $\mathscr{U}_{\rho_{\min}}$, and $\mathscr{U}_{\rho_s}$, and shows an example of the state trajectory for the closed-loop system under the sample-and-hold implementation of $u = \Phi(x)$.

Remark 4.3 Proposition 4.2 proves closed-loop stability and operational safety for the nominal system of Eq. 4.1 (i.e., $w(t) \equiv 0$). However, the stability and

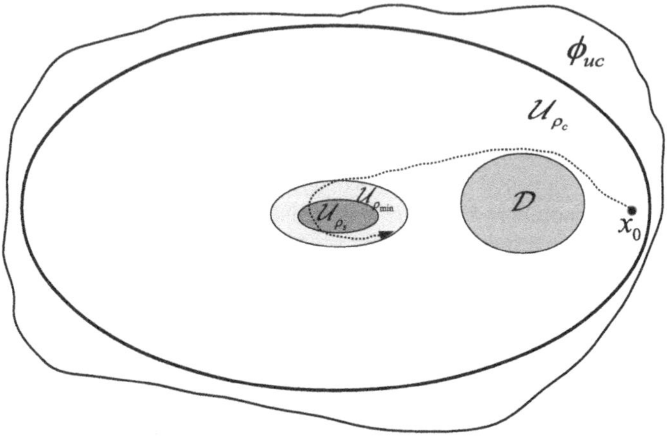

Fig. 4.4 A schematic representing the sets $\mathscr{U}_{\rho_c}$, $\mathscr{U}_{\rho_{\min}}$, and $\mathscr{U}_{\rho_s}$, where an example of the state trajectory (dotted line) for the closed-loop system under the sample-and-hold implementation of $u = \Phi(x) \in U$ is shown to ultimately enter and remain in $\mathscr{U}_{\rho_{\min}}$ while avoiding the unsafe region $\mathscr{D}$ at all times from the initial state $x_0 \in \mathscr{U}_{\rho_c}$

safety results under the sample-and-hold implementation of CLBF-based controllers are not restricted to undisturbed systems. When taking the bounded disturbance $|w(t)| \leq \theta$ into account, we can show that the results in Proposition 4.2 still hold for the nonlinear system of Eq. 4.1 with bounded disturbances. Similarly, from the local Lipschitz property of $h(\cdot)$, we can show that there exists a $k_3 > 0$, such that $|(L_h W_c(x(t)) - L_h W_c(x(t_k))| \leq k_3|x(t) - x(t_k)|$ holds. Subsequently, the following results for $\dot{W}_c(x(t), u(t))$ and $\rho'_{\min}$ that account for $w(t)$ are developed:

$$\begin{aligned} \dot{W}_c(x(t), u(t)) &\leq \dot{W}_c(x(t_k), u(t_k)) + k_4(k_1 + k_2 + k_3\theta)\Delta' \\ &< -a\rho_m + k_4(k_1 + k_2 + k_3\theta)\Delta' \\ &< -\varepsilon \end{aligned} \tag{4.25}$$

$$\rho'_{\min} = \max_{\Delta t \in [0, \Delta'')} \{W_c(x(t_k + \Delta t), u, w) \mid x(t_k) \in \mathscr{U}_{\rho_s},\ u \in U,\ |w| \leq \theta\} \tag{4.26}$$

where $\Delta' < \frac{a\rho_m - \varepsilon}{k_4(k_1+k_2+k_3\theta)}$ and $0 \leq \varepsilon < a\rho_m$, respectively. Therefore, if Δ' and ε are chosen appropriately for the sufficiently small bounded disturbance (i.e., the disturbance bound θ is sufficiently small), we can show that in the presence of disturbance, $\dot{W}_c$ remains negative within each sampling period. Additionally, if $x(t_k) \in \mathscr{B}_\delta(x_e)$, a set of feasible control actions $\bar{u}(x)$ that satisfies $W_c(x(t_{k+1})) < W_c(x(t_k)), \forall\, |w| \leq \theta$ are assumed to exist. Based on the definition of $\rho'_{\min}$ of Eq. 4.26, it is trivial to show that for any $x(t_k) \in \mathscr{U}_{\rho_s}$, the state trajectory is guaranteed to be bounded in $\mathscr{U}_{\rho'_{\min}}$, $\forall\ t \in [t_k, t_k + \Delta'')$. The above proof implies that the CLBF-based controller $u = \Phi(x) \in U$ in a sample-and-hold fashion is robust to the sufficiently small bounded disturbance.

Remark 4.4 Due to the existence of stationary points (other than the origin) in state-space, we assume that a set of feasible solutions $\bar{u}(x) \in U$ that can drive the closed-loop state away from $\mathscr{B}_\delta(x_e)$ in the direction of decreasing $W_c(x)$ exists in $\mathscr{B}_\delta(x_e)$. Such a discontinuous control law $\bar{u}(x)$ can be determined through a grid search or an optimization problem, e.g., $\bar{u}(x(t_k)) = \arg\min_{u \in U}\{W_c(x(t_{k+1})) \mid W_c(x(t_{k+1})) < W_c(x(t_k))\}$. However, if there is no input constraints for the nonlinear system of Eq. 4.1, a control action (maybe large) that decreases the value of $W_c(x)$ always exists provided that x_e is a saddle point (not a local minimum). Additionally, once the state leaves $\mathscr{B}_\delta(x_e)$ in the path with a decreasing $W_c(x)$ value, the state will move toward the origin under the CLBF-based controller $u = \Phi(x)$.

4.4.2 Formulation of CLBF-MPC

The CLBF-MPC scheme is formulated by the following optimization problem [213]:

$$\min_{u \in S(\Delta)} \int_{t_k}^{t_{k+N}} l_t(\tilde{x}(t), u(t))\mathrm{d}t \tag{4.27a}$$

$$\text{s.t.} \quad \dot{\tilde{x}}(t) = f(\tilde{x}(t)) + g(\tilde{x}(t))u(t) \tag{4.27b}$$

$$\tilde{x}(t_k) = x(t_k) \tag{4.27c}$$

$$u(t) \in U, \ \forall\, t \in [t_k, t_{k+N}) \tag{4.27d}$$

$$\dot{W}_c(x(t_k), u(t_k)) \leq \dot{W}_c(x(t_k), \Phi(x(t_k))), \quad \text{if } W_c(x(t_k)) > \rho'_{\min} \text{ and } x(t_k) \notin \mathscr{B}_\delta(x_e) \tag{4.27e}$$

$$W_c(\tilde{x}(t)) \leq \rho'_{\min}, \ \forall\, t \in [t_k, t_{k+N}), \ \text{ if } W_c(x(t_k)) \leq \rho'_{\min} \tag{4.27f}$$

$$W_c(\tilde{x}(t)) < W_c(x(t_k)), \ \forall\, t \in (t_k, t_{k+N}), \ \text{ if } x(t_k) \in \mathscr{B}_\delta(x_e) \tag{4.27g}$$

where Δ is the sampling period, $S(\Delta)$ is the set of piecewise constant functions with time interval Δ, $\tilde{x}(t)$ is the predicted state trajectory, and N is the number of sampling steps in the prediction horizon. We use $\dot{W}_c(x, u)$ to represent $\frac{\partial W_c(x)}{\partial x}(f(x) + g(x)u)$. The optimization problem of Eq. 4.27 is to minimize the object function of Eq. 4.27a subject to the constraints of Eqs. 4.27b–4.27g. Specifically, the objective function is the integral of $l_t(\tilde{x}(t), u(t))$ over the prediction horizon, in which $l_t(\tilde{x}(t), u(t))$ is developed to satisfy $l_t(\tilde{x}(t), u(t)) > 0$, $\forall(\tilde{x}(t), u(t)) \neq (0, 0)$, and $l_t(0, 0) = 0$ such that it attains the minimum value at the steady-state of the nonlinear system of Eq. 4.1. The nominal system of Eq. 4.1 with $w(t) \equiv 0$ is used in the constraint of Eq. 4.27b to predict the evolution of the closed-loop state. The initial condition for the prediction model of Eq. 4.27b is the current state measurement defined by Eq. 4.27c. The input constraints that will be applied over the entire prediction horizon are defined by Eq. 4.27d. The constraints of Eqs. 4.27e–4.27g guarantee that the closed-loop state is ultimately bounded in a small neighborhood around the origin (i.e., $\mathcal{U}_{\rho'_{\min}}$) and does not enter the unsafe region for all times. Specifically, when $x(t_k)$ is outside of $\mathcal{U}_{\rho'_{\min}}$ and $x(t_k) \notin \mathcal{B}_\delta(x_e)$, the constraint of Eq. 4.27e drives the closed-loop state into a smaller level set of $W_c(x)$ by decreasing the value of $W_c(\tilde{x})$ along the predicted state trajectory at least at the rate under the CLBF-based controller $u = \Phi(x)$. When $x(t_k)$ enters $\mathcal{U}_{\rho'_{\min}}$ (i.e., $x(t_k)$ is also bounded in a small ball around the origin $\mathcal{B}_d(0) := \{x \in \mathbf{R}^n \mid |x| \leq d\}$), the constraint of Eq. 4.27f maintains the closed-loop state inside $\mathcal{B}_d(0)$ afterwards. However, if the state is trapped in other stationary points during the path toward the origin, i.e., $x(t_k) \in \mathcal{B}_\delta(x_e)$, we activate the constraint of Eq. 4.27g to drive the state away from x_e in the direction of decreasing $W_c(x)$. It is noted that once the state leaves $\mathcal{B}_\delta(x_e)$, it will not return to $\mathcal{B}_\delta(x_e)$ because the constraint of Eq. 4.27e will be activated again to drive the state to smaller level sets of $W_c(x)$ thereafter.

The optimization problem of CLBF-MPC is solved at each sampling time based on the measured state $x(t_k)$ at $t = t_k$. After the optimal solution $u^*(t)$ is obtained from CLBF-MPC, the controller will send the first control action of $u^*(t)$ only (i.e., $u(t)$, $\forall t \in [t_k, t_{k+1})$, where $t_{k+1} := t_k + \Delta$) to the control actuators. Then, the optimization problem is resolved at the next sampling step with the horizon moving one sampling period forward.

Theorem 4.4 below shows that operational safety, closed-loop stability, and recursive feasibility of the optimization problems are guaranteed simultaneously for the nonlinear system of Eq. 4.1 under the control computed by the CLBF-MPC optimization problem of Eq. 4.27.

Theorem 4.4 *Consider the nonlinear system of Eq. 4.1 with a constrained CLBF W_c that satisfies Eq. 4.12 and has a minimum at the origin. Given any initial state $x_0 \in \mathcal{U}_{\rho_c}$, the optimization problem is guaranteed to be feasible for all times under CLBF-MPC with the sampling period $\Delta \in (0, \Delta^*]$ defined in Proposition 4.2. Additionally, operational safety and closed-loop stability are both guaranteed in the sense that for any $x_0 \in \mathcal{U}_{\rho_c}$, $x(t) \in \mathcal{U}_{\rho_c}$ holds for all $t \geq 0$, and* $\limsup_{t\to\infty} |x(t)| \leq d$.

Proof The proof consists of two parts. We first show that the optimization problem of Eq. 4.27 is recursively feasible (i.e., there always exists a feasible solution) throughout the entire operating period for any initial condition $x_0 \in \mathcal{U}_{\rho_c}$. Then, we show that the closed-loop state trajectory is bounded in $\mathcal{U}_{\rho_c}$ for all times, and can be ultimately bounded in a small region around the origin $\mathcal{U}_{\rho'_{\min}}$ under CLBF-MPC.

Part 1 : In this part, we show that the CLBF-based controller $u = \Phi(x)$ and the pre-determined discontinuous controller $u = \bar{u}(x)$ implemented in a sample-and-hold fashion are feasible solutions to the optimization problem of CLBF-MPC. Specifically, when $x(t_k) \in \mathcal{U}_{\rho_c} \backslash (\mathcal{U}_{\rho'_{\min}} \cup \mathcal{B}_\delta(x_e))$, the sample-and-hold control law $u(t) = \Phi(x(t_k + i\Delta))$, $i = 0, 1, \ldots, N-1$ satisfies both the constraint of Eq. 4.27e and the input constraint of Eq. 4.27d. Next, we show that the CLBF-based controller $u(t) = \Phi(x(t_k + i\Delta))$, $i = 0, 1, \ldots, N-1$ is also a feasible solution to CLBF-MPC when $x(t_k) \in \mathcal{U}_{\rho'_{\min}}$. Specifically, if $x(t_k) \in \mathcal{U}_{\rho_s} \subset \mathcal{U}_{\rho'_{\min}}$, from the definition of $\rho'_{\min}$ of Eq. 4.26, it is straightforward to show that any control action $u \in U$ meets the constraint of Eq. 4.27f. However, if $x(t_k) \in \mathcal{U}_{\rho'_{\min}} \backslash \mathcal{U}_{\rho_s}$, the results from Proposition 4.2 show that $\dot{W}_c(x) < -\varepsilon$ holds under the CLBF-based controller $u(t) = \Phi(x(t_k + i\Delta))$ (in sample-and-hold fashion) over one sampling period. This implies that the state is bounded in $\mathscr{U}_{\rho'_{\min}}$ within one sampling period, i.e., $W_c(x(t_{k+1})) \leq W_c(x(t_k)) \leq \rho'_{\min}$. Lastly, when the state is trapped in a neighborhood around the saddle points, i.e., $x(t_k) \in \mathscr{B}_\delta(x_e)$, $u(t) = \bar{u}(x) \in U$ is a set of feasible solutions that are pre-determined to satisfy the constraints of Eq. 4.27g and Eq. 4.27d. Therefore, it is concluded from the first part that a feasible solution to the optimization problem of Eq. 4.27 always exists when $x(t_k) \in \mathscr{U}_{\rho_c}$.

Part 2 : We now prove that if $x_0 \in \mathscr{U}_{\rho_c}$, it holds that $x(t) \in \mathscr{U}_{\rho_c}$, $\forall\, t \geq 0$. We first consider the case of a bounded unsafe region $\mathscr{D}_b$. Since the initial condition x_0 is bounded in the set $\mathscr{U}_{\rho_c}$, it is straightforward to show that $x(t) \in \mathscr{U}_{\rho_c}$ holds for all $t \geq 0$ under the constraints of Eqs. 4.27d–4.27g by letting $t_k = 0$ in the result of $W_c(x(t)) < W_c(x(t_k)) \leq \rho_c$, $\forall\, t > t_k$ obtained from Proposition 4.2. As a result, this also proves the assumption that $x(t_k) \in \mathscr{U}_{\rho_c}$ at $t = t_k$, $t_k \geq 0$ holds true in Part 1.

Finally, let $x_0 \in \mathscr{U}_{\rho_c} \backslash \mathscr{U}_{\rho'_{\min}}$, and we show that the state $x(t)$ can reach the target set $\mathscr{U}_{\rho'_{\min}}$ and remain in $\mathscr{U}_{\rho'_{\min}}$ thereafter. Based on the results from Proposition 4.2 showing that the value of W_c is continuously decreasing for every sampling step, i.e., $W_c(x(t + \Delta)) < W_c(x(t))$, when $x(t) \in \mathscr{U}_{\rho_c} \backslash (\mathscr{U}_{\rho'_{\min}} \cup \mathscr{B}_\delta(x_e))$ and $x(t) \in \mathscr{B}_\delta(x_e)$, it follows that the state trajectory will enter $\mathscr{U}_{\rho'_{\min}}$ within finite time t_s. Additionally, as shown in Part 1, once the state enters $\mathscr{U}_{\rho'_{\min}}$, i.e., $x(t) \in \mathscr{U}_{\rho'_{\min}}$, the constraint of Eq. 4.27f will be activated to maintain the state within $\mathscr{U}_{\rho'_{\min}}$ for all subsequent times $t \geq t_s$. As $W_c(\cdot)$ is a continuous function of the state, the boundedness of $W_c(x)$ in $\rho'_{\min}$ implies the existence of a positive real number d such that $\limsup_{t\to\infty} |x(t)| \leq d$ holds.

On the other hand, when the unsafe region in state-space is characterized as an unbounded set $\mathscr{D}_u$, stationary points are no longer an issue in CLBF-MPC as the origin is the unique minimum in state-space (i.e., the constraint of Eq. 4.27g remains inactive due to $\mathbb{X}_e = \emptyset$). In this case, the constraints of Eqs. 4.27e–4.27f force the state to move toward the origin and ultimately bound the state within $\mathscr{U}_{\rho'_{\min}}$.

Remark 4.5 Note that the stability and safety results established in Theorem 4.4 are not restricted to the nominal system of Eq. 4.1 (i.e., $w(t) \equiv 0$). In fact, for the nonlinear system of Eq. 4.1 subject to sufficiently small bounded disturbances, Proposition 4.2 has shown that $\dot{W}_c < -\varepsilon$ still holds for all $x \in \mathscr{U}_{\rho_c} \backslash (\mathscr{U}_{\rho'_{\min}} \cup \mathscr{B}_\delta(x_e))$ provided that the disturbances $|w(t)| \leq \theta$ and the sampling period Δ are sufficiently small. Additionally, when $x \in \mathscr{B}_\delta(x_e)$ or $x \in \mathscr{U}_{\rho'_{\min}}$, the constraints of Eq. 4.27g and Eq. 4.27f still hold since the set of control actions $\bar{u}(x)$ and the target set $\rho'_{\min}$ of Eq. 4.26 are both determined accounting for the presence of the bounded disturbances. Therefore, it is readily shown that the stability and safety results in Theorem 4.4 hold true for the disturbed system of Eq. 4.1 with $|w| \leq \theta$ under CLBF-MPC.

Remark 4.6 It is noted that by incorporating CLBFs as MPC constraints to ensure operational safety and closed-loop stability, the problem that the state may converge to the stationary points x_e instead of the origin under CLBF-based controller can be solved. Specifically, when $x(t_k) \in \mathscr{B}_\delta(x_e)$, the state is required to leave x_e under the hard constraint of Eq. 4.27g, which guarantees that the state will not be trapped in saddle points. Additionally, from the formulation of the optimization problem of CLBF-MPC, it can be inferred that a control action that avoids converging to x_e is preferred because the distances between states and the origin and also control actions are penalized in the MPC objective function (i.e., the convergence of state to any states (e.g., x_e) other than the origin will lead to a large objective function value). Therefore, the MPC optimization problem developed using CLBF-based constraints will compute the sample-and-hold control actions to drive the state toward the origin while accounting for future cost values.

4.4.3 Application to a Chemical Process Example

This section presents a chemical process example to illustrate the design of CLBF and the application of the proposed CLBF-MPC scheme. A well-mixed, non-isothermal continuous stirred tank reactor (CSTR) with an irreversible first-order exothermic reaction taking place is considered. The reactant A is converted to the product B ($A \rightarrow B$) in the reactor, for which a heating jacket that removes (supplies) heat from (to) the reactor is utilized. The following material and energy balances are used to describe the CSTR dynamics:

$$\frac{dC_A}{dt} = \frac{F}{V_L}(C_{A0} - C_A) - k_0 e^{-E/RT} C_A + w_1 \tag{4.28a}$$

$$\frac{dT}{dt} = \frac{F}{V_L}(T_0 - T) - \frac{\Delta H k_0}{\rho_L C_p} e^{-E/RT} C_A + \frac{Q}{\rho_L C_p V_L} + w_2. \tag{4.28b}$$

In Eq. 4.28, T and C_A represent the temperature of the reactor and the concentration of reactant A in the reactor, respectively. V_L is the reacting liquid volume in the reactor. Q is the heat removal/supply rate. T_0 and C_{A0} are reactant temperature and concentration in the feed stream with the volumetric flow rate F. The liquid density (ρ_L) and heat capacity (C_p) are assumed constant. ΔH, E, and k_0 are the enthalpy of the reaction, activation energy, and the reaction pre-exponential factor, respectively. Table 4.1 gives the process parameter values in Eq. 4.28. The control objective of CLBF-MPC is to operate the CSTR at the equilibrium point $(C_{As}, T_s) = (0.57\ \text{kmol/m}^3,\ 395.3\ \text{K})$, and meanwhile, to ensure that the state remains in a safe region of state-space by manipulating the inlet concentration of species A, $\Delta C_{A0} = C_{A0} - C_{A0_s}$, and the heat input rate $\Delta Q = Q - Q_s$. The input constraints for ΔC_{A0} and ΔQ are given as follows: $|\Delta C_{A0}| \leq 1\ \text{kmol/m}^3$ and $|\Delta Q| \leq 0.0167\ \text{kJ/min}$.

To make the equilibrium point of the system at the origin of the state-space, we use deviation variables to place the CSTR of Eq. 4.28 in the form of nonlinear systems of Eq. 4.1 (see, also, Sect. 3.4 for more details on representing Eq. 4.1 using deviation variables). Specifically, the state vector and the manipulated input vector in deviation variable form are denoted by $x^T = [C_A - C_{As}\ \ T - T_s]$ and $u^T = [\Delta C_{A0}\ \ \Delta Q]$, respectively. The bounded disturbance vector $w^T = [w_1\ \ w_2]$

Table 4.1 Parameter values of the CSTR with a first-order reaction

$T_0 = 310$ K	$F = 100 \times 10^{-3}$ m^3/min
$V_L = 0.1$ m^3	$E = 8.314 \times 10^4$ kJ/kmol
$k_0 = 72 \times 10^9$ min^{-1}	$\Delta H = -4.78 \times 10^4$ kJ/kmol
$C_p = 0.239$ kJ/(kg K)	$R = 8.314$ kJ/(kmol K)
$\rho_L = 1000$ kg/m^3	$C_{A0_s} = 1.0$ kmol/m^3
$Q_s = 0.0$ kJ/ min	$C_{A_s} = 0.57$ kmol/m^3
$T_s = 395.3$ K	

is designed following a Gaussian distribution with zero mean and standard deviation $\sigma_1 = 1.0$ kmol/(m^3 min), $\sigma_2 = 3.5$ K/ min. Additionally, the disturbances are bounded by $|w_1| \leq 1.0$ kmol/(m^3 min) and $|w_2| \leq 3.17$ K/ min. Finally, we develop the control Lyapunov function in a quadratic form $V(x) = x^T P x$ with $P = \begin{bmatrix} 9.35 & 0.41 \\ 0.41 & 0.02 \end{bmatrix}$.

4.4.3.1 Case Study: Bounded Unsafe Region

We first carry out the simulation study of CLBF-MPC with a bounded unsafe region $\mathscr{D}_b$ located within the set ϕ_{uc}. We define the unsafe region as an ellipse in state-space: $\mathscr{D}_b := \{x \in \mathbf{R}^2 \mid F(x) = \frac{(x_1+0.22)^2}{1} + \frac{(x_2-4.6)^2}{1\times 10^4} < 2 \times 10^{-4}\}$. Then, H is defined as a superset of $\mathscr{D}_b$: $H := \{x \in \mathbf{R}^2 \mid F(x) < 4 \times 10^{-4}\}$ such that the condition $\mathscr{D}_b \subset H \subset \phi_{uc}$ in Proposition 4.1 is satisfied. Next, we design the the control barrier function $B(x)$ as follows:

$$B(x) = \begin{cases} e^{\frac{\lambda F^2(x)}{F(x)-4\times 10^{-4}}} - e^{-2\lambda\times 10^{-4}}, & \text{if } x \in H \\ -e^{-2\lambda\times 10^{-4}}, & \text{if } x \notin H \end{cases} \tag{4.29}$$

where $\lambda > 0$ is a parameter that can be used to adjust the value of $B(x)$ in characterizing the set ϕ_{uc}. It can be seen from Eq. 4.29 that $B(x)$ remains positive in the unsafe region $\mathscr{D}_b$. Then, following the construction method in Proposition 4.1, the control Lyapunov-barrier function in the form of $W_c(x) = V(x) + \mu B(x) + \nu$ is developed with $\lambda = 0.001$, $\rho_c = 0$, $c_1 = 0.001$, $c_2 = 10$, $c_3 = \max_{x\in\partial H} |x|^2 = 34.8$, $c_4 = \min_{x\in\partial\mathscr{D}_b} |x|^2 = 16.85$, and $\nu = \rho_c - c_1c_4 = -1.685 \times 10^{-2}$. Hence, μ is chosen to be 5000 to satisfy the condition Eq. 4.19. Finally, as the unsafe region is a bounded set, we need to find all the stationary points (other than the origin) in state-space based on the above $W_c(x)$. In this example, it is shown that there exists one saddle point $x_e = (-0.235, 4.83)$ in state-space.

As the control objective is to stabilize the CSTR at its steady-state, the objective function of CLBF-MPC is developed with the following form to minimize the feed reactant concentration and the heat removal/supply rate:

$$l_t(\tilde{x}, u) = |\tilde{x}(t)|^2_{Q_L} + |u(t)|^2_{R_L} \quad (4.30)$$

where $Q_L = \begin{bmatrix} 1000 & 0 \\ 0 & 10 \end{bmatrix}$ and $R_L = \begin{bmatrix} 1 & 0 \\ 0 & 100 \end{bmatrix}$ are the weighting matrices for the states and inputs, respectively, such that the two terms in the objective function of Eq. 4.30 that penalize the state and the input deviations from the steady-state are on the same order of magnitude. In the numerical simulation, we use the explicit Euler method to numerically integrate the process model of Eq. 4.28 with a sufficiently small integration time step of $h_c = 10^{-5}$ min. Additionally, the MPC is developed with the prediction horizon $N = 10$ and the sampling period $\Delta = 2 \times 10^{-3}$ min. The MPC optimization problem is solved using a large-scale nonlinear optimizer termed IPOPT ([201]) with a 4-core CPU desktop. It is shown that under the above simulation settings, we can achieve a desired closed-loop performance with high computational efficiency and ensure that the MPC optimization problem can be done within the sampling period.

Scenario 1: We first carry out the closed-loop simulation for the nominal CSTR system (i.e., no disturbances w) in the presence of a bounded unsafe region $\mathscr{D}$. The subset $\mathscr{U}_\rho \subset \mathscr{U}_{\rho_c}$ is chosen as the safe operating region. We first choose an initial condition far away from the unsafe region $\mathscr{D}$ and show that the closed-loop state initiating from this initial condition (the green trajectory in Fig. 4.5) successfully converges to the origin. Then, we choose another three initial conditions $(-0.35, 7)$, $(-0.235, 6.5)$, and $(-0.19, 5.5)$ and show in Fig. 4.5 that all the closed-loop states can pass around the unsafe region $\mathscr{D}$ on their way and converge to the origin under CLBF-MPC.

Scenario 2: We now compare the closed-loop performance of the CSTR system under CLBF-MPC with that under a non-Lyapunov-based MPC described by the following optimization problem that uses a terminal constraint and a state constraint to guarantee closed-loop stability and operational safety, respectively.

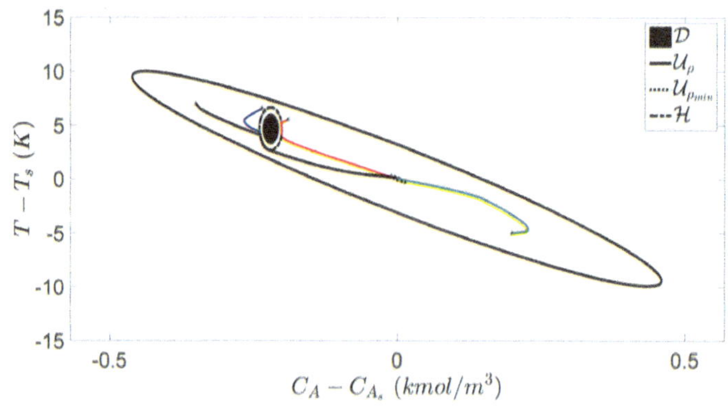

Fig. 4.5 Closed-loop state trajectories for four different initial conditions $(-0.19, 5.5)$ (red line), $(0.2, -5)$ (green line), $(-0.235, 6.5)$ (blue line), and $(-0.35, 7)$ (black line) under CLBF-MPC. The set $\mathscr{U}_\rho$ is the region between the set H and the largest ellipse, and the set of unsafe states $\mathscr{D}$ is represented by the solid black ellipse

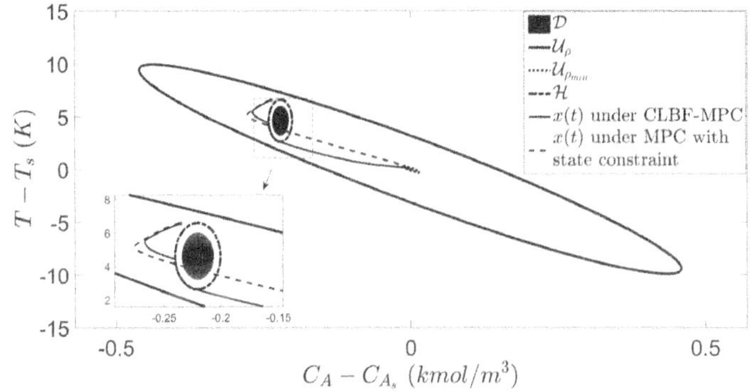

Fig. 4.6 Closed-loop state profiles under the MPC with state constraints (dashed line) and the CLBF-MPC of Eq. 4.27 (solid line) with the same initial condition $(-0.235, 6.5)$, where the unsafe region $\mathscr{D}$ is represented by the solid black ellipse

$$\min_{u \in S(\Delta)} \int_{t_k}^{t_{k+N}} l_t(\tilde{x}(t), u(t))\mathrm{d}t \tag{4.31a}$$

$$\text{s.t.} \quad \dot{\tilde{x}}(t) = f(\tilde{x}(t)) + g(\tilde{x}(t))u(t) \tag{4.31b}$$

$$\tilde{x}(t_k) = x(t_k) \tag{4.31c}$$

$$u(t) \in U, \ \forall\, t \in [t_k, t_{k+N}) \tag{4.31d}$$

$$\tilde{x}(t) \in \mathscr{U}_\rho, \ \forall\, t \in [t_k, t_{k+N}), \quad \text{if } x(t_k) \in \mathscr{U}_\rho \tag{4.31e}$$

$$\tilde{x}(t_{k+N}) \in \mathscr{U}_{\rho_{\min}}. \tag{4.31f}$$

It is demonstrated in Fig. 4.6 that starting from the same initial state $(-0.235, 6.5)$, the CLBF-MPC successfully drives the state to the origin and avoids the unsafe region, while the non-Lyapunov-based MPC drives the state across the unsafe region in the simulation. Specifically, it is seen that the closed-loop state under CLBF-MPC (black solid line) approaches the boundary of $\mathscr{D}$, moves down to pass around it, and finally, converges to the origin. However, in the non-Lyapunov-based MPC simulation (denoted by the dashed line), it is noticed that when the trajectory approaches the boundary of $\mathscr{D}$, the optimization problem becomes infeasible. Therefore, to continue the simulation, the state constraint has to be deactivated, and a feasible solution provided by the optimization problem of MPC with terminal constraint only is instead applied. As a result, the state trajectory fails to avoid the unsafe region and moves toward the origin under the terminal constraint. In this simulation study, we demonstrate that the CLBF-MPC outperforms the non-Lyapunov-based MPC that uses state constraints to avoid the unsafe region in the way that the CLBF-MPC reconciles the tasks of closed-loop stability and operational safety and with guaranteed recursive feasibility of MPC optimization problems.

Scenario 3: We carry out the closed-loop simulation subject to bounded disturbances. Figures 4.7 and 4.8 show the closed-loop state trajectory and input profiles.

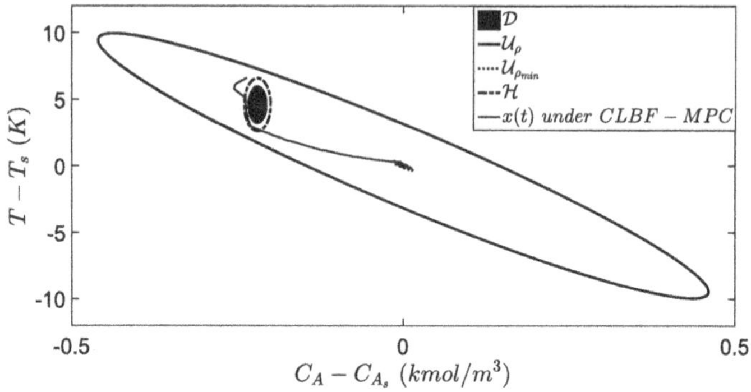

Fig. 4.7 Closed-loop state profile for the disturbed system under CLBF-MPC (solid line) with the initial condition (−0.235, 6.5)

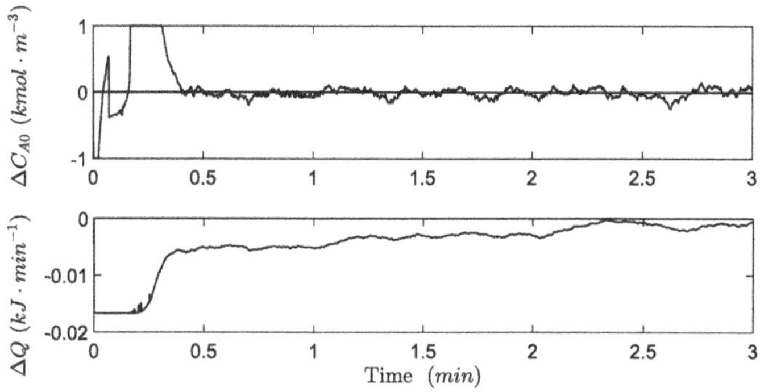

Fig. 4.8 Manipulated input profiles ($u_1 = \Delta C_{A0}$ and $u_2 = \Delta Q$) for the disturbed system under CLBF-MPC with the initial condition (−0.235, 6.5)

It is seen from Fig. 4.7 that operational safety and closed-loop stability are still guaranteed for the disturbed system under CLBF-MPC. Additionally, Fig 4.8 shows that the input profiles are varying around the steady-state after $t = 0.5$ h due to the disturbance.

Scenario 4: Lastly, we carry out the simulation studies that compare the proposed CLBF-MPC control scheme with the explicit CLBF-based controller of Eq. 4.11. Figures 4.9 and 4.10 show the closed-loop simulation results for the state trajectories and input profiles under the same initial condition (−0.235, 6.5). it is observed in Fig. 4.10 that the heat input rate ΔQ and the inlet concentration of the reactant ΔC_{A0} show significant oscillations from $t = 0.003$ min to $t = 0.2$ min under the CLBF-based control law of Eq. 4.11. As a result, the oscillation arises in the state trajectory when the state approaches the boundary of H.

The oscillation occurs under the explicit CLBF-based controller because the intrinsic dynamics of the closed-loop system attempts to drive the state toward the origin and cross the unsafe region using low energy (i.e., a small control action), while

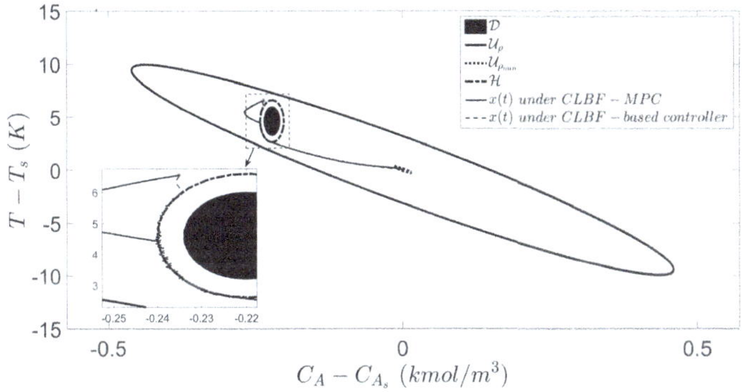

Fig. 4.9 Closed-loop state profiles under the CLBF-based controller of Eq. 4.11 (dashed line) and the CLBF-MPC of Eq. 4.27 (solid line) for the initial condition (−0.235, 6.5)

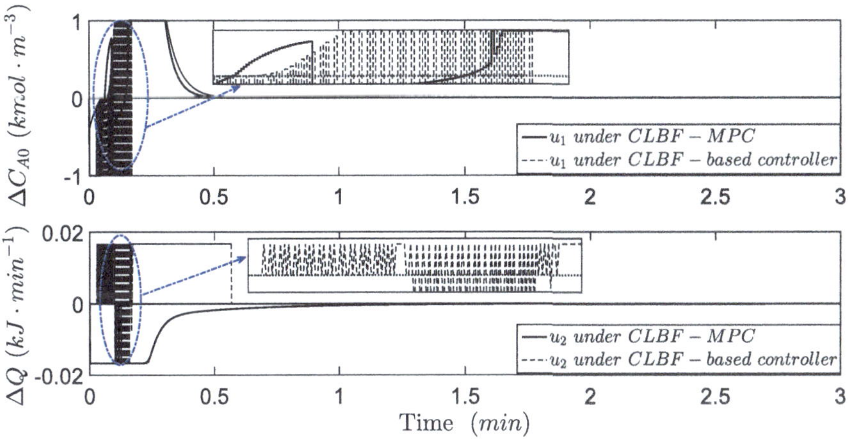

Fig. 4.10 Manipulated input profiles ($u_1 = \Delta C_{A0}$ and $u_2 = \Delta Q$) under the CLBF-based controller of Eq. 4.11 (dashed line) and the CLBF-MPC of Eq. 4.27 (solid line) for the initial condition (−0.235, 6.5)

the CLBF-based control action is to prevent this undesirable behavior by requiring W_c to be decreasing over time (i.e., W_c increases significantly for the state inside the unsafe region). Therefore, the control action shows significant oscillations, trying to balance the two opposite effects, when the state passes around the boundary of H. Overall, the dynamic performance under the proposed CLBF-MPC control scheme was shown to improve because the control actions are optimized in MPC accounting for future state behavior.

Furthermore, it is calculated based on the simulation results under CLBF-MPC that the total consumptions of energy ΔQ and of reactant ΔC_{A0} within the operating time $t_s = 3$ min are 0.006 kJ and 0.268 kmol/m^3, respectively. This shows significant improvements (i.e., 25 and 13%, respectively) compared to the numbers 0.008 kJ and 0.308 kmol/m^3 obtained from the explicit CLBF-based controller. Therefore, from the perspective of process economics, the CLBF-MPC of Eq. 4.27 outperforms the explicit CLBF-based controller of Eq. 4.11 in terms of reduced control energy consumptions and smoother control actions.

4.4.3.2 Case Study: Unbounded Unsafe Region

The closed-loop simulation for the CSTR system with an unbounded unsafe region is carried out in this section. Specifically, we define the unsafe region as an unbounded set with high concentration and temperature: $\mathscr{D}_u := \{x \in \mathbf{R}^2 \mid F(x) = x_1 + x_2 > 7.2\}$. Similarly, H is defined as a superset of $\mathscr{D}_u$: $H := \{x \in \mathbf{R}^2 \mid F(x) > 6.8\}$. Then, we design the control barrier function $B(x)$ with the following form:

$$B(x) = \begin{cases} e^{F(x)-7.2} - 2 \times e^{-0.4}, & \text{if } x \in H \\ -e^{-0.4}, & \text{if } x \notin H. \end{cases} \tag{4.32}$$

The CLBF $W_c(x) = V(x) + \mu B(x) + \nu$ is developed with the following parameters: $\rho_c = 0$, $c_1 = 0.001$, $c_2 = 10$, $c_3 = 98.78$, $c_4 = 51.99$, $\nu = \rho_c - c_1 c_4 = -1.685 \times 10^{-2}$, and $\mu = 1500$. For simplicity, we only discuss the scenario of the nominal CSTR system under CLBF-MPC. Figure 4.11 shows that all the state trajectories with initial states inside $\mathscr{U}_\rho$ converge to $\mathscr{U}_{\rho_{\min}}$ and remain outside of the unsafe region $\mathscr{D}_u$ for all times under the CLBF-MPC of Eq. 4.27.

Therefore, we conclude from the two case studies (i.e., unbounded and bound unsafe regions) in this section that simultaneous process operational safety and closed-loop stability are achieved for the nonlinear system under the CLBF-MPC

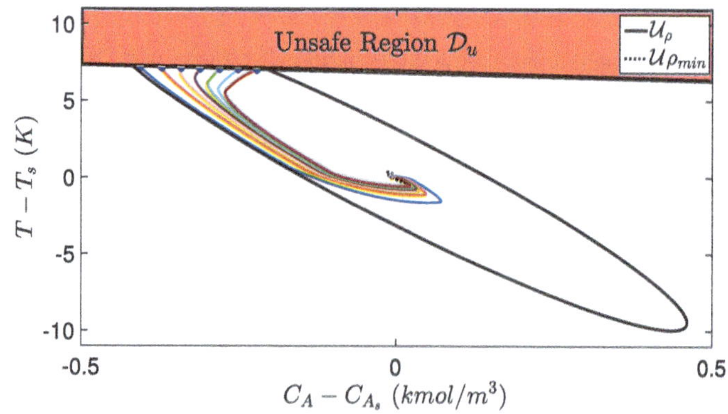

Fig. 4.11 Closed-loop state trajectories (with different initial conditions marked by stars) for the system of Eq. 4.28 under CLBF-MPC, where the unbounded unsafe region $\mathscr{D}_u$ is represented by the red area on the top

of Eq. 4.27 in the sense that the closed-loop state is guaranteed to avoid the unsafe region and remain inside $\mathcal{U}_\rho$ for all times, and will be ultimately bounded in a small neighborhood $\mathcal{U}_{\rho_{\min}}$ around the origin, for any initial state $x_0 \in \mathcal{U}_\rho \subset \mathcal{U}_{\rho_c}$.

4.5 CLBF-Based Economic Model Predictive Control

By incorporating CLBF-based constraints into tracking MPC, the state of a closed-loop nonlinear system can be driven to its set point while avoiding a bounded/unbounded unsafe region in state-space. However, given that the steady-state operation may not be optimal for industrial process operation as demonstrated in the previous chapter, economic model predictive control (EMPC) that optimizes directly in real time the economic performance of the process is an efficient method to improve process economic performance while maintaining stable operation. Therefore, based on the Lyapunov-based EMPC of Eq. 2.35 in Chap. 2, we present the design of CLBF-based EMPC (CLBF-EMPC) that ensures closed-loop stability, process operational safety, and economic optimality simultaneously. It should be noted that simultaneous closed-loop stability and operational safety now represent the boundedness of the state in a safe operating region only as EMPC does not require the convergence of state to the steady-state.

4.5.1 CLBF-Based EMPC Formulation

The CLBF-EMPC scheme is formulated by the following optimization problem [220]:

$$\max_{u(t)\in S(\Delta)} \int_{t_k}^{t_{k+N}} l_e(\tilde{x}(\tau), u(\tau))\, d\tau \tag{4.33a}$$

$$\text{s.t.} \quad \dot{\tilde{x}}(t) = f(\tilde{x}(t)) + g(\tilde{x}(t))u(t) \tag{4.33b}$$

$$\tilde{x}(t_k) = x(t_k) \tag{4.33c}$$

$$u(t) \in U,\ \forall\, t \in [t_k, t_{k+N}) \tag{4.33d}$$

$$W_c(\tilde{x}) \leq \rho_e,\quad \forall\, t \in [t_k, t_{k+N}),\quad \text{if } x(t_k) \in \mathcal{U}_{\rho_e} \tag{4.33e}$$

$$\dot{W}_c(x(t_k), u(t_k)) \leq \dot{W}_c(x(t_k), \Phi(x(t_k))),\quad \text{if } x(t_k) \in \mathcal{U}_\rho \backslash \mathcal{U}_{\rho_e} \tag{4.33f}$$

where the notation follows that for CLBF-MPC in Eq. 4.27. Unlike the CLBF-MPC that minimizes the objective function of $l_t(x, u)$ that has the minimum value 0 at the origin, the optimization problem of Eq. 4.33 maximizes the time integral of the cost function $l_e(x, u)$ of Eq. 4.33a that represents process economic benefits while satisfying the constraints of Eqs. 4.33b–4.33f. Specifically, the nominal process model of Eq. 4.33b is used as the prediction of CLBF-EMPC. The state measurement at the current time t_k is utilized as the initial condition for the optimization problem of Eq. 4.33 in Eq. 4.33c. The input constraints are defined by Eq. 4.33d. The constraints of Eqs. 4.33e–4.33f guarantee closed-loop stability and safety for the system under EMPC. Specifically, when $x(t_k)$ is inside $\mathcal{U}_{\rho_e}$, the constraint of Eq. 4.33e (Mode 1 constraint) is applied to maintain the predicted state within the set $\mathcal{U}_{\rho_e} \subset \mathcal{U}_\rho$, which is designed to make the safe operating region $\mathcal{U}_\rho$ a forward invariant set in the presence of sufficiently small disturbances (i.e., $|w(t)| \leq \theta$) and also include the states $x_e \in \mathbb{X}_e$ inside (i.e., $\mathcal{B}_\delta(x_e) \subset \mathcal{U}_{\rho_e}$).

When the state leaves $\mathcal{U}_{\rho_e}$, the contractive constraint of Eq. 4.33f (Mode 2 constraint) is activated to decrease the value of $W_c(x)$ for the next sampling step, such that within finite sampling steps, the closed-loop state will move back into $\mathcal{U}_{\rho_e}$. We implement the CLBF-EMPC in a sample-and-hold fashion and apply the first control action in the optimized input trajectory to the nonlinear system over the next sampling period.

Before we provide closed-loop stability and safety analysis for the nonlinear system under CLBF-EMPC in Theorem 4.5, a few propositions that will be used in the proof of theorem are first established. Specifically, starting from the same initial condition, the upper bound for the difference between the evolutions of the state trajectories of the disturbed system of Eq. 4.1 and of the nominal system (i.e., $w(t) \equiv 0$) are provided in Proposition 4.3. Additionally, the relationship of the sampling period, the Lipschitz constants, and the disturbance bound that are required to maintain $\dot{W}_c$ negative within one sampling period are established in Proposition 4.4. This relationship will be later used in the proof of operational safety and closed-loop stability of the CLBF-EMPC in Theorem 4.5. Also, to simplify the discussion, we only consider the initial condition $x_0 \in \mathcal{U}_\rho \subset \mathcal{U}_{\rho_c}$, for which closed-loop stability and operational safety will be demonstrated for CLBF-EMPC by showing that the state $x(t)$ is bounded in the invariant set $\mathcal{U}_\rho$ for all times.

Proposition 4.3 *Consider the nominal system $\dot{\hat{x}} = F(\hat{x}, u, 0)$ (i.e., $w(t) \equiv 0$) and the disturbed system of Eq. 4.1, i.e., $\dot{x} = F(x, u, w) := f(x) + g(x)u + h(x)w$ with the same initial conditions $x_0 = \hat{x}_0 \in \mathcal{U}_\rho \subset \mathcal{U}_{\rho_c}$. There exists a positive constant β and a class $\mathcal{K}$ function $f_w(\cdot)$ such that the inequalities below hold for all $x, \hat{x} \in \mathcal{U}_\rho$ and $w(t) \in W$:*

$$|x(t) - \hat{x}(t)| \leq f_w(t) := \frac{L_w \theta}{L_x}(e^{L_x t} - 1) \tag{4.34a}$$

$$W_c(x) \leq W_c(\hat{x}) + \alpha_4(\alpha_1^{-1}(\rho - \rho_0))|x - \hat{x}| + \beta|x - \hat{x}|^2. \tag{4.34b}$$

Proof We define the error vector as $e(t) = x(t) - \hat{x}(t)$ and obtain the time-derivative of $e(t)$ as follows:

$$|\dot{e}(t)| = |F(x(t), u(t), w(t)) - F(\hat{x}(t), u(t), 0)|. \tag{4.35}$$

Using Eq. 4.14b, the bound for $|\dot{e}(t)|$ can be obtained as follows:

$$|\dot{e}(t)| \leq L_x|x(t) - \hat{x}(t)| + L_w|w(t)| \leq L_x|e(t)| + L_w|\theta|. \tag{4.36}$$

Therefore, the upper bound of the norm of the error vector is derived for all $|w(t)| \leq \theta$ and $x(t), \hat{x}(t) \in \mathcal{U}_\rho$ with zero initial condition (i.e., $e(0) = 0$):

$$|e(t)| = |x(t) - \hat{x}(t)| \leq \frac{L_w\theta}{L_x}(e^{L_x t} - 1). \tag{4.37}$$

Then, using the Taylor series expansion of $W_c(x)$ around $\hat{x}$, we show that Eq. 4.34b holds for all $x, \hat{x} \in \mathcal{U}_\rho$ as follows:

$$W_c(x) \leq W_c(\hat{x}) + \frac{\partial W_c(\hat{x})}{\partial x}|x - \hat{x}| + \beta|x - \hat{x}|^2. \tag{4.38}$$

Finally, we substitute Eq. 4.13a and Eq. 4.13c into Eq. 4.38 and obtain the following inequality:

$$W_c(x) < W_c(\hat{x}) + \alpha_4(\alpha_1^{-1}(\rho - \rho_0))|x - \hat{x}| + \beta|x - \hat{x}|^2. \tag{4.39}$$

Proposition 4.4 *Consider the nonlinear system of Eq. 4.1 with a CLBF W_c that meets Eqs. 4.12 and 4.13, and has its minimum at the origin. Let $\varepsilon_w > 0$, $\Delta^* > 0$, $\rho > \rho_e$ satisfy*

$$-\alpha_3(\alpha_2^{-1}(\rho_e - \rho_0)) + L_x^{'} M\Delta^* + L_w^{'}\theta \leq -\varepsilon_w/\Delta^*. \tag{4.40}$$

Then, for any $x(t_k) \in \mathcal{U}_\rho \backslash \mathcal{U}_{\rho_e}$, the following inequality holds under the sample-and-hold implementation of $u = \Phi(x) \in U$:

$$W_c(x(t)) \leq W_c(x(t_k)), \ \forall t \in [t_k, t_{k+1}). \tag{4.41}$$

Proof Assuming $x(t_k) \in \mathcal{U}_\rho \backslash \mathcal{U}_{\rho_e}$, we prove that the value of $W_c(x)$ is decreasing within one sampling period under the controller $u(t) = \Phi(x(t_k)) \in U$. We first derive the time-derivative of $W_c(x)$ along the state trajectory for the nominal system of Eq. 4.1 in one sampling period, i.e., $t \in [t_k, t_{k+1})$:

$$\dot{W}_c(x(t)) = \frac{\partial W_c(x(t))}{\partial x} F(x(t), \Phi(x(t_k)), w(t)). \tag{4.42}$$

Adding $\frac{\partial W_c(x(t_k))}{\partial x} F(x(t_k), \Phi(x(t_k)), 0)$ to both sides and then using Eq. 4.13b, we obtain the following inequality:

$$\begin{aligned} \dot{W}_c(x(t)) \leq & -\alpha_3(|x(t_k)|) + \frac{\partial W_c(x(t))}{\partial x} F(x(t), \Phi(x(t_k)), w(t)) \\ & - \frac{\partial W_c(x(t_k))}{\partial x} F(x(t_k), \Phi(x(t_k)), 0). \end{aligned} \tag{4.43}$$

Then, the upper bound of $\dot{W}_c(x(t))$ for $x(t_k) \in \mathcal{U}_\rho \backslash \mathcal{U}_{\rho_e}$ can be derived using the inequalities of Eqs. 4.13a and 4.14:

$$\begin{aligned} \dot{W}_c(x(t)) &\leq -\alpha_3(\alpha_2^{-1}(\rho_e - \rho_0)) + L_x^{'}|x(t) - x(t_k)| + L_w^{'}\theta \\ &\leq -\alpha_3(\alpha_2^{-1}(\rho_e - \rho_0)) + L_x^{'} M \Delta^* + L_w^{'}\theta. \end{aligned} \tag{4.44}$$

Therefore, we prove that $\dot{W}_c(x(t)) \leq -\varepsilon_w/\Delta^*$ holds for all $x(t_k) \in \mathcal{U}_\rho \backslash \mathcal{U}_{\rho_e}$, $t \in [t_k, t_{k+1})$ if Eq. 4.40 is satisfied. Additionally, the result that $W_c(x(t_{k+1})) \leq W_c(x(t_k)) - \varepsilon_w$ and the conclusion in Eq. 4.41 are obtained by integrating $\dot{W}_c(x(t)) \leq -\varepsilon_w/\Delta^*$ over one sampling period.

Based on the results in Propositions 4.3 and 4.4, the following theorem is established to demonstrate simultaneous process operational safety and closed-loop stability for the nonlinear system of Eq. 4.1 under CLBF-EMPC. Moreover, it will be proven that the CLBF-EMPC optimization problem is recursively feasible under the sample-and-hold implementation of the control actions computed by the CLBF-EMPC of Eq. 4.33.

Theorem 4.5 *Consider the nonlinear system of Eq. 4.1 with a constrained CLBF* $W_c(x) : \mathbf{R}^n \rightarrow \mathbf{R}$ *that meets Eqs. 4.12 and 4.13 and has its minimum at the origin. Let* $\Delta \leq \Delta^*$ *and* ρ_e *be a positive real number that satisfies Eq. 4.40 and the following inequality:*

$$\rho_e \leq \rho - \alpha_4(\alpha_1^{-1}(\rho - \rho_0)) f_w(\Delta) - \beta(f_w(\Delta))^2. \tag{4.45}$$

Given any initial state $x_0 \in \mathcal{U}_\rho$, *it is guaranteed that the closed-loop state of the system of Eq. 4.1 remains inside* $\mathcal{U}_\rho$ *at all times, i.e.,* $x(t) \in \mathcal{U}_\rho$, $\forall t \geq 0$ *under the CLBF-EMPC of Eq. 4.33, where* $\mathcal{U}_\rho \subset \mathcal{U}_{\rho_c}$ *and* $\mathcal{U}_\rho \cap \mathcal{D} = \emptyset$.

Proof The proof of simultaneous operational safety and closed-loop stability of the system of Eq. 4.1 under CLBF-EMPC consists of three parts. We first prove that the closed-loop state is always bounded in the safe operating region $\mathcal{U}_\rho$ ($\mathcal{U}_\rho$ is also the stability region, where safety is implied by the fact that $\mathcal{U}_\rho \cap \mathcal{D} = \emptyset$, and stability is owing to the invariance property of the level set of $W_c(x)$) under the Mode 1 constraint of Eq. 4.33e of CLBF-EMPC. In the second part, we prove that the closed-loop state will move toward the origin, and enter $\mathcal{U}_{\rho_e}$ within finite sampling steps if the Mode 2 constraint of Eq. 4.33f is activated for $x(t_k) \in \mathcal{U}_\rho \backslash \mathcal{U}_{\rho_e}$. Finally, we show that the CLBF-EMPC optimization problem can be solved with feasible solutions for all states $x(t) \in \mathcal{U}_\rho$.

Part 1 : We prove that when $x(t_k) \in \mathcal{U}_{\rho_e}$, $t_k \geq 0$, the closed-loop state $x(t) \in \mathcal{U}_\rho$ holds for all $t \in [t_k, t_{k+1}]$. Since the state $x(t_k)$ at $t = t_k$ is assumed to be in $\mathcal{U}_{\rho_e}$, the Mode 1 constraint of Eq. 4.33e is activated and the Mode 2 constraint of Eq. 4.33f remains inactive. It is noted that in the CLBF-EMPC of Eq. 4.33, the nominal system of Eq. 4.33e is used for predicting future states. However, the actual system to which the control actions will be applied could be the nominal system (i.e., $w(t) \equiv 0$) or the disturbed system with sufficient small bounded disturbances (i.e., $|w(t)| \leq \theta$).

Therefore, we will discuss operational safety and closed-loop stability for these two scenarios. We first consider the case where the CLBF-EMPC of Eq. 4.33 is applied to the nominal system of Eq. 4.1. Since the actual process model and the prediction model are both the nominal system with $w(t) \equiv 0$, it is straightforward to show that $W_c(\hat{x}(t_{k+1})) \leq \rho_e \leq \rho$ holds for the nominal system of Eq. 4.1 from the constraint of Eq. 4.33e, where $\hat{x}$ represents the predicted state. Now we consider the case where CLBF-EMPC uses the nominal system for prediction, but is applied to the disturbed system of Eq. 4.1 (i.e., $|w(t)| \leq \theta$). In this case, it is readily shown that the predicted state is still within $\mathcal{U}_{\rho_e}$ (i.e., $W_c(\hat{x}(t_{k+1})) \leq \rho_e$) from the constraint of Eq. 4.33e. However, the true state is shown to be bounded in $\mathcal{U}_\rho$ using the results from Propositions 4.3 and 4.4 and Eq. 4.45:

$$\begin{aligned} W_c(x(t_{k+1})) &\leq +\alpha_4(\alpha_1^{-1}(\rho - \rho_0))|x(t_{k+1}) - \hat{x}(t_{k+1})| \\ &\quad + \beta|x(t_{k+1}) - \hat{x}(t_{k+1})|^2 + W_c(\hat{x}(t_{k+1})) \\ &\leq \alpha_4(\alpha_1^{-1}(\rho - \rho_0)) f_w(\Delta) + \beta(f_w(\Delta))^2 + \rho_e \\ &\leq \rho. \end{aligned} \tag{4.46}$$

Therefore, for any $x(t_k) \in \mathcal{U}_{\rho_e}$, regardless of whether the CLBF-EMPC is applied to the disturbed system with sufficiently small bounded disturbances or the nominal system, the state $x(t_{k+1})$ is always bounded in $\mathcal{U}_\rho$. Additionally, if we substitute a smaller sampling period into the monotonically increasing function $f_w(\cdot)$ in Eq. 4.46, it is straightforward to show that the above inequality holds for any time instance $t \in [t_k, t_{k+1})$.

Part 2 : In the second part, we prove that the closed-loop state $x(t)$ will move into a smaller level set of W_c (i.e., $W_c(x(t)) \leq W_c(x(t_k)), \forall t \in [t_k, t_{k+1})$), and can be bounded in $\mathcal{U}_{\rho_e}$ within finite sampling steps when $x(t_k) \in \mathcal{U}_\rho \backslash \mathcal{U}_{\rho_e}$. Specifically, when $x(t_k) \in \mathcal{U}_\rho \backslash \mathcal{U}_{\rho_e}$, the CLBF-EMPC activates the Mode 2 constraint of Eq. 4.33f. Similarly, we consider the two cases where the CLBF-EMPC is applied to the nominal system and to the disturbed system. We first consider the scenario that the actual process model and the prediction model are both the nominal system of Eq. 4.1 (i.e., $w(t) \equiv 0$). In this case, the following inequality is obtained from the constraint of Eqs. 4.33f and 4.13b:

$$\begin{aligned} \dot{W}_c(x(t_k), u(t_k)) &= \frac{\partial W_c(x(t_k))}{\partial x} F(x(t_k), u(t_k), 0) \\ &\leq \frac{\partial W_c(x(t_k))}{\partial x} F(x(t_k), \Phi(x(t_k)), 0) \\ &\leq -\alpha_3(|x(t_k)|) \end{aligned} \tag{4.47}$$

where $u(t_k)$ is the optimal solution from CLBF-EMPC that will be applied within the next sampling period $\forall t \in [t_k, t_{k+1})$. Then, we obtain the bound for $\dot{W}_c(x(t)), \forall t \in [t_k, t_{k+1})$ under the sample-and-hold implementation of CLBF-EMPC following the results in Eqs. 4.43 and 4.44 with $w(t) = 0$:

$$\begin{aligned}\dot{W}_c(x(t), u(t_k)) &= \frac{\partial W_c(x(t))}{\partial x} F(x(t), u(t_k), 0) \\ &\leq -\alpha_3(|x(t_k)|) + L_x^{'} M\Delta + L_w^{'}|0|.\end{aligned} \tag{4.48}$$

Additionally, based on Eq. 4.48, the upper bound for $\dot{W}_c(x(t), u(t_k))$ under the CLBF-EMPC that is applied to the disturbed system of Eq. 4.1 which is obtained as follows:

$$\begin{aligned}\dot{W}_c(x(t), u(t_k)) &= \frac{\partial W_c(x(t))}{\partial x} F(x(t), u(t_k), w(t)) \\ &\leq -\alpha_3(|x(t_k)|) + L_x^{'} M\Delta + L_w^{'}\theta.\end{aligned} \tag{4.49}$$

Since Eq. 4.40 in Proposition 4.4 is satisfied, it is concluded that $\dot{W}_c(x(t)) \leq -\varepsilon_w/\Delta$, $\forall t \in [t_k, t_{k+1})$ holds for both the disturbed system of Eq. 4.1 (i.e., $|w(t)| \leq \theta$) and the nominal system of Eq. 4.1 (i.e., $|w(t) \equiv 0$). As a result, it follows that $W_c(x(t)) \leq W_c(x(t_k))$ and $W_c(x(t_{k+1})) \leq W_c(x(t_k)) - \varepsilon_w, \forall t \in [t_k, t_{k+1})$. This shows that $W_c(x)$ is forced to decrease every sampling step such that the closed-loop state will be bounded in $\mathcal{U}_{\rho_e}$ within finite sampling steps.

So far, we have shown that the state at the next sampling time $x(t_{k+1})$ is guaranteed to be bounded in $\mathcal{U}_\rho$ for any initial condition $x(t_k) \in \mathcal{U}_{\rho_e}$ or $x(t_k) \in \mathcal{U}_\rho \backslash \mathcal{U}_{\rho_e}$ under the CLBF-EMPC of Eq. 4.33.

By rolling the horizon, it is straightforward to show that the state is bounded in the safe operating region $\mathcal{U}_\rho$ for all times. This completes the proof of process operational safety and closed-loop stability for the closed-loop system under CLBF-EMPC with any initial condition $x_0 \in \mathcal{U}_\rho$.

Part 3 : Lastly, we prove the existence of a feasible solution for the optimization problem of the CLBF-EMPC of Eq. 4.33 by showing that the explicit CLBF-based controller $\Phi(x)$ (in sample-and-hold fashion) provides a feasible solution to the CLBF-EMPC all the time. Specifically, when $x(t_k) \in \mathcal{U}_{\rho_e}$, the CLBF-based control law implemented in a sample-and-hold manner, i.e., $u(t) = \Phi(x(t_k + i\Delta))$, $i = 0, 1, \ldots, N-1$ satisfies both the constraint of Eq. 4.33e and the input constraint of Eq. 4.33d. As shown in the proof of the CLBF-MPC of Eq. 4.27 in the previous section, the closed-loop state may move toward the saddle points $x_e \in \mathcal{U}_{\rho_e}$ or the origin under the CLBF-based controller $u = \Phi(x)$; however, in either case, it is guaranteed that the predicted states $\tilde{x}(t_k + i\Delta)$, $i = 0, 1, \ldots, N-1$ are bounded in $\mathcal{U}_{\rho_e}$. Next, when $x(t_k) \in \mathcal{U}_\rho \backslash \mathcal{U}_{\rho_e}$, the CLBF-based controller $u(t) = \Phi(x(t_k))$ is again a feasible solution because it satisfies the constraint of Eq. 4.33f and the input constraint of Eq. 4.33d.

After we obtain the optimal solution from the CLBF-EMPC of Eq. 4.33, and apply the first control action to the system of Eq. 4.1 over the next sampling period, the time horizon will move one sampling period forward (i.e., the rolling horizon). Therefore, at the next sampling step, a feasible control action again exists for $x(t_{k+1})$ at $t = t_{k+1}$ since $x(t_{k+1}) \in \mathcal{U}_\rho$ is guaranteed. The analysis for the two scenarios: $x(t_{k+1}) \in \mathcal{U}_{\rho_e}$ or $x(t_{k+1}) \in \mathcal{U}_\rho \backslash \mathcal{U}_{\rho_e}$ follows exactly the same discussion in the last paragraph. This completes the proof of recursive feasibility of the optimization problem of the CLBF-EMPC of Eq. 4.33 for any $x(t) \in \mathcal{U}_\rho$.

Remark 4.7 The level set $\mathcal{U}_{\rho_e}$ determined by Eq. 4.45 is utilized to make $\mathcal{U}_\rho$ a forward invariant set for the disturbed system operated under the CLBF-EMPC of Eq. 4.33. Additionally, $\mathcal{U}_{\rho_e}$ is designed to include the saddle points x_e where $\partial W_c(x_e)/\partial x = 0$ such that the issue of convergence to x_e will not occur in CLBF-EMPC. Specifically, when $x(t_k) \in \mathcal{U}_\rho \backslash \mathcal{U}_{\rho_e}$, the Mode 2 constraint of Eq. 4.33f will drive the process state into $\mathcal{U}_{\rho_e}$ without having any issue of saddle points since x_e are not included in $\mathcal{U}_\rho \backslash \mathcal{U}_{\rho_e}$. Furthermore, the saddle points will not be an issue either when $x(t_k) \in \mathcal{U}_{\rho_e}$ since the state attempts to move dynamically within $\mathcal{U}_{\rho_e}$ instead of converging to a saddle point in order to maximize process economic benefits under the Mode 1 constraint of Eq. 4.33e. Therefore, saddle points do not affect system stability under CLBF-EMPC due to the nature of EMPC that process economic performance is optimized in a consistently dynamic fashion. However, if the system is required to operate at the steady-state under a tracking MPC, for example, the CLBF-MPC of Eq. 4.27, we need to carefully design the CLBF $W_c(x)$ such that x_e is a saddle point. Additionally, as shown in the CLBF-MPC of Eq. 4.27, an additional constraint that can drive the state away from x_e when the state gets trapped in the saddle point x_e is needed in the MPC layer.

4.5.2 *Application to a Chemical Process Example*

We use the same chemical process example of the non-isothermal, well-mixed, continuous stirred tank reactor (CSTR) as in Chaps. 1 and 3 to illustrate the application of CLBF-EMPC. An irreversible second-order exothermic reaction that converts reactant A to product B is taking place in the CSTR. The CSTR dynamic model and the description of process variables can be found in Sects. 1.3.1 and 3.4.

The states of the CSTR system are the concentration of A in the reactor (denoted by C_A) and the temperature of the reactor (denoted by T). The manipulated inputs are the inlet concentration of species A (denoted by C_{A0}) and the heat input rate (denoted by Q). Initially, the CSTR is operated at the steady-state $(C_{A0_s}\ Q_s) =$ $(4\,\text{kmol/m}^3,\ 0\,\text{kJ/h})$ and $(C_{As},\ T_s) = (1.22\,\text{kmol/m}^3,\ 438\,\text{K})$. Additionally, all the variables are represented in their deviation forms, i.e., the manipulated inputs and the states of the closed-loop system are $u^T = [\Delta C_{A0}\ \Delta Q]$, respectively, where $\Delta C_{A0} = C_{A0} - C_{A0_s}$ and $\Delta Q = Q - Q_s$, and $x^T = [C_A - C_{As}\ T - T_s]$. The manipulated inputs are bounded as follows: $|\Delta C_{A0}| \leq 3.5$ kmol/m^3 and $|\Delta Q| \leq 5 \times 10^5$ kJ/h. The control objective of CLBF-EMPC is to maximize the economic profit of the CSTR process, and meanwhile, to maintain the closed-loop state in the safe operating region $\mathcal{U}_\rho$ for all times. The objective function of the CLBF-EMPC is developed to optimize the production rate of B: $l_e(\tilde{x}, u) = k_0 e^{-E/RT} C_A^2$.

We define the unsafe region $\mathcal{D}$ as an open set of states with relatively high temperature inside the stability region (i.e., the level set of $V(x)$). The unsafe region is designed as an ellipse in state-space: $\mathscr{D} := \{x \in \mathbf{R}^2 \mid F(x) = (x_1 + 0.92)^2 + \frac{(x_2-42)^2}{500} < 0.06\}$. Then, we design the set H as $H := \{x \in \mathbf{R}^2 \mid F(x) < 0.07\}$, and develop the control barrier function $B(x)$ as follows:

$$B(x) = \begin{cases} e^{\frac{F(x)}{F(x)-0.07}} - e^{-6}, \text{if } x \in H \\ -e^{-6}, \text{if } x \notin H. \end{cases} \tag{4.50}$$

The control Lyapunov function $V(x) = x^T Px$ is designed in a quadratic from with $P = \begin{bmatrix} 1060 & 22 \\ 22 & 0.52 \end{bmatrix}$. Finally, the CLBF $W_c(x) = V(x) + \mu B(x) + \nu$ is developed following the construction method in Proposition 4.1 with the parameters determined as follows: $\rho_c = 0$, $c_1 = 0.1$, $c_2 = 1061$, $c_3 = \max_{x \in \partial H} |x|^2 = 2295$, $c_4 = \min_{x \in \partial \mathscr{D}} |x|^2 = 1370$, $\nu = \rho_c - c_1 c_4 = -160$, and $\mu = 1 \times 10^9$. The safe operating region $\mathscr{U}_\rho$ is then characterized with $\rho = -2.47 \times 10^6$. Given the functional form of $W_c(x)$, we do a grid search for the stationary points in state-space and find that there exists a saddle point $x_e = (-1.00, 47.5)$ in state-space. Additionally, to make the averaged reactant material available (in deviation from the steady-state value, C_{A0s}) over a given operating period $t_p = 1.0$ h to be 0, we employ the following material constraint $\frac{1}{t_p} \int_0^{t_p} u_1(\tau) d\tau = 0\,\text{kmol/m}^3$ in CLBF-EMPC. Again, the CLBF-EMPC is solved using the IPOPT software package [201] with the sampling period $\Delta = 10^{-2}$ h. The numerical simulation of the dynamic model described by Eq. 1.1 in Sect. 1.3.1 is performed using the explicit Euler method with a sufficiently small integration time step of $h_c = 10^{-4}$ h. Figures 4.12 and 4.13 show the closed-loop state trajectories and the manipulated input profiles for the CSTR system under CLBF-EMPC, where the dashed horizontal lines in Fig. 4.13 are the lower and upper bounds for ΔC_{A0} and ΔQ.

It is clearly seen from Fig. 4.12 that the closed-loop state under CLBF-EMPC remains inside the safe operating region $\mathscr{U}_\rho$ for all times, while the state under the standard LEMPC of Eq. 2.35 enters the unsafe region (the red ellipse in Fig. 4.12) in the simulation. This is consistent with the fact that the standard LEMPC can only guarantee the boundedness of the state in the stability region (i.e., $\mathscr{U}_\rho \cup \mathscr{D}$), but

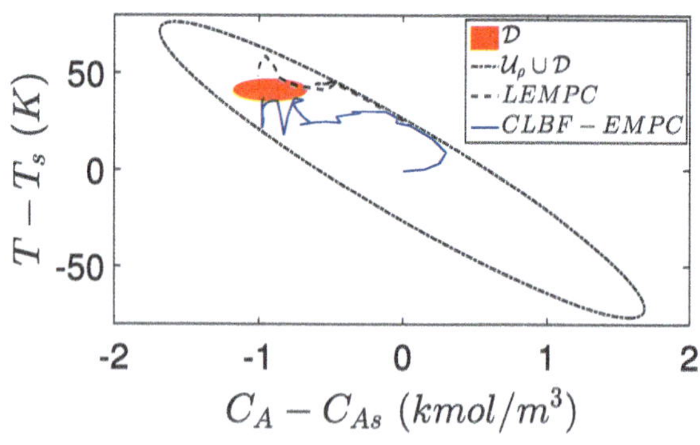

Fig. 4.12 Closed-loop state trajectories for the CSTR under the CLBF-EMPC of Eq. 4.33 and the standard LEMPC of Eq. 2.35 with the same initial condition (0, 0)

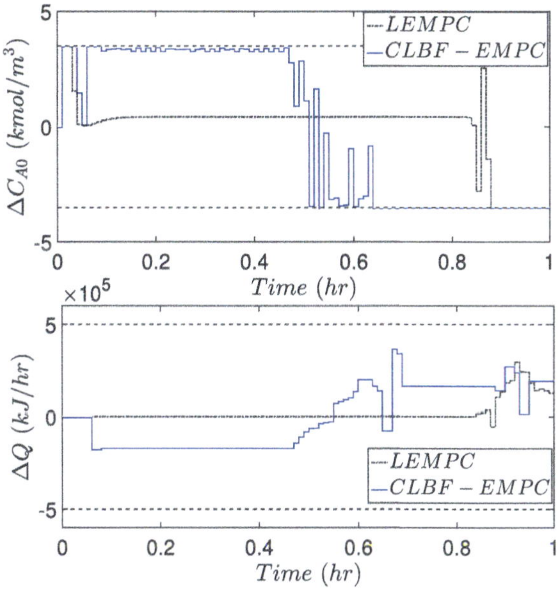

Fig. 4.13 Manipulated input profiles ($u_1 = \Delta C_{A0}$, $u_2 = \Delta Q$) for the CSTR under the CLBF-EMPC of Eq. 4.33 and the standard LEMPC of Eq. 2.35 with the same initial condition (0, 0)

not the avoidance of unsafe regions in state-space since safety considerations are not taken into account. The manipulated input profiles in Fig. 4.13 demonstrate that the material constraint and the input constraints are both satisfied by the optimized control actions from CLBF-EMPC all the time. Specifically, it can be seen that the maximum allowable reactant ΔC_{A0} is consumed during the first 0.5 h to accelerate the reaction for improving economic performance. As a result, the CLBF-EMPC has to lower the consumption at the second half-hour in order to meet the material constraint. Additionally, it is observed that the control actions from $t = 0.5$ h show an oscillatory behavior as the closed-loop state approaches the boundary of $\mathscr{D}$. The oscillation occurs because the intrinsic dynamics of the closed-loop system attempts to drive the state across the unsafe region using low energy (i.e., a small control action), while the CLBF-based constraint prevents this undesirable behavior by requiring W_c to be decreasing using a large control action.

Lastly, to demonstrate the improved economic performance, we calculate the economic benefits $L_E = \int_0^{t_p} l_e(x, u)dt$ within the entire operation period $t_p = 1$ h under the steady-state operation (i.e., the CSTR is operated at the steady-state for all times) and under the CLBF-EMPC, respectively. The economic profits are computed to be 13.9 and 16.2 for steady-state operation and the CLBF-EMPC, respectively. Therefore, through the simulation study, it is concluded that the CLBF-EMPC ensures process operational safety and economically outperforms the steady-state operation.

4.6 Conclusions

In this chapter, CLBF-based MPC and EMPC designs were developed to optimize closed-loop performance and ensure closed-loop stability and operational safety simultaneously for nonlinear systems associated with a bounded/unbounded unsafe region. CBFs were first introduced to maintain a safe operation for nonlinear systems by avoiding undesirable regions in state-space. Subsequently, a constrained CLBF was developed for input-constrained systems by combining a CLF and a CBF together following a specific construction method. Following that, CLBF-based controllers were designed with a rigorous theoretical analysis of closed-loop stability and operational safety showing that the closed-loop state is driven to the steady-state while avoiding the unsafe region for all times. Both the cases of bounded and unbounded unsafe regions were discussed. It was demonstrated that a discontinuous control action was required to address the issue of (i.e., avoid) convergence of the state to saddle points under the continuous implementation of a stabilizing controller in the presence of a bounded unsafe region.

In order to optimize closed-loop performance while accounting for closed-loop stability and operational safety, CLBF-based MPC and EMPC schemes were developed by incorporating CLBFs in the designs of stability and safety constraints. The formulations of the two MPC schemes were provided and rigorous theoretical treatments of the schemes were carried out. The effectiveness of the two MPC schemes was demonstrated using chemical process examples. Specifically, the superiority of CLBF-MPC was demonstrated through the comparison with an explicit CLBF-based controller and a standard MPC with state constraints. Additionally, the CLBF-EMPC scheme was compared with the steady-state operation, showing that closed-loop economic performance was significantly improved under EMPC. In all cases considered, closed-loop stability and operational safety were guaranteed simultaneously, and the optimization problems of MPCs were solving with recursive feasibility.

Chapter 5
Integration of Safety Systems with Control Systems

5.1 Introduction

In Chaps. 3 and 4, process operational safety has been directly incorporated into control system design to avoid unsafe operating conditions. However, it is impossible to eliminate all hazards due to disturbances and device failures in chemical plants, and therefore, a safety system, comprised of several independent layers, should be employed. Specifically, as shown in Fig. 1.1, a complete control and safety system used in industries includes basic process control systems (BPCSs), alarm systems, emergency shutdown systems (ESSs), and safety relief devices. Ideally, process variables are regulated by BPCS to their set-points while avoiding the unsafe operating conditions via Safeness Index-based constraints or control Lyapunov-barrier function-based constraints that have been discussed in Chaps. 3 and 4. When unusually large process disturbances or equipment faults occur, and the BPCS fails to maintain process variables within the desired range for safe operation, alarm systems will be triggered such that operators can take actions to prevent further unsafe operation. Additionally, the ESS and safety relief devices will be activated if process variables subsequently further exceed allowable values, which triggers an extremely dangerous operating condition. Therefore, in order to design a unified control and safety system that is able to handle various disturbances, the integration of upper-layer safety systems with BPCS needs to be studied in addition to the designs of safe BPCS.

Various methods and case studies are presented in this chapter to demonstrate the integration of safety considerations into control system design. In the first section, the dynamic interactions between feedback control and safety systems are presented, followed by a high-pressure flash drum separator and a continuous stirred tank reactor (CSTR) example to illustrate the applications of classical (i.e., proportional–integral controllers) and model-based controllers. In the second section, the Safeness Index-based MPC introduced in Chap. 3 is applied to a flash drum, an ammonia produc-

Z. Wu and P. D. Christofides, *Process Operational Safety and Cybersecurity*, Advances in Industrial Control, https://doi.org/10.1007/978-3-030-71183-2_5

tion process, and a large-scale ammonia process network for which multiple model predictive controllers were developed to improve process operational safety under process disturbances. In all the aforementioned case studies, Aspen Plus, Aspen Plus Dynamics, and MATLAB are used to build dynamic process models and control systems, and carry out the closed-loop simulations to demonstrate the applicability and effectiveness of the proposed control methods.

Aspen Plus is a widely used commercial software for process design and simulation. For example, calculation of the process steady-state can be done in Aspen Plus with appropriate thermodynamic models that are determined based on process mass and energy balances. Based on the steady-state model developed in Aspen Plus and additional detailed parameters, process dynamic simulations can be carried out in Aspen Plus Dynamics. The reader is referred to [5, 22] for further details about Aspen simulation software.

5.2 Integration of Safety and Control Systems

In the traditional process safety paradigm, the safety system is activated when the control system (BPCS) fails to operate the process in a safe operating region. However, since the process dynamics is changed after the activation of safety systems (e.g., opening of a pressure relief valve to prevent high pressure in a chemical reactor), the actions taken by the safety systems should be taken into account when calculating control actions in BPCS. In this section, we present two industrial chemical processes where safety is of significant importance. The interaction between the safety and control systems will be investigated for both the chemical reactor example using a model predictive control (MPC) scheme and the flash drum example using a proportional–integral (PI) control scheme. In the first case study, we consider a CSTR with methyl isocyanate (MIC) hydrolysis reaction, in which thermal runaway may occur due to disturbances. We will demonstrate that through appropriate integration of control and safety systems, thermal runaway can be prevented in the CSTR. The second case study focuses on a flash drum with valve malfunction that may lead to extremely high pressure. We will demonstrate that by incorporating safety system actions into control system design (i.e., the tuning of PI controller parameters account for the safety system being on or off), closed-loop performance can be much improved compared to the control system with the same parameters regardless of the activation/deactivation of safety systems.

5.2.1 *Case Study: Thermal Runaway in a CSTR System*

The first case study is the CSTR with methyl isocyanate (MIC) hydrolysis reaction, where MIC is the principal chemical involved in the Bhopal disaster [29]. The control

Table 5.1 Parameter values for the CSTR with MIC reaction

$T_0 = 293$ K	$F = 57.5$ kg/s
$m = 4.1 \times 10^4$ kg	$E_a = 6.54 \times 10^4$ J/mol
$k_0 = 4.13 \times 10^8$ s	$\Delta H = -8.04 \times 10^4$ J/mol
$C_P = 3000$ J/(kg K)	$R = 8.314$ J/(mol K)
$L = 7.1 \times 10^6$ J/(s K)	$C_{A_0} = 29.35$ mol/kg
$T_{j_s} = 293$ K	$C_{A_s} = 10.1767$ mol/kg
$T_s = 305.1881$ K	

system, i.e., Lyapunov-based MPC (LMPC), will be coordinated with the safety relief valve system to avoid unsafe operations [237].

5.2.1.1 MIC Reaction and CSTR Process Description

The exothermic hydrolysis reaction that converts the reactant, methyl isocyanate, to carbon dioxide and amine is described as follows:

$$CH_3NCO_{(l)} + H_2O_{(l)} \rightarrow CH_3NH_{2(aq)} + CO_{2(aq)}$$

The dynamic model of the process is derived based on the mass and energy balances and is of the form:

$$\begin{aligned} m\frac{dC_A}{dt} &= -mk_0 e^{\frac{-Ea}{RT}} C_A + F(C_{A0} - C_A) \\ mC_P\frac{dT}{dt} &= (-\Delta H)mk_0 e^{\frac{-Ea}{RT}} C_A + FC_P(T_0 - T) - L(T - T_j) \end{aligned} \tag{5.1}$$

where m is the total mass of the mixture in the reactor, and C_A and T are the concentration of MIC and the temperature in the reactor in units of mol/kg and K, respectively. C_{A0} and T_0 are the reactant MIC concentration and temperature in the feed stream, and F is the flow rate for both the CSTR inlet and outlet streams. The heat capacity of the reacting liquid C_P is assumed to be constant. ΔH, E_a, and k_0 represent the enthalpy, activation energy, and the reaction pre-exponential factor, respectively. L and T_j represent the heat transfer coefficient and temperature for the CSTR cooling jacket. Table 5.1 reports the process parameter values used in the simulation. In this example, we simulate the CSTR at the operating conditions reported for the Bhopal catastrophe in [190]. Note that the simulations are carried out based on the assumption that the liquid in the CSTR can vaporize. However, to allow the key aspects of the proposed method for integrating the control system (i.e., MPC) with the safety system to be explored despite the modeling approximation, we will continue to use Eq. 5.1 even when vaporization of the liquid occurs.

5.2.1.2 Lyapunov-Based MPC Design

We initially operate the CSTR at the following steady-state: jacket temperature $T_{j_s} = 293\ K$; MIC concentration and temperature $\left[C_{A_s}\ T_s\right] = \left[10.1767 \text{ mol/kg}\right.$ 305.1881 K]. The control objective is to stabilize the CSTR at its steady-state by manipulating the cooling jacket temperature T_j (i.e., manipulated inputs) accounting for limitation on control actuator (i.e., the input constraint is 280 K $\leq T_j \leq$ 300 K). We represent the input and the states in their deviation variable form from the steady-state as follows: $u = T_j - T_{j_s}$ and $x^T = [C_A - C_{A_s}\ T - T_s]$, such that the steady-states of x and u are at the origin in the state-space. Following this notation, we can rewrite the system of Eq. 5.1 in the general form of a nonlinear system: $\dot{x} = F(x, u) := f(x) + g(x)u$, where $f(x)$ and $g(x)$ are nonlinear vector functions of process state x. The Lyapunov-based MPC (LMPC) is utilized to drive the closed-loop CSTR system to a neighborhood of the origin for the state initiating from any initial condition in the stability region which is an explicitly characterizable region that ensures closed-loop stability in the state-space. The LMPC scheme is represented by the following optimization problem (see, also, the discussion in Chap. 2 for the details of LMPC):

$$\min_{u\in S(\Delta)} \int_{t_k}^{t_{k+N}} (|\tilde{x}(\tau)|^2_{Q_c} + |u(\tau)|^2_{R_c})\, \mathrm{d}\tau \tag{5.2a}$$

$$\text{s.t.} \quad \dot{\tilde{x}}(t) = f(\tilde{x}(t)) + g(\tilde{x}(t))u(t) \tag{5.2b}$$

$$\tilde{x}(t_k) = x(t_k) \tag{5.2c}$$

$$u(t) \in U,\ \forall\, t \subset [t_k, t_{k+N}) \tag{5.2d}$$

$$\frac{\partial V(x(t_k))}{\partial x} F(x(t_k), u(t_k)) \leq \frac{\partial V(x(t_k))}{\partial x} F(x(t_k), \Phi(t_k)) \tag{5.2e}$$

where Δ is the sampling period, $S(\Delta)$ is the set of piecewise constant functions, and N is the length of the prediction horizon (i.e., the number of sampling periods in the prediction horizon). $\tilde{x}$ represents the predicted process state. The notation $t_k = k\Delta$, $k = 0, 1, \ldots,$ denotes the k_{th} sampling time at which the LMPC optimization problem of Eq. 5.2 is solved with the state measurement at t_k. $\Phi(x)$ is a stabilizing controller that is assumed to exist to stabilize the CSTR system at the steady-state. The LMPC optimization problem computes the optimal input trajectory (denoted by $u^*(t|t_k)$ at t_k) and applies the first control (i.e., $u^*(t_k|t_k)$) to the CSTR for the next sampling period. Then, at the next sampling time t_{k+1}, the LMPC optimization problem will be resolved with a new state measurement at $t = t_{k+1}$. The LMPC minimizes the objective function of Eq. 5.2a, which is the time integral of the deviations of the process states x and manipulated inputs u from their steady-state: $|\tilde{x}(\tau)|^2_{Q_c} + |u(\tau)|^2_{R_c}$. $|\cdot|_{R_c}$ and $|\cdot|_{Q_c}$ represent the Euclidean norms weighted by matrices R_c and Q_c, respectively. In this example, they are chosen to be $R_c = 1$ and $Q_c = [3\ 0;\ 0\ 5]$ such that the two terms in the objective function of Eq. 5.2a are in the same order of magnitude. The CSTR model of Eq. 5.1 (represented by the deviation variables) is used to predict future states of the closed-loop system in MPC. The state measurement $x(t_k)$ at time t_k is used as the initial condition for the MPC optimization problem (i.e., Eq. 5.2c). The input constraints that are implemented for the entire prediction horizon are defined by Eq. 5.2d. Finally, the constraint of Eq. 5.2e guarantees closed-loop stability by decreasing the Lyapunov function value (at the worst-case rate under the Lyapunov-based controller $\Phi(x)$) at each sampling step such that the state x will ultimately converge to the origin.

We design the Lyapunov function in a standard quadratic form $V(x) = x^T P x$, where P is a positive definite matrix: [200 33; 33 40]. Then, we characterize the closed-loop stability region Ω_ρ as a level set of Lyapunov function $V(x)$ in state-space: $\Omega_\rho := \left\{x \in \mathbf{R}^2 \mid V(x) \leq \rho\right\}$ with $\rho = 8000$. The controller $\Phi(x)$ was designed following the Sontag control law formula in [110]:

$$\Phi(x) = \begin{cases} -\frac{L_f V + \sqrt{L_f V^2 + L_g V^4}}{L_g V^2} L_g V, & \text{if } L_g V \neq 0 \\ 0, & \text{if } L_g V = 0 \end{cases} \tag{5.3}$$

where $L_f V$ and $L_g V$ represent the Lie derivative of V along the vector fields f and g, respectively. Finally, to numerically simulate the CSTR dynamic model of Eq. 5.1, we implement the explicit Euler method to integrate Eq. 5.1 with a sufficiently small integration time step of $h_c = 10^{-2}\ s$. The nonlinear optimizer, IPOPT [201], is used to solve the LMPC optimization problem of Eq. 5.2 with the following parameters: prediction horizon $N = 10$ and sampling period $\Delta = 1\ s$.

5.2.1.3 Simulation Results Under Disturbances

We consider a small feed disturbance that changes feed concentration from 29.35 mol/kg to 35 mol/kg at the beginning of the closed-loop simulation of the CSTR of Eq. 5.1 under LMPC. The simulation results are shown in Fig. 5.1. It is demonstrated that the LMPC is robust to the small feed disturbance in the sense that the closed-loop state is driven to another steady-state, but still maintained within the stability region under LMPC. Then, we consider a large disturbance in feed concentration due to, for example, a device failure in feed distribution. In this case, the feed concentration varies from 29.35 mol/kg to 70 mol/kg at the beginning of the closed-loop simulation (i.e., $t = 0$ s). Figure 5.2 shows that the temperature T_j drops quickly at $t = 200$ s when the closed-loop state leaves the stability region under disturbances. Meanwhile, to cool down the reactor under a large disturbance in feed concentration, the manipulated input has reached its lower bound since $t = 200\ s$. However, it is shown in Fig. 5.2 that after implementing the maximum cooling for around 600 s (i.e., from $t = 200$ s to $t = 800$ s), the reactor temperature in the state-space plot starts increasing significantly, which implies that the thermal runaway occurs in the reactor. The occurrence of thermal runaway in this example is because the increase of the reactor temperature under the feed concentration disturbance accelerates the

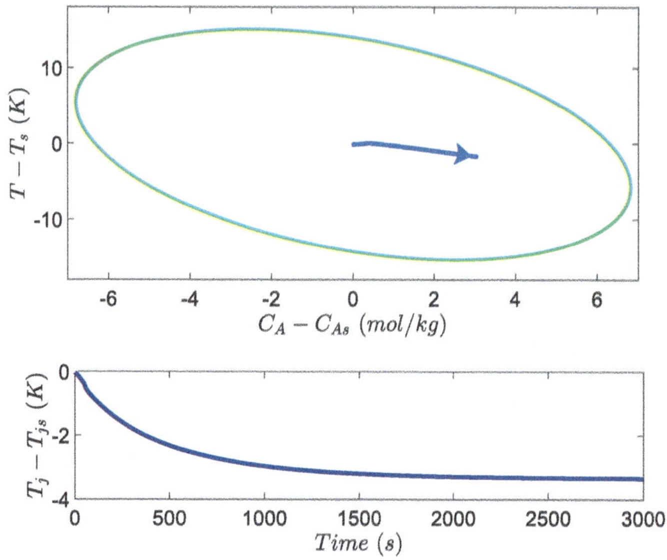

Fig. 5.1 State-space profile (top) and input trajectory (bottom) under a small disturbance

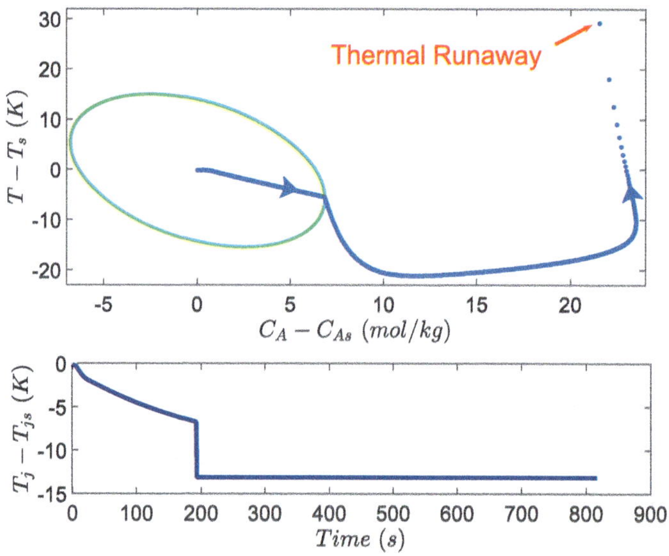

Fig. 5.2 State-space profile (top) and input trajectory (bottom) under a large disturbance

exothermic reaction in the reactor even when using the maximum cooling, which further increases the reactor temperature, and eventually drives the system to an unsafe operating condition. Additionally, it should be mentioned that all the data points shown in Fig. 5.2 are sampled at the same time interval during the entire simulation period. This implies that the reactor temperature varies slowly before the occurrence of thermal runaway, and increases rapidly at the end of the simulation when thermal runaway is approached.

From the two simulation studies of the CSTR with a small and a large disturbance on feed concentration, we have demonstrated that thermal runaway may occur in the reactor in the presence of a large disturbance, and explained that the unsafe operation is due to the restriction of the control actuator. Therefore, it motivates us to incorporate the safety system in the control system design to maintain reactor safety.

5.2.1.4 Integration of MPC with Safety System

We use two different safety mechanisms to design the safety system for the CSTR with the MIC hydrolysis reaction: (a) a safety relief valve and (b) cold water injection. We first consider using an outlet valve in the reactor (termed safety relief valve in this example) that can discharge material to reduce the reactor temperature. Therefore, the opening of the safety relief valve will be triggered by the safety system when the temperature is high in the reactor to prevent thermal runaway. There are various types of device failures that can cause thermal runaway in industrial chemical plants, for example, the cooling system failures that change the flow rate or the temperature of the coolant, and the device failure in feed distribution. Since in general it is challenging to predict and control the aforementioned unsafe scenarios, to prevent the thermal runaway that can vaporize the liquid in a reactor, a safety relief system is generally used to prevent fatal accidents [73]. Specifically, to ensure process operational safety, the relief valve should be designed with a suitable size. On the one hand, if the relief valve is oversized, we may waste too much material during its opening and the process itself may become unstable [56]. On the other hand, if the relief valve is undersized, then equipment failure may occur, which could lead to high pressure in the reactor. In this example, we design the relief valve to open once the reactor temperature exceeds 320 K to prevent high temperatures that may cause thermal runaway. To simplify the discussion, all the relief discharge flows are assumed to be in the liquid phase. Additionally, through closed-loop simulations, the relief valve size is determined to be 4×10^{-3} m^2 such that the closed-loop state can be driven back into the stability region upon the activation of the safety system. The relief flow G_{relief} is computed using the following equation [73]:

$$G_{\text{relief}} = 0.9 \times 144 \times \frac{\mathrm{d}P}{\mathrm{d}T} \times \left(\frac{32.2}{778.16} \times \frac{T}{C_P} \right) \tag{5.4}$$

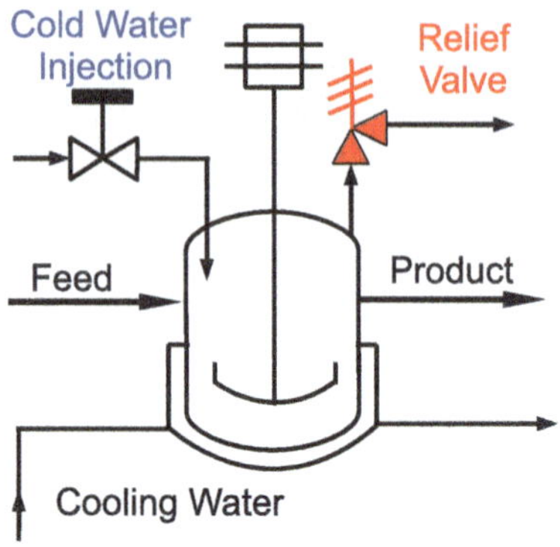

Fig. 5.3 Safety system for the CSTR with an MIC hydrolysis reaction

where G_{relief} (kg/m^2) represents the mass of the mixture per area for flow through the relief valve, and C_p (J/kg K), T (K), and P (Pa) are the heat capacity, the temperature of the relief flow, and the pressure in the reactor. The parameter values are estimated using the process simulation data from Aspen Plus.

Additionally, since the cold water injection is demonstrated in both the experiments and simulations [200] as an efficient approach that can reduce reactor temperature significantly and rapidly in the presence of an exothermic reaction, it is also used in the safety system in this example to reduce reaction mixture's temperature in the MIC hydrolysis reaction.

Specifically, when the reactor temperature exceeds 320 K, we inject the cool water with a temperature of 280 K and the same mass flow rate as that of the material being discharged through the relief valve such that the total mass in the reactor remains unchanged upon the activation of cool water injection. Figure 5.3 is a schematic of the CSTR with the safety system (i.e., the relief valve and cold water injection) under the control system that manipulates the cooling water temperature. Additionally, we assume that the positions of the cooling water inlet and outlet valves are fixed in this example.

5.2.1.5 Integration of Control and Safety Systems

This subsection presents the methodology for integrating the safety system (i.e., relief valve and cold water injection) with the control system (i.e., LMPC). The goal of the integrated control and safety system is to prevent thermal runaway when the LMPC fails to maintain the closed-loop state within the safe operating region (i.e., the closed-loop stability region Ω_ρ) in the presence of process disturbances. Specifically, we divide the entire state-space into three regions in which control and safety systems take different actions. Figure 5.4 shows a schematic of the three regions in state-space and gives an example of a state trajectory under the proposed integrated safety and control scheme. The implementation strategy of the safety and control systems is presented as follows:

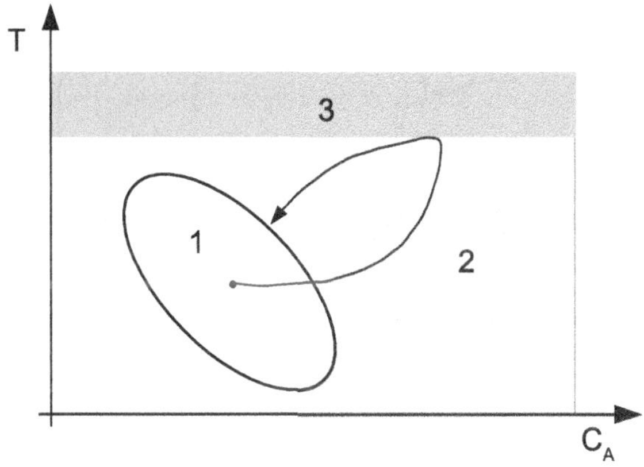

Fig. 5.4 A schematic showing, in the $C_A - T$ state-space, the stability region (white region), unsafe operating region (light gray region), and the thermal runaway region (dark gray region), together with an example trajectory starting from the origin

In Region 1 (termed stability region), the LMPC is utilized to stabilize the CSTR at its steady-state for any initial condition within this stability region. It has been demonstrated in the previous section that the LMPC is robust to a small disturbance but may not be able to handle a relatively large disturbance. Therefore, for the nominal system or the disturbed system with a sufficiently small disturbance, the closed-loop state will be maintained in the stability region under LMPC, while the safety system is not activated in this case.

In Region 2 (termed unsafe operating region), the states leave the stability region due to a large disturbance. Therefore, we set the manipulated input (i.e., the cooling water temperature T_j) to its lower bound (i.e., the lowest cooling jacket temperature) to prevent any unsafe operations. However, as shown in the previous simulation study, the LMPC may not be able to drive the state into the stability region due to the limitation of control actions, and thus, further safety system actions may be needed to prevent thermal runaway.

In Region 3 (termed thermal runaway region), the reactor temperature increases rapidly reaching a high value (i.e., the lower boundary of Region 3) due to the failure of maximum cooling for handling large disturbances. As a result, the safety relief valve will open the moment that the state enters Region 3, and will remain open until the state re-enters Region 1. In the meantime, we inject cold water and set the cooling jacket temperature to its lower bound to further cool down the reactor until the state returns to Region 1. The temperature value for the boundary between Region 3 and Region 2 in this example is determined from the closed-loop simulation to be the point from which the temperature starts increasing rapidly.

5.2.1.6 Closed-Loop Simulation Results

Figure 5.5 shows the closed-loop simulation results under the integration of safety and control systems. It is observed that the closed-loop state leaves the stability region due to a large disturbance at the beginning of the simulation (i.e., the feed concentration

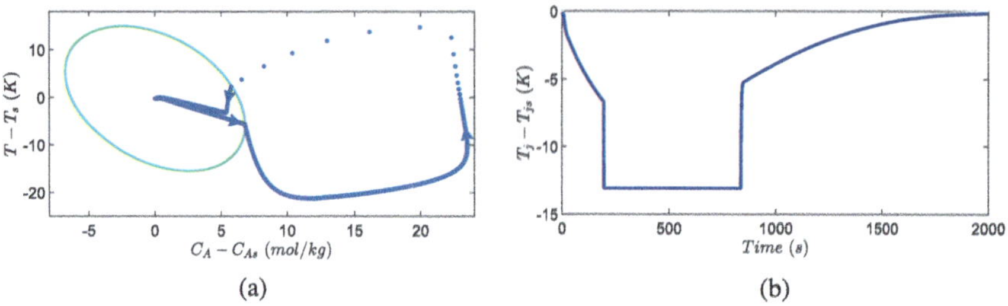

Fig. 5.5 State-space plot and input plot of LMPC integrated with the safety system for the MIC hydrolysis reaction in a CSTR

varies from 29.35 mol/kg to 70 mol/kg as discussed in Sect. 5.2.1.3). As a result, the LMPC fails to maintain the state within the stability region, and after 600 s, the concentration of the reactant increases to such an extent that the temperature shows a rapid increase and reaches 320 K (i.e., the safety limit for temperature) because the maximum cooling power is not able to fully remove the heat generated from the exothermic reaction. Once the reactor temperature reaches the boundary of Region 3, we open the relief valve to discharge hot fluid from the reactor, and inject cold water into the reactor to further decrease the temperature. Since the cold water promptly dilutes the reactant and lowers the reactor temperature, and the liquid relief flow rapidly decreases the reactant concentration and the total internal energy in the reactor, it is seen from Fig. 5.5 that the activation of safety systems quickly drives the closed-loop state into the stability region (Region 1) (it takes around 10 seconds). The coolant temperature is set to the lower bound when the state is outside of the stability region. Additionally, after the state re-enters the stability region, we assume that the disturbances disappear (i.e., the device failure is fixed), and we turn off the relief valve and water injection, and use LMPC again to stabilize the CSTR at its steady-state by manipulating T_j. It is guaranteed that the CSTR can be stabilized at the origin under LMPC in the absence of process disturbances. Additionally, it is noted that the data points in Fig. 5.5 are sampled with the same time interval, which implies the reactor temperature shows a rapid change when the CSTR approaches the thermal runaway region.

Remark 5.1 We assume the device failure is fixed after a certain period of time such that the disturbances no longer exist when the state returns to the stability region. However, if the disturbance still exists when the closed-loop state is driven into the stability region, it is straightforward to show that the state will leave it again and activate the safety system as we have shown in this example. In this case, beyond implementing the safety and control systems proposed in this section, engineers may need to perform process diagnostics and maintenance to fix the device failures to prevent frequent activations of the safety system.

In this case study, we demonstrate the integration of a safety system with MPC for the methyl isocyanate hydrolysis reaction in a CSTR subject to disturbances that could lead to reactor thermal runaway. The closed-loop system state was demonstrated to remain in the stability region under the LMPC in the presence of small disturbances, while the state left the stability region but process operational safety was still guaranteed in the presence of large disturbances. It was demonstrated that the thermal runaway was avoided and the process state was driven back into the stability region in finite time under the integration of LMPC and safety systems. The effectiveness of the proposed safety systems was also demonstrated by the quick movement of the state into the stability region upon the activation of safety systems.

5.2.2 *Case Study: High Pressure in a Flash Drum*

In the second case study, we present a high-pressure flash drum process that is often used in the chemical industry for separating a typical mixture, and demonstrate the dynamic interaction between safety systems and classical feedback control systems. Specifically, two PI controllers are utilized to regulate the temperature and the liquid level in the flash drum, and a pressure relief valve is used for safety consideration. We consider an unsafe scenario where the outlet vapor stream valve experiences a fault (e.g., the valve is blocked) that can lead to a significant pressure rise in the flash drum. We will demonstrate that the PI controller with tuning parameters varying upon the activation of the safety system achieves improved closed-loop performance than the PI controller with the same set of parameters regardless of the status of safety systems.

5.2.2.1 Flash Drum Process Description and Relief Valve Design

Figure 5.6 shows a schematic of a flash process [118] that is used to separate a mixture of ethane (20%), methane (10%), pentane (5%), butane (35%), and propane (30%) for the downstream distillation towers. Specifically, a liquid feed stream goes through a heat exchanger with heating duty Q and is heated up to a temperature T_{in} with a corresponding pressure P_{in}. The feed flow rate, temperature, mole fraction of component i, and pressure are denoted by F, T_f, z_i, and P_f, respectively. In this simulation study, the feed pressure P_f and the feed temperature T_f are set to 45 bar and 40 °C. The mole fractions of i-butane, ethane, n-butane, methane, propane, and n-pentane in the feed stream (i.e., the z_i) are 0.15, 0.2, 0.2, 0.1, 0.3, and 0.05, respectively. The flash drum is developed with height to be 4 ft and diameter to be 1 ft. The mole fractions of ethane, i-butane, methane, n-butane, n-pentane, and propane in the feed stream (i.e., the z_i) are 0.2, 0.15, 0.1, 0.2, 0.05, and 0.3, respectively. The heated stream then goes through a throttling valve and is adiabatically separated into a vapor stream of flow rate V with composition y_i and a liquid stream of flow rate L with composition x_i in the flash drum. Both the vapor and liquid streams exiting the flash drum have pressure P and temperature T. The flash drum separates five components based on different vapor pressures.

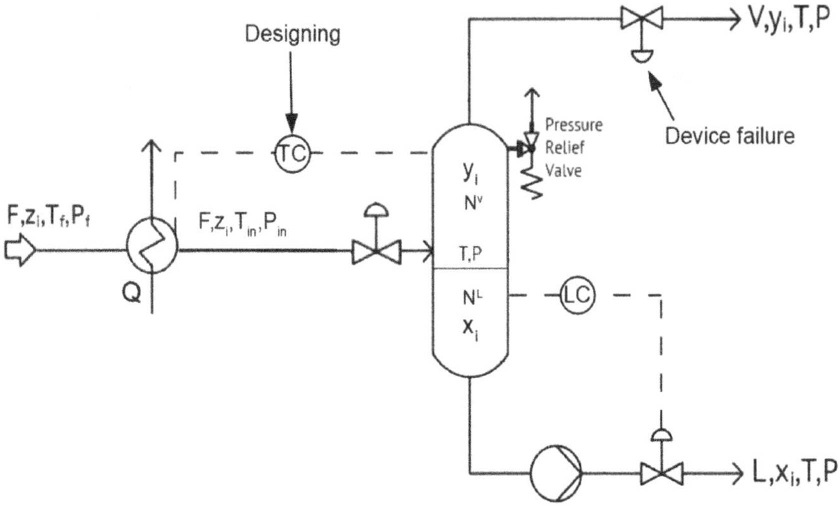

Fig. 5.6 A schematic of the flash process with a heat exchanger, flash drum, pump (from left to right), valves, and controllers that control the temperature and liquid level. The temperature controller (marked by "Designing") is designed to account for the safety system activation for handling vapor effluent valve failure (marked by "Device failure")

We use energy balance, component molar balances, and phase equilibrium equations to represent the flash drum process as a nonlinear dynamic system that can be described by a system of first-order nonlinear ordinary differential equations. The following process state variables are accounted for in the process model: mole fractions x_i and y_i of component i in liquid and vapor phases, number of moles N_i of component i in the flash drum, the total number of moles N^L and N^V in the liquid and vapor phases, respectively, drum temperature T, and drum pressure P. We develop and simulate the dynamic model in Aspen Plus Dynamics following the schematic in Fig. 5.6. Specifically, in Fig. 5.6, two controllers (i.e., the level controller (LC) and the temperature controller (TC)) are utilized to control the liquid level and the drum temperature T to their desired values by manipulating the liquid effluent valve and the heating duty Q, respectively. PI controllers are used to calculate the control actions in this example. Since the drum pressure and temperature are related through thermodynamics, the drum pressure P will be affected by controlling the drum temperature. The process is initially operated normally with all process equipment working properly. As shown in [118], the level and temperature controllers are able to track the liquid level and the drum temperature to their desired values under normal operation. However, an unsafe operation may occur due to various device failures. For example, a common process fault that leads to extremely high pressure in the drum is the valve failure in, for example, the bottom liquid effluent valve and the top vapor effluent valve. If the valve gets blocked, in order to prevent a potential dangerous high-pressure operation, a pressure relief valve is often used in the flash drum. Specifically, we utilize a pressure-actuated relief valve in this example to protect a pressurized vessel during an overpressure event. Note that this is different from the safety relief valve that we introduced in the first case study (i.e., MIC hydrolysis example), for which the valve was temperature-actuated through electrical signals. Aspen Plus is used to design the pressure relief (i.e., size and opening/reseating pressure) valve for the flash drum. We consider the worst-case scenario that the top vapor valve is fully closed due to device failure, and determine the pressure relief valve parameters for this case. Specifically, to quickly lower the drum pressure in such an unsafe scenario, the required mass flow rate is calculated to be 523 kg/h in Aspen Plus such that the pressure can be maintained below the maximum pressure that the drum can sustain. Correspondingly, a standardized orifice size of 8.303 cm^2 is utilized to satisfy the obtained relief flow rate based on operating conditions, fluid properties, and relieving conditions. Additionally, considering that the normal operating pressure and the maximum allowable drum pressure for flash drum are 10 and 12 bar, we choose the opening pressure of the relief valve to be 10.5 bar. To ensure that the relief valve remains open until the process equipment faults causing a high pressure are fixed, the reseating pressure (i.e., the pressure that triggers the close of a relief valve) is set to 9 bar. Moreover, the following assumptions/settings are made in the simulation: (1) the flash calculation is based on constant enthalpy, (2) the discharge flow is considered to be vapor only, and (3) the relief flow is considered to be a compressible fluid with the discharge coefficient being 0.96.

5.2.2.2 Feedback Controller Design

The flash drum is initially operated at the steady-state for 0.002 h. Then, we assume that a device failure occurs such that the opening of the vapor effluent valve varies from 50% to 0% (i.e., the valve becomes fully closed). As a result, the drum pressure increases rapidly and quickly reaches the opening pressure of the pressure relief valve. After the pressure relief valve opens, high-pressure vapor is discharged, and the drum pressure and temperature are both reduced. However, since the system dynamic is changed due to the opening of the pressure relief valve, a more effective PI controller should be developed to account for the activation of the safety relief system by changing the tuning of PI parameters instead of leaving the tuning unchanged. Therefore, in this case study, we develop sets of PI control parameters for controlling the drum temperature when the pressure relief valve is open and closed, respectively. The control objective is to stabilize the drum temperature at the set-point, and to ensure process operational safety in terms of maintaining drum pressure below its maximum operating pressure of 12 bar at all times in the presence of a vapor effluent valve failure. Additionally, it is noted that we vary the tuning of PI parameters in the temperature controller only to clearly analyze the effectiveness of the proposed controller design that accounts for safety system actions. As a result, the level controller parameters ($K_c = 10$ and $\tau_I = 3600$ s) remain unchanged throughout the simulation period.

To determine the PI tuning parameters for the cases of an open and closed relief valve, respectively, empirical linear models are first developed to capture the rela-

Table 5.2 Parameter values of the empirical model of Eq. 5.5 when the pressure relief valve is open and closed, respectively

Relief valve closed	Relief valve open
$b = 0.0202$	$b = 0.0206$
$a = 0.105$	$a = 0.113$

tionship between feed heating duty and drum temperature. Specifically, a first-order transfer function model is developed using extensive open-loop simulation of the drum temperature T under a step change in feed heating duty Q in Aspen Plus Dynamics. Subsequently, we develop a single-input-single-output model and implement the maximum likelihood estimation (MLE) method to the dataset collected from open-loop simulations to identify the unknown coefficients:

$$y(s) = \frac{b}{s+a}u(s) \tag{5.5}$$

where y (°C) and u (kW) are the drum temperature and heat duty in deviation variable form, respectively. Table 5.2 reports the model coefficients a and b for the two cases: (1) the relief valve is closed when the vapor effluent valve works properly (denoted by "relief valve closed") and (2) the relief valve is opened after the vapor effluent valve closes (denoted by "relief valve open").

It should be mentioned that the two sets of model parameters obtained from the simulation data are specific to the process disturbances resulting from the vapor valve failure (i.e., the vapor effluent valve closes from 50% to 0% open due to a device failure). This means that the PI parameters that we will tune are also specific to the unsafe scenario that we considered in this example. If the proposed method is to be implemented in an industrial system, then a variety of process fault scenarios need to be accounted for in order to develop a set of PI parameters for different unsafe scenarios. In this example, we develop the PI controller based on the linear model with two sets of parameters as follows:

$$\begin{aligned} e(t_k) &= T^{set} - T(t_k) \\ u_{PI}(t_k) &= K_c\left(e(t_k) + \frac{1}{\tau_I}\int_0^{t_k} e(\tau)\,d\tau\right) \\ Q(t_k + \Delta t) &= Q(t = 0) + u_{PI}(t_k) \\ 0 \le Q(t_k &+ \Delta t) \le Q_{\max} \end{aligned} \tag{5.6}$$

where t_k and Δt are the current time and the sampling period, respectively. $e(t_k)$ represents the error between the temperature measurement $T(t_k)$ at time t_k and its set-point $T^{set} = 25$ °C, and is updated at every sampling step. $u_{PI}(t_k)$ is the control action obtained from the PI controller at $t = t_k$. $Q(t = 0) = 87.2625$ kW and $Q(t_k + \Delta t)$

Table 5.3 Parameter values of a PI temperature controller for the cases when the relief valve is open and closed, respectively

Relief valve closed	Relief valve open
$K_c = 4$	$K_c = 6$
$\tau_I = 14\ s$	$\tau_I = 10\ s$

are the heat duty at the initial steady-state and at the next sampling time. The lower and upper bounds of the heat duty are 0 and $Q_{\max} = 160$ kW, respectively. The PI parameters in Eq. 5.6 (i.e., the gain K_c and the time constant τ_I) for the two cases (i.e., "relief valve closed" and "relief valve open") are developed based on the two sets of model parameters in Eq. 5.5, and are reported in Table 5.3.

5.2.2.3 Closed-Loop Simulation Results

Closed-loop simulation of the flash drum process under the PI controller with varying parameters is carried out using Aspen Plus Dynamics. In Fig. 5.7, it is shown that the drum temperature increases rapidly after the vapor effluent valve is closed at $t = 0.002$ h. Then, the temperature controller adjusts the heat duty to reduce the difference between the set-point of the drum temperature and the current measurement. However, as shown in Fig. 5.8, the temperature controller is not able to prevent the drum pressure from increasing rapidly, and as a result, the pressure relief valve is triggered once the drum pressure reaches its opening (set) pressure, 10.5 bar. In the meantime, the PI controller switches to the other set of parameters after the opening of the pressure relief valve as discussed in the previous section. It can be seen in Fig. 5.8 that the drum pressure and temperature drop quickly under a decreasing heating duty and the opening of the relief valve. Finally, the drum temperature is re-stabilized at its set-point under the temperature controller after the pressure relief valve remains open for a while.

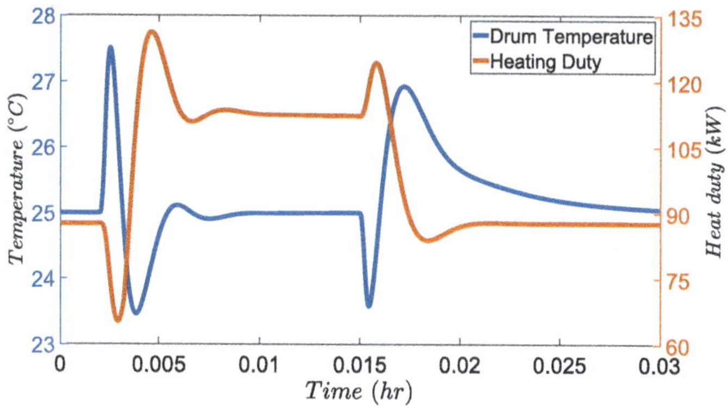

Fig. 5.7 Manipulated input and controlled output profiles for the temperature controller with varying tuning parameters to account for the activation of the safety system in a flash drum

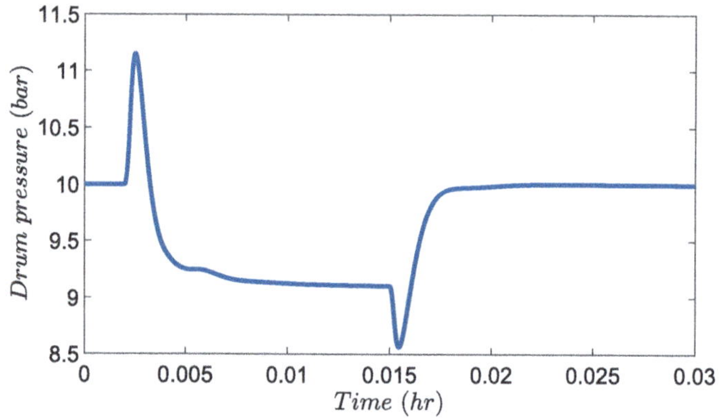

Fig. 5.8 Drum pressure profile under the temperature controller with varying tuning parameters to account for the activation of the safety system in a flash drum

The device failure that results in the vapor effluent valve closure is assumed to be resolved at $t = 0.015$ h (i.e., the vapor effluent valve returns to its normal operating condition, 50% opening). However, since the vapor valve is fixed and opens abruptly, the drum pressure experiences a sudden drop and reaches the reseating pressure (i.e., 9 bar). As a result, the relief valve is closed and the PI parameters switch to the original set of values for the remaining time of the simulation. It is observed that shortly after 0.015 h, the drum temperature increases, shows an overshoot above its set-point, and eventually stabilizes at the set-point under the temperature controller.

To demonstrate the improved closed-loop performance under the proposed PI control scheme with varying tuning parameters to account for the activation of the safety system, we compare it with the standard PI control scheme with a set of fixed parameters throughout the entire operation. The closed-loop flash drum temperature profiles under these two PI control schemes are shown in Fig. 5.9. It is clearly seen

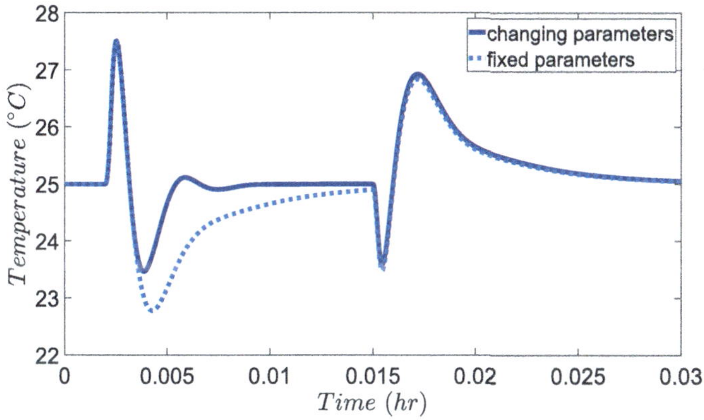

Fig. 5.9 Flash drum temperature profile under the temperature controllers with fixed parameters, and varying tuning parameters to account for the activation of the safety system, respectively

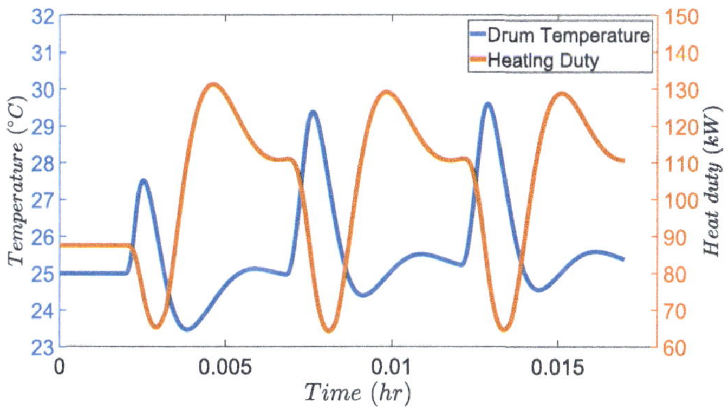

Fig. 5.10 Drum temperature and heating duty profiles under the temperature controller with varying tuning parameters to account for the activation of the relief valve with the reseating pressure of 9.2 bar in a flash drum

in Fig. 5.9 that the drum temperature varies in a smaller range under the PI controller with varying tuning parameters compared to that with a fixed tuning regardless of the status of the safety system. Also, the drum temperature returns to its set-point more quickly under the proposed PI controller than that with the fixed tuning. Therefore, it is concluded that the closed-loop performance is improved under the PI controller that accounts for the safety system activation.

It should be noted that the reseating pressure also plays an important role in the proper use of the relief valve. Specifically, the reseating pressure of the relief valve should be sufficiently low such that the relief valve can remain open until the process fault that causes the vapor effluent valve closure is fixed. Otherwise, if we choose a large reseating pressure, the relief valve will be closed while the vapor effluent valve failure is still blocked. As a result, the drum pressure will rise up again, and

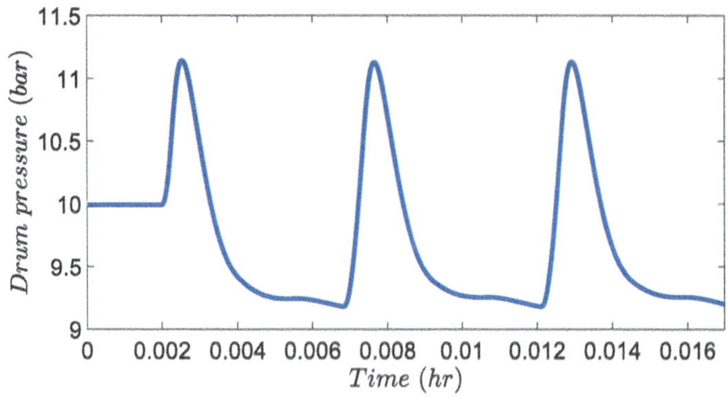

Fig. 5.11 Drum pressure profile under the temperature controller with varying tuning parameters to account for the activation of the relief valve with the reseating pressure of 9.2 bar in a flash drum

lead to a re-opening of the pressure relief valve, which is undesirable and should be avoided in practical implementation. To demonstrate the importance of the reseating pressure, we carry out the simulation studies that use a higher value, 9.2 bar, than its normal value 9 bar that we utilized in the previous simulations (Figs. 5.7, 5.8, and 5.9). Figures 5.10 and 5.11 show the simulation results with a reseating pressure of 9.2 bar, from which the opening and closure of the pressure relief valve show a periodic pattern because the drum pressure increases rapidly again after the relief valve is closed at 9.2 bar. Therefore, by using a lower reseating pressure in closing the safety relief valve, we can avoid undesired opening and closure of the relief valve in a short time. This indicates that it is necessary to design control and safety systems together in order to coordinate them effectively. Without analyzing the safety system actions, a high reseating pressure may be chosen, which leads to undesirable control performance. Additionally, since in general the vessel pressure may vary differently under various types of disturbances which are unknown a priori, it is also important to perform closed-loop simulations to determine a reasonable reseating pressure for the disturbance case under consideration. If possible, it would be of great help to allow for manual relief valve opening and closure in designing the pressure relief valve to improve the initial selection of the parameters.

To conclude, this case study is focused on a flash drum process under a PI controller that is developed accounting for the activation of a pressure relief valve. It was demonstrated that the closed-loop performance was much improved by modifying the PI parameters based on the safety system being on or off compared to the standard PI controller with fixed parameters regardless of the actions of the safety system.

5.3 Safeness Index-Based MPC

As discussed in Chap. 3, Safeness Index function is developed to provide thresholds as triggers for activating the safety system based on system-level safety considerations. Additionally, it can be used as a constraint in model predictive control (MPC) design to provide some coordination between safety and control systems to avoid unnecessarily abrupt changes in operating conditions and reduce safety system activation [8, 214, 221, 238, 240]. Please see, also, the discussions of Safeness Index-based MPC and EMPC in Chap. 3. In this section, we illustrate applications of the Safeness Index-based MPCs to a safety-critical chemical process to demonstrate the improved process operational safety.

5.3.1 Case Study: Flash Drum

In this subsection, the process and potential failures of a flash drum process are first introduced. Subsequently, we develop the Safeness Index function and the Safeness Index-based MPC that accounts for safety considerations of the flash drum process operation. The Safeness Index-based MPC is then applied to the closed-loop simulations of the flash drum process in the presence of disturbances with different magnitudes using Aspen Dynamics Plus simulators to demonstrate improved operational safety.

5.3.1.1 Flash Drum Process Description and Control Objective

We consider the same flash drum process that is used to separate a mixture of ethane (20%), methane (10%), propane (30%), pentane (5%), and butane (35%) to a level that can be used by distillation towers in the downstream. The notations and parameter values follow those in Sect. 5.2.2.1 and the schematic of the flash drum process is shown in Fig. 5.6. The flash process is modeled by a dynamic system with state variables of drum temperature T, drum pressure P, mole fractions x_i and y_i of component i in liquid and vapor phases, respectively, the total number of moles N^L and N^V in the liquid phase and the vapor phase, respectively, and the number of moles N_i of component i in the drum. This flash drum using the model in Fig. 5.6 is dynamically simulated in Aspen Plus Dynamics. In a safe scenario where the process equipment such as pressure sensor and valves operate properly, the liquid controller and the temperature controller can stabilize the liquid level and drum temperature at their steady-states such that the drum pressure is also stabilized at the desired level [118]. However, in an unsafe scenario where the broken pressure sensor leads to improper control actions, or the bottom (liquid) and top (vapor) effluent valves are accidentally closed, an extremely high pressure might occur in the drum. In this case, to prevent unsafe operation under high drum pressure, a pressure relief valve is often utilized in the industrial flash drum process.

The design of the pressure relief valve follows the same steps as performed in Sect. 5.2.2.1. Specifically, to guarantee device safety, we use Aspen Plus to compute the required relief flow (i.e., the minimum flow) to be 523 kg/h. The standardized orifice size is determined to be 8.303 cm^2 to satisfy the requirement of relief flow accounting for the operating conditions, fluid properties, and relieving conditions. Additionally, considering that the highest device durable pressure is 12 bar and the flash drum is normally operated at 10 bar, we set the reseating pressure of the relief valve to be 9 bar, and the opening pressure to be 10.5 bar.

Unlike the previous case study that uses two PI controllers, the flash drum in this example is initially operated at the steady-state under a model predictive controller. Then, we consider the same unsafe scenario (i.e., a device failure occurs in the top vapor valve) that leads to extremely high pressure in the drum. However, in this case, we assume that the top vapor valve closes from 50% opening to a smaller opening instead of fully closing (i.e., to 0% opening) under device failure. After the device failure occurs, the temperature T and pressure P in the drum rise up quickly, and drive the system to an unsafe operating condition. We have demonstrated in the previous case study that the drum pressure and temperature can be re-stabilized at their set-points using the two PI controllers and the pressure relief valve with appropriate settings of reseating and opening pressures when there is a severe device failure causing extremely high pressure in the drum. In this study, however, we will develop the control system to maintain the drum temperature T and pressure P within the desired range (i.e., safe operating region), and prevent the relief valve from frequently opening in the presence of small disturbances. Specifically, the controller aims to control the temperature T at its set-point and maintain the drum pressure P below its maximum allowable value 10.5 bar by manipulating the heating duty of the feed Q. In this example, we consider the worst-case scenario where the top vapor valve is originally operated at 50% opening and is reduced to 35% opening due to a device failure. It is noted that if a larger disturbance occurs (e.g., the opening of the top vapor valve becomes less than 35%), then the pressure relief valve will have to be activated following the strategy we presented in the previous case study. Meanwhile, the control system needs to account for the activation of the relief valve to safely control the process while maintaining the drum pressure below the device's maximum operating pressure of 12 bar.

5.3.1.2 Safeness Index-Based MPC

To develop the Safeness Index-based MPC, we first build a data-driven process model that will be used as the MPC prediction model based on extensive Aspen Plus Dynamics open-loop simulation results. Then the Safeness Index function is developed to characterize the safe and unsafe operating regions for the flash drum process, and the Safeness Index-based MPC is developed by incorporating the data-driven process model for prediction and using the Safeness Index function as constraints to maintain process operational safety.

The dynamic simulation of the flash drum is done in Aspen Plus Dynamics. The system is initially operated at the steady-state pressure and temperature, $P_s = 10$ bar and $T_s = 25\ ^\circ$C, and the steady-state heating duty $Q_s = 87.6$ kW. We use deviation variables to represent the manipulated input and the process states, i.e., $u = Q - Q_s$ and $x^T = [T - T_s\ P - P_s]$, such that the system has an equilibrium point at the origin of the state-space. The following linear dynamic model is developed using extensive open-loop simulation data for predicting future states in MPC:

$$\frac{\mathrm{dx}}{\mathrm{dt}} = Ax + Bu \tag{5.7}$$

where $A \in \mathbf{R}^{2\times 2}$ and $B \in \mathbf{R}^{2\times 1}$. Aspen open-loop simulation data is used to identify the coefficient matrices A and B. Specifically, we run extensive open-loop simulation under pseudorandom binary sequence (PRBS) signal in heating duty Q to generate the dataset of drum pressure P and temperature T. Then, we implement the Multivariable Output Error State Space (MOSEP) algorithm to identify the matrices A and B from the simulation dataset: $A = \begin{bmatrix} -0.047453 & -0.22548 \\ -0.001111 & -0.097369 \end{bmatrix}$ and $B = \begin{bmatrix} 0.01488 \\ 0.002277 \end{bmatrix}$.

We design the Safeness Index function based on the fundamental process knowledge that high pressure and high temperature are safety-critical process variables in the flash drum. Specifically, a high pressure P and temperature T should be considered unsafe by the Safeness Index, while a pressure P and temperature T below the steady-state values are considered as a safe operation. In this example, the Safeness Index function is designed to be zero when both the drum temperature and pressure are below the steady-state values (i.e., both x_1 and x_2 are negative), and to be positive when either the drum temperature or pressure is above the steady-state values (i.e., at least one of x_1 and x_2 is positive). Based on the function $f^+(x)$ of Eq. 5.8, the Safeness Index $S(x)$ is developed in the following form:

$$f^+(x) = \begin{cases} x, & \text{if } x \geq 0 \\ 0, & \text{if } x < 0 \end{cases} \tag{5.8}$$

$$S(x) = k_T \left[f^+\left(\frac{x_1}{T_s}\right)\right]^2 + k_P \left[f^+\left(\frac{x_2}{P_s}\right)\right]^2 \tag{5.9}$$

where k_P and k_T are the weights for drum pressure and temperature, respectively. We normalize the pressure and temperature terms in Eq. 5.8 by dividing their values (in deviation form) by their steady-state values P_s and T_s such that they are in the same order of magnitude in the Safeness Index function. Additionally, $S(x)$ is designed with a quadratic form of Eq. 5.9 such that $S(x)$ is nonnegative for all x_1 and x_2, and grows faster when pressure P and temperature T are far above the steady-state. Additionally, the weights $k_T = 1000$ and $k_P = 3000$ are chosen due to the consideration that high pressure is considered more dangerous than high temperature in this example. To avoid triggering the safety relief valve, the threshold S_{TH} for the Safeness Index function $S(x)$ should be designed lower than the threshold (i.e., the opening pressure) utilized in the pressure relief valve. In consideration of the sample-and-hold implementation of MPC and potential model mismatch, the actual threshold in the control system is determined to be more conservative such that some overshoot of Safeness Index is allowed (but it should not exceed the threshold that triggers the opening of pressure relief valve). Based on the value of $S(x)$ at the opening pressure 10.5 bar that triggers the relief valve (i.e., $S([0\ 0.5]^T) = 7.5$ when $P = 10.5$ bar and $T = 25$ °C), the threshold in the controller is finally chosen to be $S_{TH} = 6$.

5.3.1.3 Formulation of Safeness Index-Based MPC

Following the formulation of the Safeness Index-based MPC in Sect. 3.3, the Safeness Index-based MPC designed for the flash drum process is formulated as the following optimization problem:

$$\min_{u \in S(\Delta), y} \int_{t_k}^{t_{k+N}} (|\tilde{x}_1(\tau)|^2_{Q_c} + |u(\tau)|^2_{R_c})\, d\tau + \sum_{i=1}^{N} k_1 e^{-k_2 y(i)} \tag{5.10a}$$

$$\text{s.t.} \quad \dot{\tilde{x}}(t) = A\tilde{x}(t) + Bu(t) \tag{5.10b}$$

$$\tilde{x}(t_k) = x(t_k) \tag{5.10c}$$

$$u(t) \in U, \ \forall\, t \in [t_k, t_{k+N}) \tag{5.10d}$$

$$S(\tilde{x}\,(t_{k+i})) + y(i) = S_{TH}, \quad i = 1, 2, \ldots, N \tag{5.10e}$$

$$y(i) \geq 0, \ i = 1, 2, \ldots, N, \quad \text{if } S(x(t_k)) \leq S_{TH} \tag{5.10f}$$

where $\Delta, \tilde{x}, S(\Delta)$, and N are the sampling period, the predicted state trajectory, the set of piecewise constant functions, and the length of the prediction horizon, respectively. The objective function of Eq. 5.10a is minimizing the integral of $|\tilde{x}_1(\tau)|^2_{Q_c} + |u(\tau)|^2_{R_c}$ over the prediction horizon and the penalty term $\sum_{i=1}^{N} k_1 e^{-k_2 y(i)}$ of slack variables $y(i)$. It is noted that the objective function penalizes the inputs and the state x_1 only (instead of full state x) because only the drum temperature T will be stabilized at its set-point by the Safeness Index-based MPC of Eq. 5.10, while the drum pressure P will be maintained in a desired range by the Safeness Index constraints in Eq. 5.10e. The linear model of Eq. 5.7 is used in the constraint of Eq. 5.10b to predict the closed-loop states over the prediction horizon. The state measurement $x(t_k)$ at $t = t_k$ is used as the initial condition $\tilde{x}(t_k)$ for the optimization problem in Eq. 5.10c. The input constraints are applied in Eq. 5.10d over the prediction horizon. The manipulated input u (i.e., the heating duty Q) is bounded by $U = [-87.6,\ 72.4]$ (i.e., $0 \leq Q \leq 160$ kW). The Safeness Index function is used in the soft constraint of Eq. 5.10e, where $y(i)$ is a slack variable. Specifically, the input will be affected by the soft constraint of Eq. 5.10e as the Safeness Index value $S(x)$ approaches the threshold value S_{TH}; however, if the slack variables are not used in the constraint of Eq. 5.10e, we will see an abrupt input change when $S(x)$ hits the threshold value S_{TH}. When $S(x(t_k)) > S_{TH}$, a negative slack variable $y(i)$ can be used to satisfy the constraint of Eq. 5.10e, under which the Safeness Index $S(x)$ can stay above the threshold S_{TH} to avoid infeasibility issue in solving the optimization problem. However, in this case, the objective function of Eq. 5.10a will play a dominant role in decreasing the Safeness Index value by minimizing the penalty term $\sum_{i=1}^{N} k_1 e^{-k_2 y(i)}$ in Eq. 5.10a. On the other hand, when $S(x(t_k)) \leq S_{TH}$, Eq. 5.10f requires a nonnegative slack variable $y(i)$ to ensure process operational safety by maintaining $S(x)$ below S_{TH}. Additionally, we need to carefully choose the parameters k_1 and k_2 in the objective function of Eq. 5.10a, such that the control actions are barely affected by the slack variables $y(i)$ when $S(x(t_k))$ is below S_{TH}, and are significantly affected when $S(x(t_k))$ exceeds S_{TH}. In this example, k_1 and k_2 are determined to be 90 and 1.6, receptively, to achieve the desired performance.

The dynamic model of Eq. 5.7 is numerically integrated using the explicit Euler method with a sufficiently small integration time step of $h_c = 10^{-3}$ s. The solver FilterSD on OPTI Toolbox in MATLAB is used to solve the nonlinear optimization problem of the Safeness Index-based MPC of Eq. 5.10 with the following settings: prediction horizon $N = 10$ and sampling period $\Delta = 0.5$ s. $R_c = 0.0005$ and $Q_c = 1$ are chosen to balance the magnitude of the two terms with respect to the manipulated input and the process state, respectively, in the objective function of Eq. 5.10a.

5.3.1.4 Closed-Loop Simulation Results

In this section, we implement the Safeness Index-based MPC to a flash drum process to control the drum pressure in a safe operating region in the presence of disturbances with different magnitudes. Specifically, we first discuss a scenario in which process operational safety is ensured under Safeness Index-based MPC without activating the safety system in the presence of small disturbances. The second scenario demonstrates that the integration of the Safeness Index-based MPC and the safety system is able to bring the process state to the safe region in the presence of large disturbances that cannot be handled by the control system only.

Scenario 1: Closed-loop simulation without safety system activation

We first consider a device failure that reduces the opening of the top vapor valve from 50% to 45 %. In Figs. 5.12 and 5.13, it is demonstrated that the temperature, pressure, and Safeness Index value immediately increase after the device failure occurs. As a result, the Safeness Index-based MPC of Eq. 5.10 reduces the input Q in order to minimize both the terms $|\tilde{x}_1(\tau)|^2_{Q_c}$ and $\sum_{i=1}^{N} k_1 e^{-k_2 y(i)}$ in the objective function of Eq. 5.10a. Specifically, $\sum_{i=1}^{N} k_1 e^{-k_2 y(i)}$ can be minimized with a large slack variable, which leads to a decrease of the Safeness Index function value over the prediction horizon and eventually leads to a decreasing input Q. Additionally, it is noted that the closed-loop state does not reach the original steady-state ($x = 0$) at the end of the simulation because of the model mismatch between the actual disturbed process

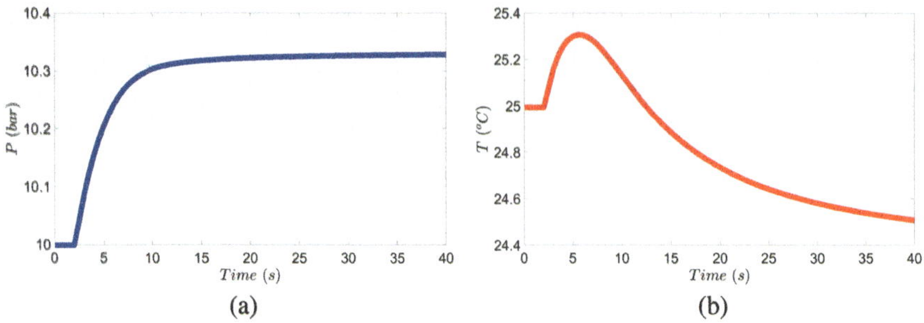

Fig. 5.12 **a** Drum pressure and **b** temperature profiles under Safeness Index-based MPC with a device failure that changes the top vapor valve opening from 50% to 45%

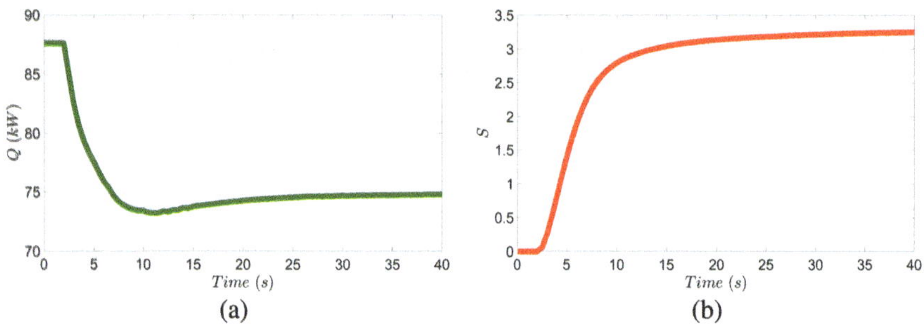

Fig. 5.13 **a** Manipulated input and **b** Safeness Index profiles under Safeness Index-based MPC with a device failure that changes the top vapor valve opening from 50% to 45%

and the prediction model (i.e., the nominal process model) used in Safeness Index-based MPC. To eliminate the offset, an offset-free MPC can be employed [239], or an integral control term can be added to the MPC control action if required.

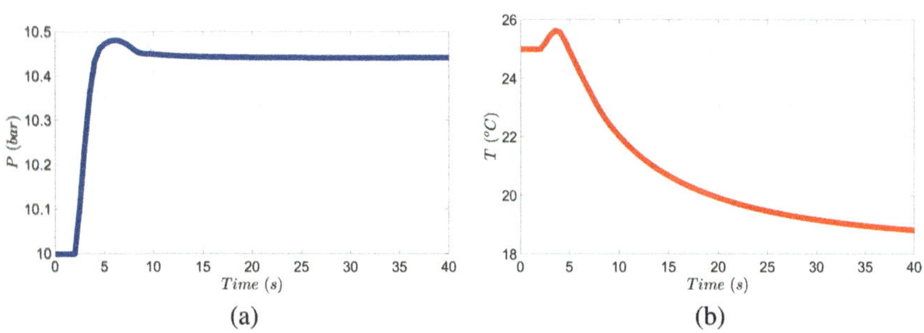

Fig. 5.14 **a** Drum pressure and **b** temperature profiles under Safeness Index-based MPC with a device failure that changes the top vapor valve opening from 50% to 35%.when the top vapor valve is closed from 50% to 35%

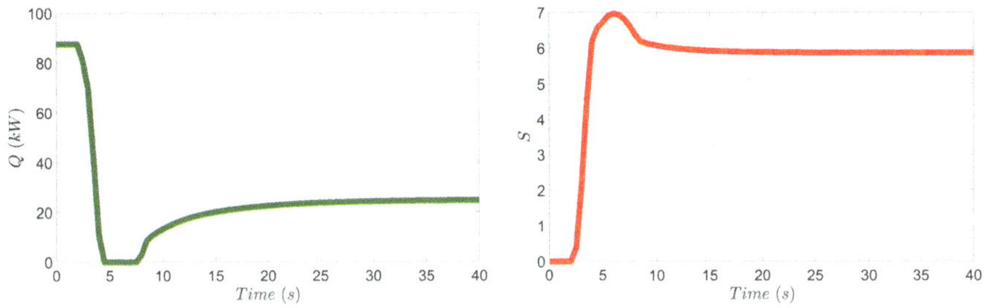

(a) (b)

Fig. 5.15 **a** Manipulated input and **b** Safeness Index profiles under Safeness Index-based MPC with a device failure that changes the top vapor valve opening from 50% to 35%

When the opening of the top vapor valve is further reduced to 35 %, Figs. 5.14 and 5.15 show that the drum temperature and pressure increase rapidly, and the Safeness Index value approaches the threshold value S_{TH}. As a result, the Safeness Index-based MPC computes aggressive control actions Q to prevent the drum pressure from exceeding the threshold value 10.5 bar. When $S(x(t_k))$ exceeds S_{TH}, a negative slack variable $y(i)$ is used by the Safeness Index-based MPC to meet the constraint of Eq. 5.10e. It is noted that a relatively large slack variable $y(i)$ is preferred to decrease future Safeness Index $S(\tilde{x}\ (t_{k+i}))$ when $y(i)$ is small, and the term $\sum_{i=1}^{N} k_1 e^{-k_2 y(i)}$ dominates the objective function of Eq. 5.10a. It is shown in Figs. 5.14 and 5.15 that when the Safeness Index is approaching or exceeding the threshold S_{TH}, the MPC computes an aggressive control action (i.e., the input Q is at the lower bound) to stop Safeness Index increasing. Additionally, due to the change of process dynamics under disturbances, the steady-state values of pressure and temperature are also changed under the same input. Also, the model mismatch may cause the Safeness Index based on the actual states to exceed the threshold S_{TH}. To alleviate the adverse effect of model mismatch, a smaller k_2 is preferred in the objective function of Eq. 5.10a to obtain a conservative Safeness Index value. Specifically, when the slack variables are positive, a large value of the slack variable will be utilized under a small k_2. From Eq. 5.10e, we can clearly see that a larger slack variable can lead to a conservative Safeness Index value. Therefore, the resulting lower value of the Safeness Index might help alleviate the adverse effect of model mismatch.

Additionally, parameters k_1 and k_2 should be carefully chosen to account for the conservatism of the threshold S_{TH} used in Safeness Index-based MPC. Specifically, a large k_2 is preferred for the case of a conservative threshold S_{TH} to allow a desired closed-loop performance with the Safeness Index not exceeding the threshold S_{TH} under all disturbances. However, when the threshold S_{TH} is less conservative, to maintain the Safeness Index below the threshold S_{TH} for all times, a small value of k_2 is preferred. Additionally, based on the value of k_2, the parameter k_1 can be chosen such that the term $\sum_{i=1}^{N} k_1 e^{-k_2 y(i)}$ in the objective function of Eq. 5.10a is

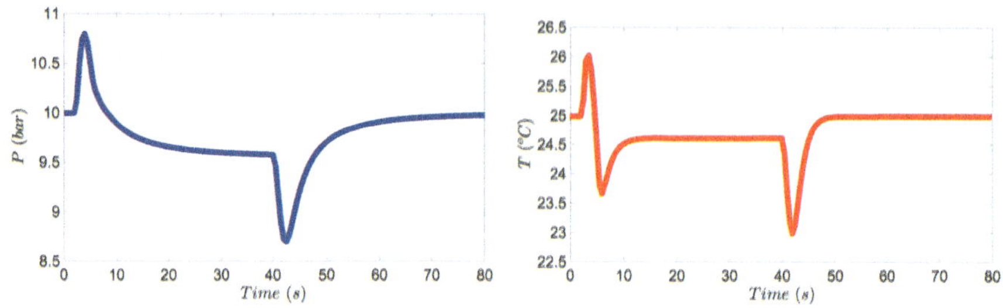

Fig. 5.16 **a** Drum pressure and **b** temperature profiles under Safeness Index-based MPC with a device failure that changes the top vapor valve opening from 50% to 10%

smaller than the term $\int_{t_k}^{t_{k+N}} (|\tilde{x}_1(\tau)|^2_{Q_c} + |u(\tau)|^2_{R_c})\, d\tau$ under a large but handleable disturbance, and larger than that under a relatively small disturbance.

Scenario 2: Closed-loop simulation with safety system activation
In the presence of a large disturbance that reduces the top vapor valve opening from 50% to 10%, the Safeness Index-based MPC system is unable to prevent the occurrence of extremely high pressure in the flash drum due to actuator constraints. As shown in Figs. 5.16 and 5.17, even using the minimum heating duty Q, the drum pressure P still exceeds 10.5 bar quickly, which triggers the pressure relief valve opening to allow the pressurized fluid to flow out of the drum. As a result, the drum pressure and temperature decrease, and settle to a new steady-state under MPC. It is noted that the new steady-state is slightly different from the initial steady-state (e.g., the temperature T is 0.2 °C below the set-point 25 °C at $t = 40$ s) due to the model mismatch from relief valve opening and vapor valve disturbance. We assume that the device failure is fixed after $t = 40$ s such that the top vapor valve opening returns to 50%. Then, as the pressure relief valve and the vapor valve are open at the same time, it is seen that the drum pressure P decreases heavily. When the drum pressure P reaches the reseating pressure 9 bar, the safety relief valve is turned off, and the process states (i.e., temperature and pressure) are re-stabilized at the initial steady-state under Safeness Index-based MPC.

In this case study, control systems were integrated with a pressure relief valve as a safety system to maintain the safe operation of a flash drum process in the presence of various disturbances. We developed the Safeness Index function and its threshold using the process and safety system information to characterize the safeness of process operating conditions. The Safeness Index-based MPC was developed using a data-driven linear model that was developed using extensive Aspen simulation data to carry out the closed-loop simulation of the flash drum separator in MATLAB and Aspen. In the presence of a small disturbance, the drum pressure was demonstrated to remain below the triggering pressure of the relief valve under Safeness Index-based MPC. However, when there exists a large disturbance that cannot be handled by Safeness Index-based MPC, the relief valve will be activated to work with the controller to further improve process operational safety.

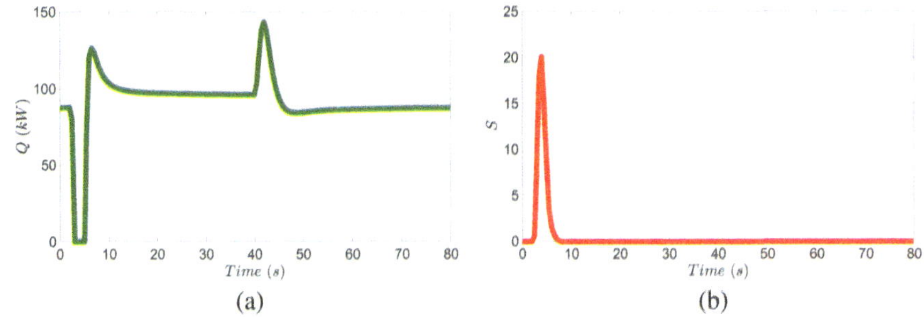

Fig. 5.17 **a** Manipulated input and **b** Safeness Index profiles under Safeness Index-based MPC with a device failure that changes the top vapor valve opening from 50% to 10%

5.3.2 *Case Study: Ammonia Process*

In this section, the Safeness Index-based MPC is applied to a multi-unit process to improve process operational safety. We use the ammonia process as an example since it has been intensely studied by researchers and engineers over the past century in order to meet the ammonia demand. The ammonia process has been demonstrated to be one of the greatest risks in a survey of major accidents over a period of 70 years. Ojha and Dhiman [143] have reviewed the main problems and accidents in the ammonia process. For example, various types of disturbances in the high-temperature shift converter have been studied in simulation studies to prevent poison of the catalyst and thermal runaway in the methanator [10]. Additionally, the failures such as piping failure, drop in activity of the catalyst, and oil leak might also occur in ammonia processes. In this case study, we develop the ammonia plant model in Aspen Plus software and carry out the closed-loop simulation under the Safeness Index-based MPC to investigate the impact of potential disturbances such as feed temperature change and catalyst deactivation in an ammonia plant.

5.3.2.1 Ammonia Process Descriptions and Simulations

We focus on the shift conversion, carbon dioxide removal, and methanation parts of the ammonia process in this case study. Figure 5.18 shows a schematic of an ammonia process, in which the three parts we study are used to remove carbon dioxide and carbon monoxide produced in the previous steam reformer. Figure 5.19 shows the schematic of all simulated units in this example. Specifically, the low-temperature shift reactor and the high-temperature shift reactor are the two adiabatic tube reactors that convert water and carbon monoxide into hydrogen and carbon dioxide. A two-bed operation under the temperatures 400 and 200 °C, respectively, is performed using a different catalyst in each bed. The carbon monoxide can be reduced to 2∼4% in the high-temperature shift reactor, and an output of carbon monoxide around 0.1∼0.3% can be obtained from the low-temperature shift reactor under normal

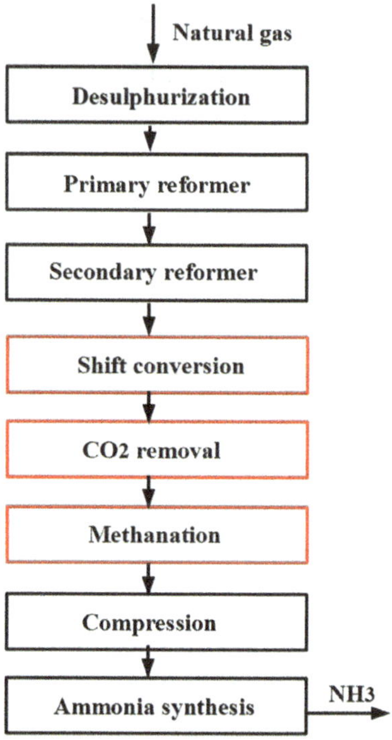

Fig. 5.18 A schematic of an ammonia process

operating conditions [18, 63, 193]. Then, the adsorption column removes the carbon dioxide and water vapor in the gas, in order to prevent ammonia synthesis catalysts from being poisoned [18]. However, since a small amount of carbon dioxide and carbon monoxide in the syngas is poisonous to ammonia synthesis catalysts, trace amounts of carbon dioxide and carbon monoxide are removed in the methanation unit next to the removal unit. The concentrations of carbon dioxide and carbon monoxide can be reduced to less than 5 ppm catalytically by exothermic methanation reaction in the methanator [143, 193].

We use Aspen Plus and Aspen Plus Dynamics to perform high-fidelity dynamic simulation of the gas purification process accounting for the interaction among units in the ammonia plant. The components in the simulation were carefully chosen with the Redlich–Kwong–Soave–Boston–Mathias (RKS-BM) model being used as a reference to calculate thermodynamic properties of all involved chemical components. Based on the example provided by Aspen [23], we first build a steady-state model in Aspen Plus, and a dynamic model in Aspen Plus Dynamics. Specifically, the model configuration is validated using the Pressure Checker Tool. Then, using the Dynamic Mode Tool, the steady-state model is exported to a pressure-driven dynamic file. The reaction rate equations from [23, 63] are used to model the gas phase reactions in all units. Due to the limitation of the kinetic models provided in Aspen Plus, user-defined routine is adopted for complex kinetic modeling. In this study, we program the reaction rate equations in a FORTRAN user-kinetics subroutine file, compile the

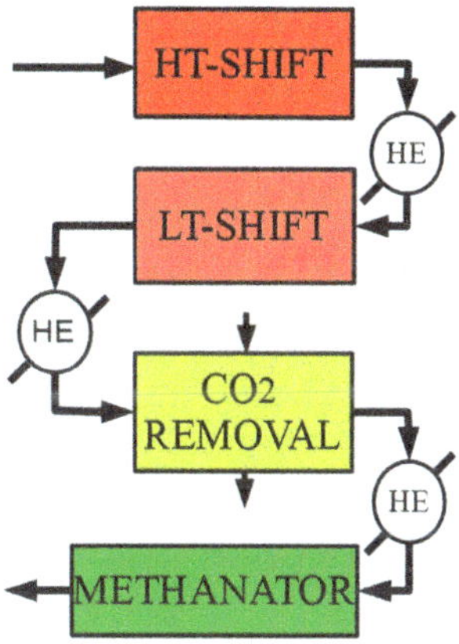

Fig. 5.19 A schematic of all simulated units, where the high-temperature shift reactor, heat exchanger, low-temperature shift reactor, CO_2 removal, and methanator are denoted by HT-SHIFT, HE, LT-SHIFT, CO_2 REMOVAL, and METHANATOR, respectively

program into an objective file, and finally use it as a dynamic link library file in Aspen Plus software. The rate equations for all units are shown as follows [23, 63]:

(a) High-temperature shift reaction: $CO + H_2O \rightleftarrows CO_2 + H_2$, $\Delta H = -41.2$ kJ/mol:

$$r_{CO} = -A_c \exp\left(-\frac{300.69}{T} + 8.02\right)(P)^{1/2}\left(y_{CO} - \frac{y_{H_2} y_{CO_2}}{K_{eq} y_{H_2O}}\right),\ K_{eq} = \exp\left(\frac{8240}{T} - 4.33\right). \tag{5.11}$$

(b) Low-temperature shift reaction: $CO + H_2O \rightleftarrows CO_2 + H_2$, $\Delta H = -41.2$ kJ/mol:

$$\begin{aligned} r_{CO} &= -A_c \frac{513.15}{T} \frac{K_L y_{CO} y_{H_2O}^{1/2}\left(1 - \frac{K}{K_{eq}}\right)}{\frac{1}{P} + K_A y_{CO} + K_B y_{CO_2}},\ K = \frac{y_{H_2} y_{CO_2}}{y_{CO} y_{H_2O}},\ K_{eq} = \exp\left(\frac{8240}{T} - 4.33\right) \\ K_L &= 68.4 \exp\left(-3620\left(\frac{1}{513.15} - \frac{1}{T}\right)\right),\ K_A = 4.31 \exp\left(-4580\left(\frac{1}{513.15} - \frac{1}{T}\right)\right), \\ K_B &= 1.35 \exp\left(-1500\left(\frac{1}{513.15} - \frac{1}{T}\right)\right). \end{aligned} \tag{5.12}$$

(c) Methanation reaction 1: $CO + 3H_2 \rightleftarrows CH_4 + H_2O$, $\Delta H = -206$ kJ/mol:

$$r_{CO} = -A_c\, 3.119 \exp\left(1300\left(\frac{1}{T} - \frac{1}{513}\right)\right)\left(\frac{P}{y_{H_2}}\right)^{1/2}\left(y_{CO} - \frac{y_{CH_4}\, y_{H_2O}}{y_{H_2}^3\, P^2 \exp\left(-38.4523 + \frac{2627}{T}\right)}\right). \tag{5.13}$$

(d) Methanation reaction 2: $CO_2 + 4H_2 \rightleftarrows CH_4 + 2H_2O$, $\Delta H = -164$ kJ/mol:

$$r_{CO} = -A_c\, 3.119 \exp\left(1300\left(\frac{1}{T} - \frac{1}{513}\right)\right)\left(\frac{P}{y_{H_2}}\right)^{1/2}\left(y_{CO_2} - \frac{y_{CH_4}\, y_{H_2O}^2}{y_{H_2}^4\, P^2 \exp\left(-38.4523 + \frac{2627}{T}\right)}\right) \tag{5.14}$$

where r_{CO} (gmol/m^{3}s), A_c, T (K), P (atm) and y_i are the reaction rate of CO, catalyst activity, temperature, total pressure, and the mole fraction of component i, respectively. In the simulation, all the three tube reactors are adiabatic, and the outlet temperature of all heat exchangers is fixed. We simulate the CO_2 removal unit as a

Table 5.4 Parameter values of the ammonia process simulation

	Parameter	Value
Feed	Temperature	980 °C
	Pressure	29 bar
	Mole flow rate	3435 mol/s
	Mole fraction y_{CO}	0.0839
	Mole fraction y_{CO_2}	0.0507
	Mole fraction y_{H_2}	0.355
	Mole fraction y_{H_2O}	0.353
	Mole fraction y_{N_2}	0.152
HT-shift	Reactor length	15.8 m
	Reactor diameter	4.4 m
	Loaded catalyst	9.61×10^4 kg
	Voidage	0.5
	Catalyst heat capacity	900 J/kg K
	Feed temperature	360 °C
LT-shift	Reactor length	7.7 m
	Reactor diameter	3.7 m
	Loaded catalyst	3.48×10^4 kg
	Voidage	0.5
	Catalyst heat capacity	850 J/kg K
	Feed temperature	210 °C
CO_2 Removal	Volume	49.09 m^3
	Temperature	30 $°C$
	Pressure	26.9 bar
	CO_2 remove rate	98.6%
	H_2O remove rate	99.7%
Methanator	Reactor length	4 m
	Reactor diameter	2.5 m
	Loaded catalyst	1.57×10^4 kg
	Voidage	0.5
	Catalyst heat capacity	900 J/kg K
	Feed temperature	280 °C

flash drum operated at $T = 30$ °C to condense water and remove CO_2 with feeding ammonia solution. Detailed reaction kinetic and electrolyte solution chemistry in CO_2 removal unit is discussed in [23]. Table 5.4 reports the process parameter values and the steady-state values.

The catalyst properties and the size of the high-temperature shift reactor are designed following [10, 23, 63, 193]. It is noted that an optimal feed temperature exists in this process since a low temperature might lead to a better equilibrium in an exothermic reversible reaction, while a high temperature can accelerate reaction rate [163, 175]. Therefore, we determine the optimal feed temperature to be the one at which the highest conversion of carbon monoxide is achieved among many simulations with various feed temperatures. The results are demonstrated to be consistent with the industrial data. The size (i.e., diameter and length) of the low-temperature shift reactor as well as the catalyst properties are designed following [10, 23, 63, 193]. Note that the optimal feed temperature for the low-temperature shift reactor could be low due to a low feed concentration. However, the dew point of the gas is the limiting condition since condensed water is poisonous to the catalyst. We use the Aspen analysis tool (in the Properties sheet) for the mixture to find the dew point 169 °C for the specific high temperature and high pressure gas, and finally choose 210 °C according to industrial data [10, 23, 63]. Carbon dioxide is removed and the water from the gas phase is condensed in the carbon dioxide removal unit with an aqueous ammonia solution. The adsorption column unit is represented by a flash drum in the simulation. A stream of water (85%) and ammonia (15%) as well as the gas from the shift reactor that has been cooled down to 40 °C are fed into the flash drum. Water and carbon dioxide can be removed up to 99.7 and 98.6% in the gas leaving the flash drum. [23, 100] discussed the detailed reaction kinetic and electrolyte solution chemistry properties in CO_2 removal. After the carbon dioxide removal unit, the gas is heated up and fed into the methanation unit to prepare for the final purification step. We choose the feed temperature to be 280 °C such that the mole fraction of CO_2 and CO in the final outlet is below 0.0005%. The key parameter values in the methanator are carefully chosen based on the data in [95, 171, 193].

We assume that all units are initially operated at the steady-states. When catalyst activity decreases (i.e., considered as a process disturbance in this study) in the first high-temperature shift reactor, the consumption of CO becomes less in the shift reactor. Then, more CO goes into the methanator since the CO_2 removal unit does not remove CO. This will lead to a temperature increase in the adiabatic tube reactor due to the exothermic reaction of methanation. The open-loop simulation result for the process with a disturbance that decreases catalyst activity from 1 to 0.1 within 300 s is shown in Fig. 5.20a. Additionally, when the feed temperature drops (i.e., considered as another type of process disturbance) in the first high-temperature shift reactor, less CO will be reacted in the shift reactor. Similarly, more CO enters the methanator, causing an increase of temperature in the methanator as CO is not removed by the CO_2 removal unit. The open-loop simulation result for the process with a disturbance that decreases the feed temperature from 380 °C to 280 °C within 300 s for the high-temperature shift reactor is shown in Fig. 5.20b. To improve process operational safety for the ammonia plant when these two types of disturbances occur, we will develop a controller in the following section to regulate the methanator outlet temperature with the methanator inlet feed temperature as the manipulated inputs.

5.3.2.2 Safeness Index-Based MPC

Initially, the methanator is operated at the steady-state where the outlet temperature and the feed temperature are $T_{\text{out}} = 327.27$ °C and $T_{\text{in}} = 280$ °C, respectively. Since

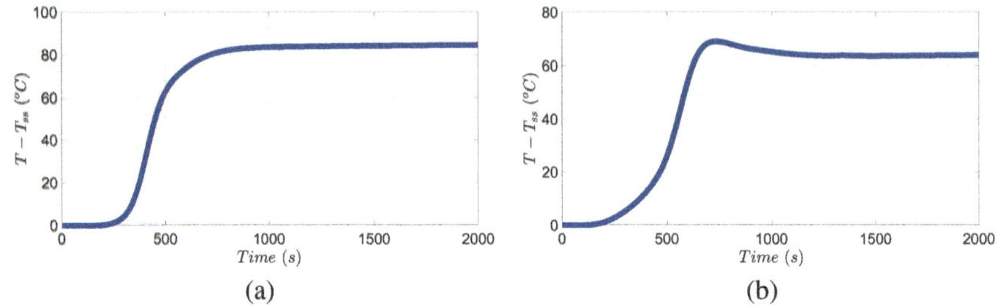

Fig. 5.20 Methanator outlet temperature profiles, from which it is shown that $T - T_{ss}$ increases more than **a** 80 °C after the catalyst activity decreases from 1 to 0.1 in 300 s, and **b** 60 °C after the feed temperature decreases from 380 °C to 280 °C in 300 s, respectively, in the high-temperature shift reactor

the feed concentration of CO has a dominating impact on the produced heat, we consider CO mole fraction y_{CO} as a measurable disturbance with a steady-state value of 3.55×10^{-3}. The manipulated input, process states, and disturbance are represented as $u = T_{\mathrm{in}} - T_{\mathrm{in,ss}}$, $x = T_{\mathrm{out}} - T_{\mathrm{out,ss}}$, and $d = y_{\mathrm{CO}} - y_{\mathrm{CO,ss}}$ (i.e., in deviation variable form), such that the system has an equilibrium point at the origin. As a time delay between the outlet temperature T_{out} and the feed temperature T_{in} exists in this example, we introduce the time delay t_d (in seconds) into the following linear dynamic model to represent the Aspen Plus model:

$$\frac{\mathrm{d}x(t)}{\mathrm{d}t} = Ax(t) + Bu(t - t_d) + Kd(t - t_d). \tag{5.15}$$

It is noted that the disturbance time delay and the input time delay are the same because it takes the same amount of time for the outlet temperature T_{out} of a tube reactor to be affected by the CO mole fraction of y_{CO} and the feed temperature T_{in}. Aspen open-loop simulations are performed to generate the dataset for developing the model of Eq. 5.15. Specifically, we run extensive open-loop simulations to collect time-series data of outlet temperature T_{out} under various step changes in feed temperature T_{in}. The matrices A and B are then identified from the dataset using the Multivariable Output Error State Space (MOSEP) algorithm in MATLAB. Subsequently, the gain K of the disturbance term is obtained following the same steps for A and B based on another dataset with various step changes in CO mole fraction of y_{CO}. The model parameter values are reported as follows: $t_d = 100\ s$, $K = 32.887$, $A = -0.005136$, and $B = 0.01207$.

Since temperature control is of significant importance in safe operation of the methanator, to avoid a high outlet temperature that could lead to unsafe operations, the Safeness Index in this example is designed to treat high temperature T_{out} as unsafe operating conditions and all the temperatures T_{out} below the steady-state value as safe operating conditions. Specifically, Safeness Index is designed in the following form:

$$S(x) = [f^{+}(x)]^2 \tag{5.16}$$

where $f^+(x)$ is the same function as shown in Eq. 5.8. The Safeness Index threshold S_{TH} is carefully chosen to avoid an extremely high temperature in the methanator and also to account for the large time delay, sample-and-hold implementation of the controller, and the model mismatch. Due to the above considerations, a more conservative value, i.e., to be $S_{TH} = 25$, is finally chosen for the threshold of the Safeness Index function. Then, the controller u is developed by integrating a feedforward control action with Safeness Index-based MPC as shown in Eq. 5.17:

$$u(t_k) = u_{MPC}(t_k) + u_{\text{feedforward}}(t_k) \tag{5.17}$$

where $u_{\text{feedforward}}(t_k)$ and $u_{\text{MPC}}(t_k)$ represent the control actions computed by the feedforward controller and MPC, respectively. We use Eq. 5.18 to calculate $u_{\text{feedforward}}(t_k)$ and solve the optimization problem of Eq. 5.19 to obtain the optimal solution $u^*(t)$, from which the first control $u_{\text{MPC}}(t_k)$ will be applied to the ammonia process.

$$u_{\text{feedforward}}(t_k) = -\frac{K}{B}\mathrm{d}(t_k) \tag{5.18}$$

$$\min_{u\in S(\Delta),y} \int_{t_k+t_d}^{t_{k+N}+t_d} (|\tilde{x}(\tau)|^2_{Q_c})\, d\tau + \int_{t_k}^{t_{k+N}} (|u(\tau)|^2_{R_c})\, \mathrm{d}\tau + \sum_{i=1}^{N} k_1 e^{-k_2 y(i)} \tag{5.19a}$$

$$\text{s.t.}\quad \dot{\tilde{x}}(t) = A\tilde{x}(t) + Bu(t - t_d) \tag{5.19b}$$

$$\tilde{x}(t_k) = x(t_k) \tag{5.19c}$$

$$u(t) = u_{pre}(t),\ \forall\, t \in [t_k - t_d, t_k) \tag{5.19d}$$

$$u(t) \in U,\ \forall\, t \in [t_k, t_{k+N}) \tag{5.19e}$$

$$S(\tilde{x}(t_{k+i} + t_d)) + y(i) \le S_{TH},\ i = 1, 2, \ldots, N \tag{5.19f}$$

$$y(i) \ge 0,\ i = 1, 2, \ldots, N,\ \text{ if }\ S(\tilde{x}(t_k + t_d)) \le S_{TH} \tag{5.19g}$$

$$y(i) \in R,\ i = 1, 2, \ldots, N,\ \text{ if }\ S(\tilde{x}(t_k + t_d)) > S_{TH} \tag{5.19h}$$

where the notation follows that in Eq. 5.10 and $k_1,\ k_2 > 0$. The optimal solution to the MPC $u^*(t)$ is obtained for the entire prediction horizon $t \in [t_k,\ t_{k+N})$. However, we only apply the first control action $u(t_k)$ to the next sampling period. The objective function of Eq. 5.19a is minimizing the deviation of the current states and inputs from their steady-state values (i.e., the integral terms $\int_{t_k+t_d}^{t_{k+N}+t_d} (|\tilde{x}(\tau)|^2_{Q_c})\, \mathrm{d}\tau$ and $\int_{t_k}^{t_{k+N}} (|u(\tau)|^2_{R_c})\, \mathrm{d}\tau$) as well as the penalty term $\sum_{i=1}^{N} k_1 e^{-k_2 y(i)}$ of slack variables $y(i)$. It is noted that due to the time delay in process dynamics, the states from t_k to $t_k + t_d$ have been already determined by the previous control actions, and thus, the state is integrated from $t_k + t_d$ to $t_{k+N} + t_d$ only. The nominal linear model of Eq. 5.15 is used to predict the closed-loop states in the prediction horizon in Eq. 5.19b. The feedforward control action $u_{\text{feedforward}}(t_k)$ is to mitigate the impact of disturbances, and the MPC utilizes the nominal system of Eq. 5.19b as the prediction model to optimize the control action u_{MPC}. The initial condition $\tilde{x}(t_k)$ of the optimization problem uses the measured state $x(t_k)$ at $t = t_k$ in Eq. 5.19c. The input trajectories from previous time steps are provided in Eq. 5.19d for predicting the state from t_k to $t_k + t_d$ at the current time step. The input constraints are imposed over the entire prediction horizon in Eq. 5.19e. In this example, the bounds for the manipulated input (i.e., feed temperature T_{in}) are $U = [-100,\ 100]$ (i.e., $180\,°\text{C} \leq T_{\text{in}} \leq 380\,°\text{C}$). The Safeness index constraint is developed as a soft constraint in Eq. 5.19f with slack variables $y(i)$. Specifically, to maintain $S(x)$ below the threshold S_{TH} $S(x(t_k + t_d)) > S_{TH}$, a nonnegative slack variable $y(i)$ is required by Eq. 5.19g; when $S(x(t_k + t_d))$ exceeds S_{TH}, negative slack variables can be used to allow the Safeness Index to exceed the threshold S_{TH} such that a feasible solution exists for the optimization problem. Additionally, $k_1 = 10^5$ and $k_2 = 0.2$ are chosen in this example to balance the effects when $S(x(t_k + t_d))$ is far below S_{TH} and when it is close to S_{TH}.

The dynamic model of Eq. 5.15 is numerically integrated with a sufficiently small integration time step of $h_c = 10^{-1}$ s using the explicit Euler method. The solver FilterSD on OPTI Toolbox in MATLAB is used to solve the nonlinear optimization problem of Safeness Index-based MPC with the following settings: prediction horizon $N = 30$ and sampling period $\Delta = 20$ s. The coefficients $Q_c = 1$ and $R_c = 0.5$ are chosen to balance the order of magnitude of the two terms with respect to the input and the state, respectively, in the objective function of Eq. 5.19a.

5.3.2.3 Closed-Loop Simulation Results

We implement the proposed controller to the ammonia process in the presence of various disturbances in this subsection.

Disturbance 1: Variation of catalyst activity

We consider a process disturbance in catalyst activity. Specifically, we assume that the catalyst activity decreases from 1 to 0.1 within 300 s in the high-temperature shift reactor. As a result, less CO is reacted in the high-temperature shift reactor, and a higher concentration of CO reaches the methanator, leading to a temperature increase in the methanator. At each sampling time t_k, we measure the feed mole fraction of CO, i.e., y_{CO}, and send it to the feedforward controller to compute the control actions that can compensate the effect of disturbance. However, it should be noted that the feedforward control action may not be able to fully mitigate the impact of disturbances because the process dynamics has been changed due to catalyst deactivation, and it cannot be fully characterized by the change of feed mole fraction of CO only. Additionally, since the process model used in MPC is developed as a linear model, there exists a model mismatch between the actual ammonia process and the MPC prediction model. Therefore, in the presence of a disturbance in catalyst activity, the methanator outlet temperature may be stabilized at another steady-state that is different from the initial steady-state. Figure 5.21 shows the methanator outlet temperature and feed temperature profiles in the presence of such a disturbance. It can be seen that the increase of the outlet temperature T_{out} is less than 30 °C under the MPC

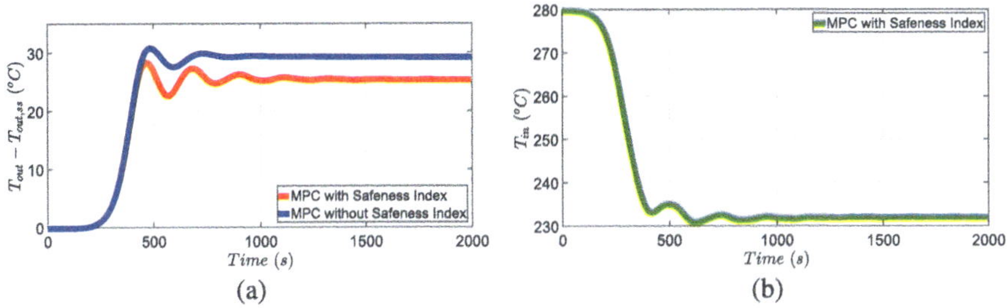

Fig. 5.21 Close-loop methanator **a** outlet temperature and **b** feed temperature profiles when the catalyst activity decreases from 1 to 0.1 within 300 s in the high-temperature shift reactor

that uses Safeness Index constraints, while the temperature increase exceeds 30 °C using the standard MPC that does not account for safety considerations. Therefore, an improved process operational safety for an ammonia plant that is demonstrated under the Safeness Index-based MPC is demonstrated in this simulation study.

Disturbance 2: Variation of feed temperature

We consider another type of process disturbance, under which the feed temperature drops from 380 °C to 280 °C in the high-temperature shift reactor. Since it takes time for a large amount of catalyst to cool down, the temperature in the high-temperature shift reactor will slowly decrease after the disturbance occurs. Correspondingly, the feed mole fraction of CO, i.e., y_{CO}, in the methanator will gradually increase, and meanwhile, the methanator temperature will also increase but will be slower than that in the disturbance of catalyst activity. Figure 5.22 shows the methanator outlet temperature and feed temperature profiles in the presence of a feed temperature disturbance. Specifically, it is observed that the increase of the methanator outlet temperature T_{out} exceeds 40 °C under the standard MPC that does not use Safeness Index constraints; however, the outlet temperature increase is maintained below 40 °C when the Safeness Index-based MPC is utilized. Additionally, if we compare the closed-loop simulation results (i.e., Figs. 5.22 and 5.21) for the two types of dis-

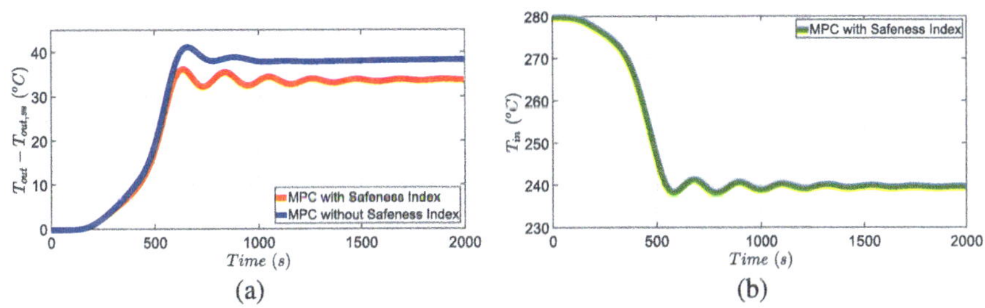

Fig. 5.22 Close-loop methanator **a** outlet temperature and **b** feed temperature profiles when the feed temperature decreases from 380 °C to 280 °C within 300 s in the high-temperature shift reactor

turbances, we can see that the temperature increases more in the case of disturbance 2. This can be explained by the fact that the CO concentration in the feed stream of the methanator is much less under the second disturbance than that under the first disturbance. Therefore, a sufficiently small feedforward control action $u_{\text{feedforward}}$ is utilized in the case of disturbance 2, and the overall control action obtained from the integration of the feedforward controller and MPC, i.e., $u = u_{\text{MPC}} + u_{\text{feedforward}}$, becomes less than that under disturbance 1. Additionally, it is noticed that under both disturbances, the closed-loop state shows an offset. This is because not all process disturbances in a multi-unit process are measurable, and thus, cannot be fully compensated by feedforward control actions. To eliminate the offset, an offset-free control can be used, which will be demonstrated in the next case study.

Safeness Index-based MPC was applied to the ammonia process with four units in this case study. Specifically, to ensure process operational safety for the ammonia process with a significant propagated disturbance in the methanation unit, we developed a Safeness Index function to characterize process safeness based on the state measurement of process variables in an adiabatic methanation tube reactor. Subsequently, based on the linear dynamic model that was identified for the methanator with time delay and disturbance terms, we developed the Safeness Index-based MPC with the integration of feedforward control that compensates the disturbance to improve process operational safety under the propagated disturbances.

5.3.3 Case Study: Ammonia Process Network

The case study in this section develops a safety control scheme for integrating process control and operational safety for a multi-unit ammonia process network, and investigates its performance through the simulation of the multi-unit ammonia process under a practical disturbance that is often encountered by engineers. Specifically, we consider a disturbance of the decrease of catalyst activity in the first unit of the proposed ammonia plant network. It is shown in the previous case study that an unsafe operation may occur since the gas concentration and temperature in the methanation unit are affected by the variation of catalyst activity. Unlike the control system we designed in the previous case study, in this example, to further improve process operational safety, we develop a tracking model predictive controller (MPC) for the high-temperature shift reactor and a Safeness Index-based MPC that is similar to that in Sect. 5.3.2 for the methanator, respectively. MATLAB and Aspen Plus are coordinated to simulate the closed-loop ammonia plant under the proposed control systems.

5.3.3.1 Ammonia Process Description

In this section, we present a simplified description of the ammonia process with the key elements for each unit in the ammonia process network. As shown in Fig. 5.23, the

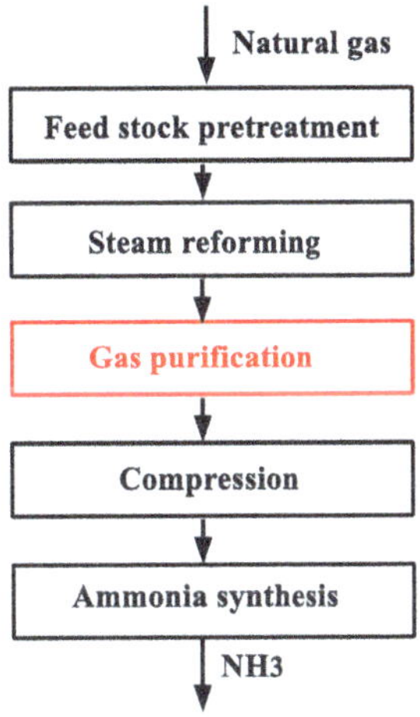

Fig. 5.23 A schematic of the entire ammonia process network

ammonia process network consists of the following units: feedstock pre-treatment, steam reforming, gas purification, compression, and ammonia synthesis [18]. The case study in this section focuses on the same gas purification step as in the previous case study, which includes shift conversion, carbon dioxide removal, and methanation processes. All three sub-processes are developed to remove carbon dioxide and carbon monoxide generated in the previous steam reforming step. A schematic of the ammonia process network implemented in this section is shown in Fig. 5.24, where the high-temperature shift reactor, low-temperature shift reactor, and heat exchanger are denoted by HT-shift, LT-shift, and HE, respectively. It is noted that the difference between Fig. 5.19 and Fig. 5.24 is that two controllers, i.e., C_1 and C_2, are employed in this case study of an ammonia process network while only the controller for the methanator was utilized in the case study in Sect. 5.3.2.

We use the same two-bed adiabatic operation as that in Sect. 5.3.2.1 (i.e., the beds operate at 400 and 200 °C, respectively, with different catalyst components) for eliminating the carbon monoxide in the shift reactors. An exothermic reaction takes place in each reactor using water and carbon monoxide to yield hydrogen and carbon dioxide. Specifically, a large amount of carbon monoxide is removed in the high-temperature shift reactor, followed by the low-temperature shift reactor that is used to further remove carbon monoxide. A surplus amount of water vapor is introduced into the gas stream in both the high-temperature and low-temperature shift reactors, and a large amount of carbon dioxide is produced. To protect the ammonia synthesis catalyst, carbon dioxide and water vapor are removed afterwards in an adsorption column [18]. Additionally, the adsorption column purifies the gas by using various types of solvents. Finally, since trace amounts of carbon dioxide and carbon monoxide still exist in the stream after the gas leaves the adsorption column, a catalytic methanation reaction unit is needed. The remaining carbon dioxide and carbon monoxide with hydrogen are converted into water and methane in the adiabatic tubular methanation reactor where two exothermic catalytic reactions take

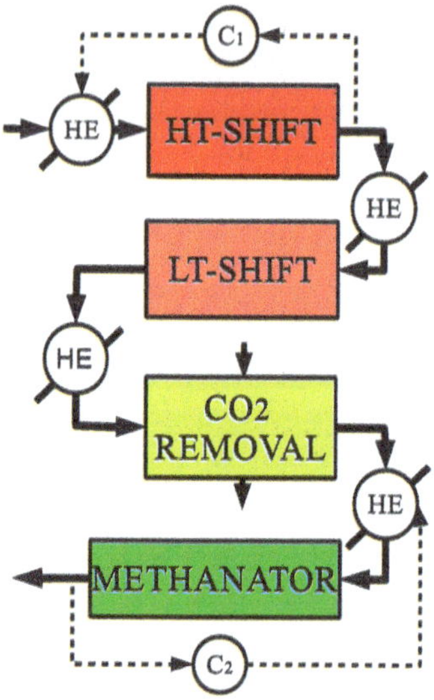

Fig. 5.24 A schematic of the control structure that uses two control loops, where C_1 and C_2 represent controller 1 and controller 2

place [122, 171]. The concentrations of carbon dioxide and carbon monoxide are expected to be reduced below $0.0005 - 0.001\%$ in the methanation unit [143, 193]. The detailed description of the ammonia process network including its parameter values and reaction rate equations can be found in Sect. 5.3.2.1 and is omitted here. However, the safety control of this methanation unit has been challenging due to many reasons such as potential thermal runaway and sensitivity to the catalyst [171]. For example, thermal runaway may occur in the methanator when there is a disturbance in the upstream shift conversion reactors that can cause high heat generation of reaction in the methanator [10].

5.3.3.2 Disturbance and Process Operational Safety

Runaway reactions resulting from the catalyst deactivation in shift reactors are one of the most common safety issues in an ammonia plant (see, for example, [10, 19, 202]). Figure 5.25 shows how the disturbance of catalyst deactivation affects the operation of an ammonia process. Specifically, in Fig 5.25, it is demonstrated that less CO is consumed after the catalyst becomes less active in the high-temperature shift reactor. Although a small amount of the increasing CO content is buffered in the low-temperature shift reactor (no CO is removed in the CO_2 removal unit), there are still more CO going into the methanator as the reactant, leading to a drastic temperature increase due to the exothermic reaction in the methanator. Figure 5.26 shows an example of the methanator outlet temperature profile under the open-loop simulation with a decrease of the catalyst activity from 1 to 0.2 within 300 s. It is observed that the outlet temperature is initially at 330 °C and increases to 390 °C in the methanator under the above disturbance.

5.3.3.3 Feedback Controller Design

In the previous case study in Sect. 5.3.2, we have demonstrated improved process operational safety by incorporating Safeness Index functions into model predictive control scheme and utilizing a feedforward control action to compensate the impact of disturbances. In order to further improve operational safety, two controllers are developed in this example for the high-temperature shift reactor and the methanator, respectively. The first controller is designed with reactor outlet temperature as the controlled variable and reactor inlet temperature as the manipulated variable. The second controller is designed with methanator outlet temperature as the controlled variable and methanator inlet temperature as the manipulated variable. Since the gas temperature in both the methanator and the high-temperature shift reactor increases monotonously from the inlet to the outlet, outlet temperature measurement can be used to indicate the safeness of the reactor operation. The schematic for the two control loops implemented in the ammonia process network is shown in Fig. 5.24. The controller design for the high-temperature shift reactor and the methanator, respectively, are detailed as follows.

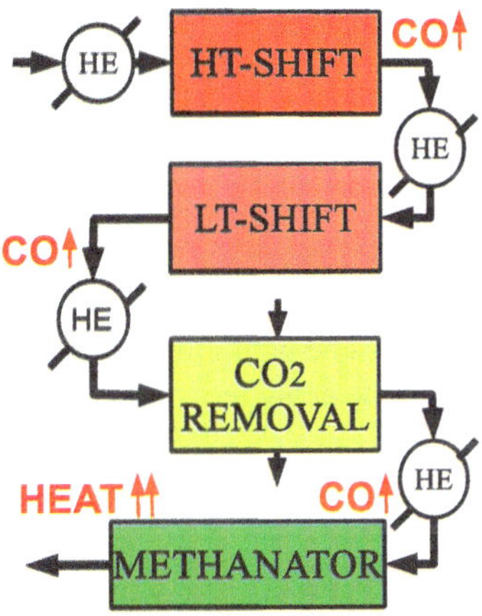

Fig. 5.25 A schematic of disturbance propagation showing that a reaction thermal runaway may occur due to the increasing concentration of CO in high-temperature shift reactor

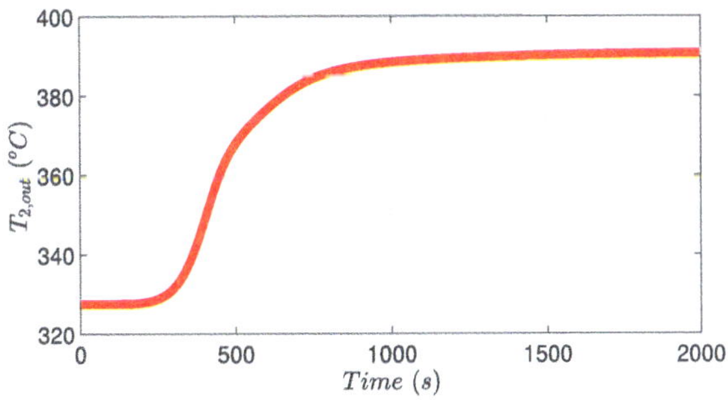

Fig. 5.26 Open-loop methanator outlet temperature profile for the ammonia process under a decrease of catalyst activity from 1 to 0.2 within 300 s in the high-temperature shift reactor

High-temperature shift reactor controller

The reaction rate in the tube decreases after the catalyst becomes less active in the high-temperature shift reactor, which also leads to the decrease of outlet temperature in the reactor. To avoid the unsafe operation resulting from the increasing amount of carbon monoxide, we need to manipulate the heat exchanger to increase the inlet temperature such that more carbon monoxide is consumed with a more active reaction. To efficiently regulate the outlet temperature, we develop a model predictive controller (MPC) that controls the outlet temperature by manipulating the inlet temperature in the high-temperature shift reactor. Since the MPC uses a nominal process model that does not account for any process disturbances, an integral control term will be added to the control action calculated by MPC to eliminate the offset due to model mismatch. To develop an MPC for the high-temperature shift reactor, we first need to develop a process model following the same steps of system identification as performed in the previous case studies. Specifically, the high-temperature shift reactor is initially operated at the steady-state, i.e., outlet temperature $T_{1,\text{out,ss}} = 429.18$ °C and feed temperature $T_{1,\text{in,ss}} = 360$ °C. Both the manipulated input and the process state are represented in deviation variable form as $u_1 = T_{1,\text{in}} - T_{1,\text{in,ss}}$ and $x_1 = T_{1,\text{out}} - T_{1,\text{out,ss}}$, such that the system has an equilibrium point at the origin. An inverse response is observed between the outlet temperature $T_{1,\text{out}}$ and the feed temperature $T_{1,\text{in}}$ for the high-temperature shift reactor, and it can be explained by the fact that a decrease of inlet temperature in an adiabatic exothermic tube reactor would reduce the reaction rate at the cooler upstream part of the tube reactor, which leads to a higher CO concentration in the gas stream entering the downstream part of the tube reactor, and also a temporary increase of heat generation at the reactor outlet. However, since the inlet temperature is decreased, the outlet temperature will eventually decrease below its original value [153, 194]. In this study, the inverse response is treated as a time delay in dynamic response since its impact is demonstrated to be negligible. Therefore, we develop the following linear dynamic model with time delay term $t_{d,1}$ (s) to represent the dynamics of the high-temperature shift reactor:

$$\frac{\mathrm{d}x_1(t)}{\mathrm{d}t} = A_1 x_1(t) + B_1 u_1(t - t_{d,1}). \tag{5.20}$$

The model is developed based on the time-series dataset generated from extensive Apsen open-loop simulations that collect the outlet temperature $T_{1,\text{out}}$ under various step changes in the feed temperature $T_{1,\text{in}}$. Similarly, we identify the matrices A_1 and B_1 by implementing the Multivariable Output Error State Space (MOSEP) algorithm in MATLAB.

$$A_1 = -0.015; \quad B_1 = 0.0142; \quad t_{d,1} = 360 \text{ s}.$$

The first controller (i.e., C_1 in Fig. 5.24) is the MPC with an integral control term, for which the control action $u_1(t_k)$ consists of an integral control action $u_{1,\text{integral}}(t_k)$ and an MPC control action $u_{1,\text{MPC}}(t_k)$. $u_{1,\text{integral}}(t_k)$ is calculated by Eq. 5.22 and $u_{1,\text{MPC}}(t_k)$ is the first element in the optimal solution $u^*(t)$ to the optimization problem of Eq. 5.23 as follows:

$$u_1(t_k) = u_{1,\mathrm{MPC}}(t_k) + u_{1,\mathrm{integral}}(t_k) \tag{5.21}$$

$$\begin{aligned} e(t_k) &= T_{1,\mathrm{out,set}} - T_{1,\mathrm{out}}(t_k) \\ u_{\mathrm{integral}}(t_k) &= \frac{1}{\tau_I} \int_0^{t_k} e(\tau)\, d\tau \end{aligned} \tag{5.22}$$

and

$$\min_{u \in S(\Delta), y} \int_{t_k + t_{d,1}}^{t_{k+N} + t_{d,1}} (|\tilde{x}_1(\tau)|^2_{Q_c})\, d\tau + \int_{t_k}^{t_{k+N}} (|u_1(\tau)|^2_{R_c})\, d\tau \tag{5.23a}$$

$$\text{s.t.} \quad \dot{\tilde{x}}_1(t) = A_1 \tilde{x}_1(t) + B_1 u_1(t - t_{d,1}) \tag{5.23b}$$

$$\tilde{x}_1(t_k) = x_1(t_k) \tag{5.23c}$$

$$u_1(t) = u_{1,pre}(t),\ \forall\, t \in [t_k - t_{d,1}, t_k) \tag{5.23d}$$

$$u_1(t) \in U_1,\ \forall\, t \in [t_k, t_{k+N}). \tag{5.23e}$$

The optimization problem of Eq. 5.23 minimizes the sum of $\int_{t_k+t_{d,1}}^{t_{k+N}+t_{d,1}} (|\tilde{x}_1(\tau)|^2_{Q_c})\, d\tau$ and $\int_{t_k}^{t_{k+N}} (|u_1(\tau)|^2_{R_c})\, d\tau$ in the objective function of Eq. 5.23a to stabilize the system at the steady-state. The nominal linear model of Eq. 5.20 is used in Eq. 5.23b to predict future states. The state measurement $x(t_k)$ is used as the initial condition $\tilde{x}(t_k)$ for the optimization problem at $t = t_k$ in Eq. 5.23c. To predict the states from t_k to $t_k + t_{d,1}$, the input trajectories calculated from the previous time steps are provided in Eq. 5.23d. Equation 5.23e is the constraint on the control actions. Specifically, the manipulated input (i.e., feed temperature $T_{1,\mathrm{in}}$) is bounded by: $U_1 = [-50,\ 50]$ (i.e., 310 °C $\leq T_{1,\mathrm{in}} \leq$ 410 °C) in this example. The dynamic model of Eq. 5.23b is numerically integrated with a sufficiently small integration time step of $h_c = 10^{-1}\ s$ using the explicit Euler method. The solver FilterSD on OPTI Toolbox in MATLAB is used to solve the nonlinear optimization problem of MPC of Eq. 5.23 with the following settings: prediction horizon $N = 30$ and sampling period $\Delta = 20$ s. $R_c = 0.2$ and $Q_c = 1$ are chosen to balance the order of magnitude of the two terms in the objective function. The integral control term with the time constant $\tau_I = 33.3$ s is used to eliminate the offset due to model mismatch.

Methanator controller

Although the increase of CO concentration from the high-temperature shift reactor is mitigated by the low-temperature shift reactor, the feed stream to the methanator may still contain a higher concentration of CO. To avoid potential high temperature due to more CO coming into the methanator, another MPC is designed to regulate the methanator outlet temperature with the inlet temperature as manipulated input. Similarly, open-loop simulations are performed to collect data for the development of a data-driven process model for the methanator. The methanator is initially operated at the steady-state with outlet temperature $T_{2,\text{out,ss}} = 327.98$ °C and feed temperature $T_{2,\text{in,ss}} = 280$ °C. The mole fraction of CO, i.e., y_{CO}, is treated as a measurable disturbance since the feed concentration of CO plays a dominant role in heat production. We represent the disturbance in deviation form, i.e., $\text{d}_2 = y_{\text{CO}} - y_{\text{CO,ss}}$ with $y_{\text{CO,ss}} = 3.55 \times 10^{-3}$. Additionally, it should be noted that the steady-state of the methanator varies when feed mole fraction of CO is changed because working condition of the methanator is significantly changed in the presence of an increasing CO in the feed stream. Therefore, we calculate offline a set of steady-state values as a function of the feed mole fraction of CO for the methanator. The future steady-states for the outlet and inlet temperature can be represented as the following function of disturbance d_2:

$$\begin{aligned}
\Delta\text{d} &= \text{d}_2(t_k) - \text{d}_2(t_{k-1}) \\
T(t_{k+N})_{2,\text{in,ss}} &= 280 - 4080.7(\text{d}_2 + \gamma\ \Delta\text{d}\ N) \\
T(t_{k+N})_{2,\text{out,ss}} &= 327.27 + 1616.3(\text{d}_2 + \gamma\ \Delta\text{d}\ N)
\end{aligned} \tag{5.24}$$

where $\gamma = 0.5$ is used to estimate the variation of future disturbance. For example, the future disturbance is anticipated to increase at the rate of $\gamma\ \Delta d$ at each subsequent sampling time if the current disturbance $d_2(t_k)$ shows an increase of Δd compared to the previous disturbance $d_2(t_{k-1})$. It is noted that all the steady-states that we obtain offline are reasonable working conditions, under which the corresponding outlet temperature is below 340 °C and the outlet CO content is below 5×10^{-6} in the presence of a small disturbance d_2.

The process input and states are represented using deviation variables, $u_2 = T_{2,\text{in}} - T_{2,\text{in,ss}}$ and $x_2 = T_{2,\text{out}} - T_{2,\text{out,ss}}$ such that the system has the equilibrium point at the origin. Since under a step change of the feed temperature $T_{2,\text{in}}$, the dynamic response of the outlet temperature $T_{2,\text{out}}$ shows a time delay, the following linear dynamic model with a time delay term $t_{d,2}$ (s) is developed to represent the methanator model in Aspen Plus:

$$\frac{\text{d}x_2(t)}{\text{d}t} = A_2 x_2(t) + B_2 u_2(t - t_{d,2}). \tag{5.25}$$

Again, we identify the matrices A_2 and B_2 by implementing the MOSEP algorithm to the dataset from extensive Aspen open-loop simulations under various step changes in feed temperature $T_{2,\text{in}}$. It is demonstrated that the disturbance in feed CO content changes the methanator steady-state only but barely affects the process dynamics (i.e., gain, time delay, and time constant). Therefore, the model of Eq. 5.25 is able to capture the process dynamics well for all the steady-states corresponding to different feed CO content with the following parameter values:

$$A_2 = -0.005136; \quad B_2 = 0.01207; \quad t_{d,2} = 100 \text{ s}.$$

The methanator controller (i.e., C_2 in Fig. 5.24) is developed to improve operational safety of the ammonia process to avoid unsafe operations due to a high outlet temperature above the steady-state. Specifically, Safeness Index function $S(x)$ is developed and incorporated as a constraint in the second MPC (i.e., the controller C_2) due to safety considerations in the methanator; however, the controller C_1 is a tracking MPC without explicitly accounting for any safety considerations as no critical safety issues are encountered in the high-temperature shift reactor. The Safeness Index function should be developed such that all the methanator outlet temperature values $T_{2,\text{out}}$ below the steady-state operating condition $T_{2,\text{out,ss}}$ are considered safe, and all the temperature values above the steady-state value are considered unsafe. To develop the Safeness Index function that satisfies the above requirement, we first define a function $f^+(x)$:

$$f^+(x) = \begin{cases} x, & \text{if } x \geq 0 \\ 0, & \text{if } x < 0. \end{cases} \tag{5.26}$$

Then, the Safeness Index function is designed as follows:

$$S(T) = [f^+(T_{2,\text{out}} - T_{2,\text{out,initial}})]^2 \tag{5.27}$$

where the initial steady-state value of outlet temperature is at $T_{2,\text{out,initial}} = 327.98\,°\text{C}$. Considering that the steady-state value $T_{2,\text{out,ss}}$ of the outlet temperature is varying over time due to the variation of disturbances d_2 (see Eq. 5.24), we rewrite Eq. 5.27 using the deviation variable x_2 as follows:

$$S(x_2(t_{k+N})) = [f^+(x_2(t_{k+N}) + 1616.3(d_2 + \gamma\ \Delta d\ N))]^2. \tag{5.28}$$

The Safeness Index threshold S_{TH} needs to be carefully chosen to prevent an operating condition of high temperature in the methanator. Specifically, the threshold for $S(x_2)$ is calculated to be $(340 - 327.98)^2 = 144.48$ when the methanator outlet temperature $T_{2,\text{out}}$ remains below 340 °C. Additionally, a conservative value of $S_{TH} = 121$ is used for the actual system due to the same considerations as discussed in Sect. 5.3.2.2. Based on the Safeness Index function of Eq. 5.28, we develop the Safeness Index-based MPC as the following optimization problem:

$$\min_{u \in S(\Delta), y} \int_{t_k + t_{d,2}}^{t_{k+N} + t_{d,2}} (|\tilde{x}_2(\tau)|^2_{Q_c})\, d\tau + \int_{t_k}^{t_{k+N}} (|u_2(\tau)|^2_{R_c})\, d\tau + \sum_{i=1}^{N} k_1 e^{-k_2 y(i)} \tag{5.29a}$$

$$\text{s.t. } \dot{\tilde{x}}_2(t) = A_2 \tilde{x}_2(t) + B_2 u_2(t - t_{d,2}) \tag{5.29b}$$

$$\tilde{x}_2(t_k) = x_2(t_k) \tag{5.29c}$$

$$u_2(t) = u_{2,pre}(t),\ \forall\, t \in [t_k - t_d, t_k) \tag{5.29d}$$

$$u_2(t) \in U_2,\ \forall\, t \in [t_k, t_{k+N}) \tag{5.29e}$$

$$S(\tilde{x}_2(t_{k+i} + t_d)) + y(i) \leq S_{TH},\ i = 1, 2, \ldots, N \tag{5.29f}$$

where the notation follows that in Eq. 5.10 and $k_1,\ k_2 > 0$. It is noted that in the objective function of Eq. 5.29a, the state is integrated from $t_k + t_{d,2}$ to $t_{k+N} + t_{d,2}$ because the states from t_k to $t_k + t_{d,2}$ are already determined by the previously implemented control actions. The nominal linear model of Eq. 5.25 is used as the prediction model in the constraint of Eq. 5.29b. The state measurement $x_2(t_k)$ at $t = t_k$ is used as the initial condition for MPC in Eq. 5.29c. The input trajectories obtained from the previous time steps are provided in Eq. 5.29d for the current prediction from t_k to $t_k + t_{d,2}$. Equation 5.29e defines the constraints on the control actions, which is $U_2 = [-100,\ 100]$ (i.e., 180 °C $\leq T_{2,in} \leq$ 380 °C) on the manipulated input (i.e., feed temperature) in this example. Equation 5.29f is the soft constraint based on Safeness Index function with slack variables $y(i)$. It can be seen from the objective function that $y(i)$ is maximized by the penalty term of $\sum_{i=1}^{N} k_1 e^{-k_2 y(i)}$ to maintain the Safeness Index below its threshold S_{TH} as much as possible. Additionally, we carefully choose the parameter values for k_1 and k_2 in the objective function of Eq. 5.29a ($k_1 = 10^3$ and $k_2 = 0.2$ in this example) to achieve the desired closed-loop performance under Safeness Index-based MPC. Specifically, when $S(x_2(t_k + t_d))$ is far below the threshold S_{TH}, the slack variables $y(i)$ should not have a major impact on the optimization of control actions; however, when $S(x_2(t_k + t_d))$ approaches the threshold, the Safeness Index constraint with the slack variables $y(i)$ should play a dominant role in the calculation of control actions.

The dynamic model of Eq. 5.25 is numerically integrated with a sufficiently small integration time step of $h_c = 10^{-1}$ s using the explicit Euler method. The FilterSD solver on OPTI Toolbox in MATLAB is used to solve the nonlinear optimization problem of the Safeness Index-based MPC of Eq. 5.29 with the following settings: prediction horizon $N = 30$ and sampling period $\Delta = 20$ s. Similarly, $R_c = 2$ and $Q_c = 1$ are determined to balance the contributions of states and of inputs in the MPC objective function.

5.3.3.4 Closed-Loop Simulation Results

In this subsection, closed-loop simulations are carried out for the following four scenarios: (a) both C_1 and C_2 controllers are used, (b) only C_2 controller is applied, (c) PI controllers are applied to replace the MPC and the Safeness Index-based MPC for C_1 and C_2, and (d) the MPC is applied without Safeness Index constraints, to demonstrate the benefits of the proposed Safeness Index-based MPC.

(a) Simulation results using both C_1 and C_2
Figures 5.27, 5.28, and 5.29 show the closed-loop simulation results of the entire ammonia process under the proposed controllers C_1 and C_2 when a disturbance of catalyst deactivation occurs. As a result of catalyst deactivation from 1 to 0.2 within the first 300 s in the high-temperature shift reactor, less CO is consumed in it. Figure 5.27a shows that after a small inverse response, the outlet temperature of the high-temperature shift reactor drops at around 400 s below its steady-state value. Figure 5.27b shows that the first controller C_1 measures the decreasing outlet

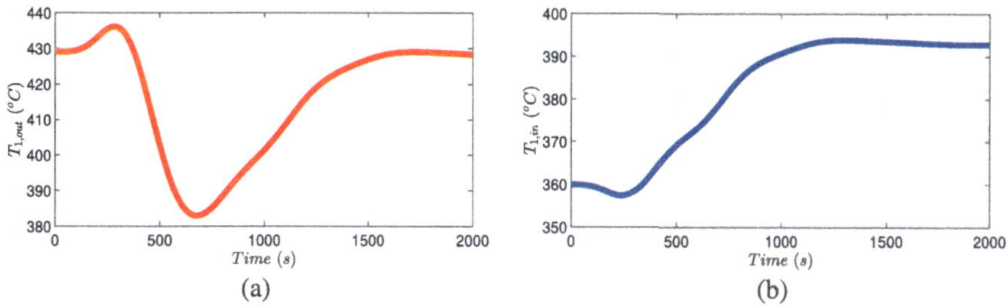

Fig. 5.27 Closed-loop **a** outlet temperature and **b** inlet temperature profiles of the high-temperature shift reactor using the proposed MPC and Safeness Index-based MPC for C_1 and C_2, respectively

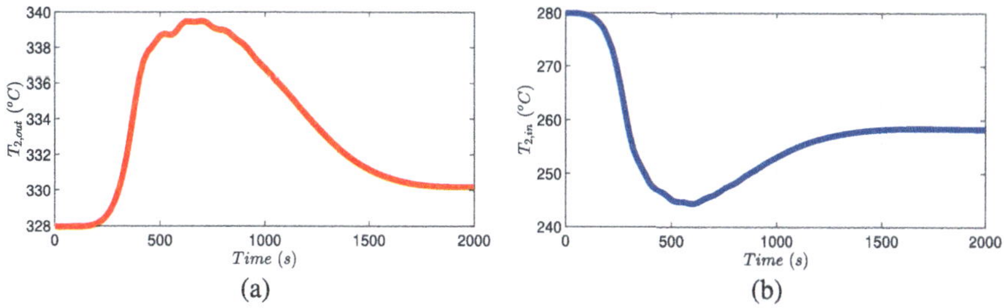

Fig. 5.28 Closed-loop **a** outlet temperature and **b** inlet temperature profiles of the methanator using the proposed MPC and Safeness Index-based MPC for C_1 and C_2, respectively

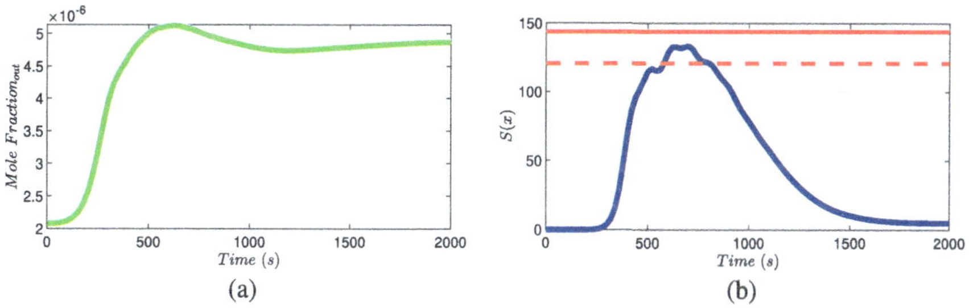

Fig. 5.29 Closed-loop **a** outlet mole fraction of carbon monoxide of the methanator, and **b** Safeness Index profiles using the proposed MPC and Safeness Index-based MPC for C_1 and C_2, respectively, where the solid line is the actual process threshold, and the dashed line is the threshold used in the controller

temperature $T_{1,\text{out}}$, and then increases the inlet temperature $T_{1,\text{in}}$ in order to consume more CO in the high-temperature shift reactor. It is demonstrated that the outlet temperature $T_{1,\text{out}}$ returns to its steady-state within 1500 s under the MPC of Eq. 5.23. However, as the catalyst has become inactive in the high-temperature shift reactor, less CO will be reacted compared to the initial nominal condition. Therefore, there is more residual unreacted CO entering the low-temperature shift reactor, which might cause an unsafe operation.

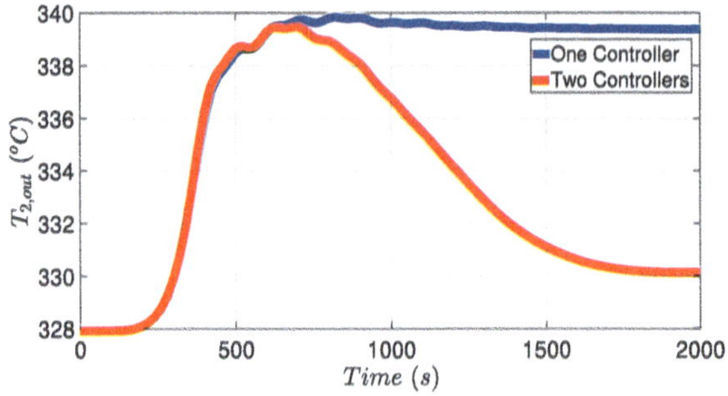

Fig. 5.30 Methanator outlet temperature profiles under C_2 only, and under both C_1 and C_2

As the next unit of the process network, the low-temperature shift reactor mitigates a portion of the increased CO content; however, the outlet stream leaving the low-temperature shift reactor still contains a higher concentration of CO compared to the nominal case without catalyst deactivation. As a result, in Fig. 5.28a, the methanator outlet temperature $T_{2,\text{out}}$ starts to increase at around 200 s. Since the process dynamics is changed under disturbances, we measure the mole fraction of CO in the feed stream to the methanator at each sampling time and calculate the new steady-state following Eq. 5.24. Then, the Safeness Index-based MPC (i.e., C_2 controller) drives the methanator outlet temperature $T_{2,\text{out}}$ to the new steady-state by adjusting the inlet temperature $T_{2,\text{in}}$ based on the current measurement of the outlet temperature $T_{2,\text{out}}$. Figure 5.29a shows that the steady-state values calculated offline work well since the outlet carbon monoxide mole fraction is maintained in a range (i.e., 5×10^{-6}) that meets the process requirements.

(b) Comparison with use of C_2 only

In this simulation study, we run closed-loop simulation for the ammonia process under the same disturbance but with the controller C_2 only. Figure 5.30 compares the simulation results under a single controller C_2 and under both C_1 and C_2 controllers. Specifically, when C_1 and C_2 are both used, to mitigate the impact of reduced catalyst activity, the first controller C_1 increases the feed temperature $T_{1,\text{in}}$ and the reaction rate in the high-temperature shift reactor. Then, the CO content in the feed stream to the methanator will first increase, and gradually decrease in some time (i.e., time delay) when the controller C_1 is used. Figure 5.30 shows that the controller C_1 mitigates the impact of disturbance d_2, and ultimately stabilizes the system at a new steady-state corresponding to the disturbance d_2. However, when we use the controller C_2 only in the ammonia process, more CO is produced from the high-temperature shift reactor, and thus the feed stream to the methanator contains a high level of CO concentration, which represents a large disturbance to the methanator under C_2. Therefore, the system will be ultimately driven to a new steady-state that corresponds to the large disturbance d_2 in this case. Figure 5.30 shows that when only the controller C_2 is utilized, the methanator outlet temperature is stabilized at around 340 °C, which is higher than the case of two controllers. Additionally, it is seen from Fig. 5.30 that the two controllers take longer time (i.e., around 2000 s) than the single controller to stabilize the methanator outlet temperature $T_{2,\text{out}}$ at the

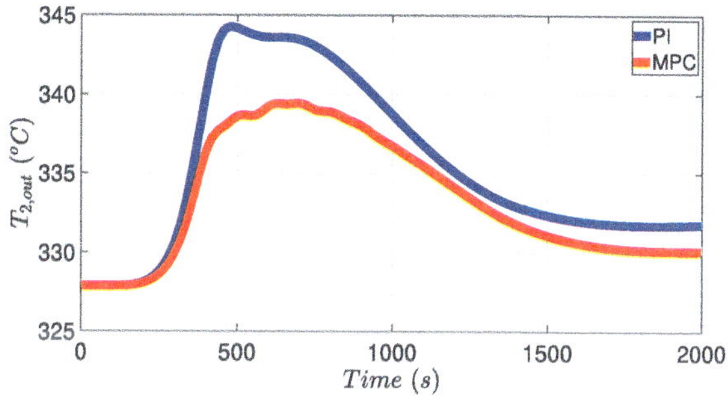

Fig. 5.31 Comparison of methanator outlet temperature under MPC (both C_1 and C_2) and under PI (both C_1 and C_2) control schemes

new steady-state due to the existence of a large time delay (i.e., $t_{d,1} = 360$ s) in the high-temperature shift reactor.

(c) Comparison with PI controllers

Since proportional–integral (PI) control is one of the most popular control algorithms used in chemical industries, we also perform the closed-loop simulation of the ammonia process under the same disturbance using the PI controllers for both C_1 and C_2. Figure 5.31 compares the simulation results between the PI controllers and the two MPCs in Part a. The PI controller C_2 also measures the current disturbance $d_2(t_k)$ and changes the set-point as discussed in the previous section. It is noted that one of the shortcomings of PI controllers is that the state constraints (e.g., the Safeness Index-based constraint of Eq. 5.29f in MPC) cannot be employed. Moreover, the PI controller does not take anticipated variation of future disturbances from $d_2(t_{k+1})$ to $d_2(t_{k+N})$ into account, while the MPC can consider the future steady-state corresponding to the future disturbances in Eq. 5.24. Therefore, the closed-loop performance under the PI controllers cannot avoid extreme states in the presence of disturbance. In Fig. 5.31, it is shown that the methanator outlet temperature $T_{2,\text{out}}$ exceeds 340 °C for around 500 s under the PI controllers, while $T_{2,\text{out}}$ remains below 340 °C at all times under the MPCs with Safeness Index-based constraints in Part a.

(d) MPC without Safeness Index constraints

Lastly, we simulate the ammonia process under the same disturbance using the standard MPC (i.e., without the Safeness Index constraint of Eq. 5.29f) for both C_1 and C_2. Figure 5.32 compares the simulation results between the MPCs with and without Safeness Index constraints, respectively. The objective of the controllers is to avoid a potential runaway reaction by maintaining the methanator outlet temperature $T_{2,\text{out}}$ below 340 °C. However, Fig. 5.32 shows that the methanator outlet temperature $T_{2,\text{out}}$ exceeds 340 °C for 400 s under the standard MPC without Safeness Index constraints; however, the temperature is maintained below 340 °C for the entire operating period under the MPC using the Safeness Index constraints of Eq. 5.29f. This implies that by incorporating the Safeness Index constraints within MPC, process operational safety is much improved for the ammonia plant compared to the standard MPC that does not account for safety considerations.

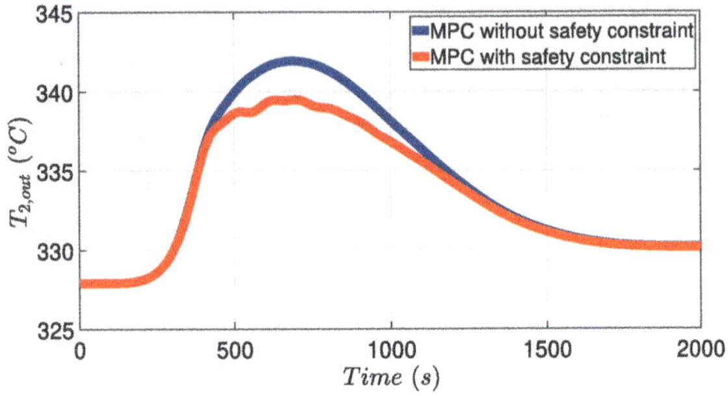

Fig. 5.32 Methanator outlet temperature profiles under the MPC with and without Safeness Index constraints

In this case study, process operational safety for a multi-unit ammonia network was improved by incorporating Safeness Index-based constraints within multiple model predictive controllers. We studied a common and problematic disturbance, catalyst deactivation, in the dynamic operation of the ammonia plant. Specifically, catalyst deactivation occurred in the high-temperature shift reactor, affected the downstream units, and finally caused an unsafe operation with a dramatic temperature increase in the methanation unit. To improve process operational safety, we developed an MPC with an integral control term for the high-temperature shift reactor, and developed a Safeness Index-based MPC for the methanator. Through closed-loop simulations, it was demonstrated that the proposed controllers were able to prevent extremely high temperatures for an ammonia plant subject to significant disturbances.

5.4 Conclusions

This chapter presented a number of methods and case studies to demonstrate the control system designs that account for safety considerations. Dynamic interaction between safety systems and feedback control systems was first discussed and illustrated using the examples of the MIC reaction in a CSTR and a flash drum process under model-based controllers and classical controllers, respectively. Subsequently, Safeness Index-based MPC was applied to a high-pressure flash drum separator, and an ammonia plant for improving process operational safety under various disturbances. Additionally, an ammonia process network with multiple controllers (i.e., the Safeness Index-based MPC) was simulated to demonstrate the improvement of process operational safety in the sense of avoidance of an extremely high temperature when significant disturbances occur.

Chapter 6
Machine Learning in Process Operational Safety

6.1 Introduction

In the previous chapters, Safeness Index-based MPCs and control Lyapunov-barrier function (CLBF)-based MPCs were developed with theoretical treatments of guaranteed process operational safety for nonlinear systems based on the assumption that an accurate first-principles model is available. However, this assumption is typically not satisfied in practical applications, and therefore, in industry linear empirical models are often used in controllers to operate the system at the steady-state. The case studies of large-scale chemical processes in Chap. 5 discussed the applications of data-driven modeling using Aspen simulation data, and demonstrated the improvement of process operational safety under the Safeness Index-based MPCs using linear state-space models. Despite the successful applications of linear empirical modeling in process industries, modeling nonlinear systems is always valuable to address systems-level task since chemical processes are inherently nonlinear, and thus, require nonlinear process models to improve the closed-loop performance of model-based controllers. Motivated by the above, machine learning, a method of data analysis that can be utilized to model nonlinear systems for model-based controllers, has received an increased level of attention in model identification in recent years. Specifically, recurrent neural networks (RNNs) are an efficient approach for modeling a general class of nonlinear dynamical systems using time-series data. While feedforward neural networks use a one-way connectivity between units to model nonlinear systems, RNN architectures include feedback loops that introduce the past information stored in hidden neurons at earlier time steps to the current network. Thus, the RNN internal states preserve the past network information and can be considered as the memory of an RNN. It allows RNN models to capture process dynamic behavior in a way conceptually similar to nonlinear dynamical systems that can be described by state-space ordinary differential equations. The history of recurrent neural networks can be traced back to the 1980s, at which time researchers

Z. Wu and P. D. Christofides, *Process Operational Safety and Cybersecurity*,
Advances in Industrial Control, https://doi.org/10.1007/978-3-030-71183-2_6

created the Hopfield networks for pattern recognition [80]. Since then, modern RNN structures (e.g., Bidirectional recurrent neural networks, gated recurrent unit (GRU), and long short-term memory (LSTM) networks) and a variety of learning algorithms have been developed for various applications including natural language processing and human visual pattern recognition. In the meantime, due to the rapid development of open-source machine learning libraries such as Keras and Tensorflow and of the availability of significant computational resources, machine learning techniques have become popular in tackling problems with a massive amount of data outside the field of computer science and engineering. For example, machine learning techniques have been successfully implemented to model nonlinear systems in classical engineering fields, e.g., [102, 177, 191, 207, 222, 226, 227]. Moreover, considering that a process with complex dynamics may not be able to be well represented by a single data-driven model, ensemble learning, a multi-model approach, has been proposed to combine the results of multiple machine learning models to improve the overall approximation performance. By training multiple models at a learning step to approximate particular outputs, ensemble learning is demonstrated to have improved accuracy and robustness in solving regression and classification problems than a single machine learning model, (e.g., [33, 127, 142, 155, 169, 189, 225, 234]).

In this chapter, the concept of recurrent neural networks and a general framework for developing RNN models for nonlinear dynamical systems are introduced. Subsequently, CLBF-MPC and CLBF-EMPC schemes that use RNN models for prediction are presented with guaranteed operational safety and closed-loop stability, followed by the discussion of ensemble learning of multiple RNN models in MPCs to improve prediction accuracy and the use of parallel computing to address the resulting computational implementation issues. Online learning of RNN models is also discussed to update machine learning models in real-time implementation of controllers to capture the most recent process dynamics subject to time-varying disturbances. The applications of machine learning-based control schemes to a chemical reactor example demonstrate the ability of RNNs to model nonlinear dynamical systems and the effectiveness of the control schemes in stabilizing systems with guaranteed safety.

6.1.1 Class of Nonlinear Systems

The following system of nonlinear first-order ordinary differential equations is used to represent the class of nonlinear systems that we consider

$$\dot{x} = F(x, u, w) := f(x) + g(x)u + h(x)w, \ x(t_0) = x_0 \tag{6.1}$$

where $x \in D \subset \mathbf{R}^n$, $u \in U \subset \mathbf{R}^m$, and $w \in W$ are the state vector, the manipulated input vector, and the disturbance vector, respectively. The control action is constrained by the following bounds: $u \in U := \{u_{\min} \leq u \leq u_{\max}\} \subset \mathbf{R}^m$, where $u_{\max}$ and $u_{\min}$ are the upper and lower bounds for the input vector, respectively.

The disturbance is assumed to be constrained by a sufficiently small bound, i.e., $W := \{w \in \mathbf{R}^l \mid |w| \le w_m,\ w_m \ge 0\}$. The vector and matrix functions $f(\cdot)$, $g(\cdot)$, and $h(\cdot)$ of dimensions $n \times 1$, $n \times m$, and $n \times l$, respectively, are assumed to be sufficiently smooth with $f(0) = 0$. Therefore, the origin is a steady-state of the nominal system of Eq. 6.1 with $w(t) \equiv 0$. The measurement of $x(t)$ is assumed to be available for feedback at each sampling time $t_k = t_0 + k\Delta,\ k = 0, 1, \ldots$, where Δ is the sampling period.

6.1.2 Stabilizability Assumption

Throughout this chapter, we will make the following assumptions.

Assumption 6.1 We assume that there exists a stabilizing feedback controller $u = \Phi(x) \in U$ for the nominal system of Eq. 6.1 with $w(t) \equiv 0$ that renders the origin of the closed-loop system under continuous implementation of the controller exponentially stable in the sense that a $\mathscr{C}^1$ Lyapunov function $V : D \to \mathbf{R}_+$ exists such that for all states x in D (a small region around the origin), the following inequalities hold:

$$c_1|x|^2 \le V(x) \le c_2|x|^2, \tag{6.2a}$$

$$\frac{\partial V(x)}{\partial x} F(x, \Phi(x), 0) \le -c_3|x|^2, \tag{6.2b}$$

$$\left|\frac{\partial V(x)}{\partial x}\right| \le c_4|x|, \tag{6.2c}$$

where c_i, $i = 1, 2, 3, 4$ are positive real numbers.

The stability region Ω_ρ is designed as a level set of the Lyapunov function $V(x)$ within D, from which Eq. 6.2 is satisfied: $\Omega_\rho := \{x \in D \mid V(x) \le \rho,\ \rho > 0\}$. Using Lyapunov function properties, it is straightforward to show that Ω_ρ is an invariant set since it holds that $\dot{V} \le -c_3|x|^2$ under $u = \Phi(x) \in U$ for all $x \in \Omega_\rho$. The following control law is used to render the origin of the nominal system of Eq. 6.1 exponentially stable.

$$k_i(x) = \begin{cases} -\dfrac{p + \sqrt{p^2 + \gamma |q|^4}}{|q|^2} q_i & \text{if } q \neq 0 \\ 0 & \text{if } q = 0 \end{cases} \tag{6.3a}$$

$$\Phi_i(x) = \begin{cases} u_{\min} & \text{if } k_i(x) < u_{\min} \\ k_i(x) & \text{if } u_{\min} \le k_i(x) \le u_{\max} \\ u_{\max} & \text{if } k_i(x) > u_{\max} \end{cases} \tag{6.3b}$$

where p denotes $L_f V(x)$, q_i, $(i = 1, 2, \ldots, m)$ denotes $L_{g_i} V(x)$, $q = [q_1 \ldots q_m]^T$, $f = [f_1 \ldots f_n]^T$, $g_i = [g_{i1} \ldots g_{in}]^T$ and $\gamma > 0$. $k_i(x)$ of Eqs. 6.3a and 6.3b represent the original Sontag control law without saturation, and the i_{th} component of the saturated control law $\Phi(x)$ that accounts for the constraints on manipulated inputs, i.e., $u \in U$, respectively.

6.2 Recurrent Neural Network Modeling

A recurrent neural network (RNN) model of the following form is developed for the nonlinear system of Eq. 6.1:

$$\dot{\hat{x}} = F_{nn}(\hat{x}, u) := A\hat{x} + \Theta^T y \tag{6.4}$$

where $\hat{x} \in D \subset \mathbf{R}^n$, $u \in \mathbf{R}^m$, and $y = [y_1, \ldots, y_n, y_{n+1}, \ldots, y_{m+n}] = [\sigma(\hat{x}_1), \ldots, \sigma(\hat{x}_n), u_1, \ldots, u_m] \in \mathbf{R}^{n+m}$ are the RNN state vector, the manipulated input vector, and a vector of both the input u and the network state $\hat{x}$, respectively. To introduce nonlinearity into the RNN model of Eq. 6.4, a nonlinear activation function $\sigma(\cdot)$ (e.g., a sigmoid function $\sigma(x) = 1/(1 + e^{-x})$) is used in y. A and Θ are the coefficient matrices defined as follows: $A = diag\{-a_1, \ldots, -a_n\} \in \mathbf{R}^{n\times n}$, and $\Theta = [\theta_1, \ldots, \theta_n] \in \mathbf{R}^{(m+n)\times n}$ where $\theta_i = b_i[w_{i1}, \ldots, w_{i(m+n)}], i = 1, \ldots, n$, and a_i, b_i are constants. Specifically, w_{ij} represents the weight between the jth input and the ith neuron where $i = 1, \ldots, n$ and $j = 1, \ldots, (m + n)$. We assume that all the diagonal elements a_i in A are positive such that each RNN state $\hat{x}_i$ is bounded-input bounded-state stable. It is noted that the bias term is not included in the notation to simplify the discussion as it can be considered an additional constant input, and thus, does not affect the formulation of RNN of Eq. 6.4. Throughout this chapter, we use $\hat{x}$ and x to represent the state of the RNN model of Eq. 6.4 and the state of the actual nonlinear system of Eq. 6.1, respectively, unless stated otherwise. Additionally, it is noted that the RNN model of Eq. 6.4 is an input-affine system, and therefore, we can rewrite Eq. 6.4 in the form that is similar to Eq. 6.1:

$$\dot{\hat{x}} = \hat{f}(\hat{x}) + \hat{g}(\hat{x})u \tag{6.5}$$

where $\hat{f}(\cdot)$ and $\hat{g}(\cdot)$ are nonlinear vector functions that can be derived from the coefficient matrices A and Θ in Eq. 6.4. Similarly, $\hat{f}(\cdot)$ and $\hat{g}(\cdot)$ are assumed to be sufficiently smooth. Unlike the one-way connectivity used in feedforward neural networks (FNNs), the signals travel in both directions in RNN models due to the feedback loops in the hidden layer. Figure 6.1 shows the structure of an RNN model, from which we can see that the model exhibits a dynamic behavior because the past network information (i.e., the hidden state value in earlier time steps) are fed into the current network. Therefore, RNN model provides a solution to model the nonlinear dynamical systems of Eq. 6.1 using time-series data. In fact, it can be shown via the universal approximation theorem, e.g., [102, 182], that the RNN model with a sufficient number of neurons can approximate any dynamic nonlinear system on

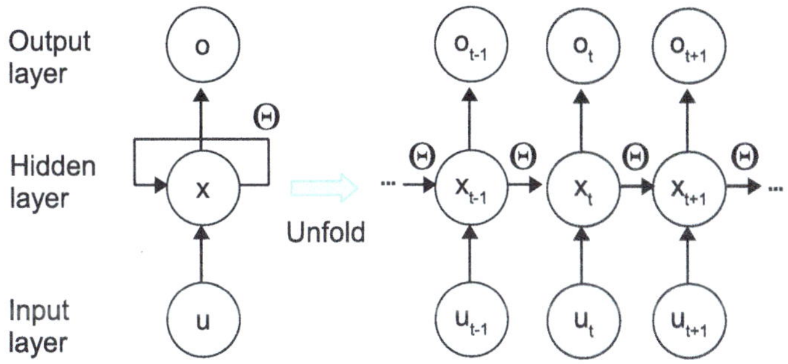

Fig. 6.1 A recurrent neural network (left) and its unfolded structure (right), where x, u, o, and Θ are the input vector, the state vector, the output vector, and the weight matrix, respectively

compact subsets of the state-space for finite time. The approximation property of the RNN model is summarized in the following proposition:

Proposition 6.1 (Universal Approximation Theorem, c.f. [182]) *Consider the RNN model of Eq. 6.4 and the nonlinear system of Eq. 6.1 with the same initial condition $x(0) = \hat{x}(0) = x_0 \in \Omega_\rho$. For any $\varepsilon > 0$ and $T > 0$, an optimal weight matrix Θ^* exists such that the following equation is satisfied by the RNN state $\hat{x}$ under $\Theta = \Theta^*$:*

$$\sup_{t\in[0,T]} |x(t) - \hat{x}(t)| \leq \varepsilon. \tag{6.6}$$

Remark 6.1 As shown in Eq. 6.4, a continuous-time RNN model is utilized to approximate the nonlinear dynamic system of Eq. 6.1. By writing RNN models in continuous-time form, it allows us to derive desirable properties of RNNs, for example, the well-defined derivative of the internal state with respect to time [150]. However, it should be noted that the discrete-time RNNs can be equally well applied to model the nonlinear system of Eq. 6.1. Similar stability analysis and learning procedures can be derived for discrete-time RNNs through numerical approximation methods. In this chapter, we will discuss the stability properties for model-based controllers that use continuous-time RNN models for prediction; however, in the simulation study, the RNN model is actually developed based on a training dataset consisting of sampled data (e.g., sampled sensor measurement) and simulated in a sample-and-hold fashion with a sufficiently small sampling period Δ to be consistent with industrial practice.

Remark 6.2 To simplify the discussion, the RNN model of Eq. 6.4 is formulated as a one-hidden-layer RNN with n states to approximate the nonlinear system of n first-order ODEs of Eq. 6.1. However, the development of RNN models for approximation of the nonlinear system of Eq. 6.1 is not restricted to n-state, one-hidden-layer RNN model. Instead, to achieve a desired approximation performance of the nonlinear system of Eq. 6.1, a multi-layer RNN with a sufficient number of neurons is generally utilized. In that case, the RNN states $\hat{x} \in \mathbf{R}^n$ in Eq. 6.4 will be the last hidden layer or the output layer of an RNN.

6.2.1 RNN Learning Algorithm

In this section, we present the RNN learning algorithm that computes the optimal weight matrix Θ^* by minimizing the error between the RNN states $\hat{x}(t)$ and the actual state $x(t)$ of the nominal system of Eq. 6.1 with $w(t) \equiv 0$. Although Proposition 6.1 states the universal approximation ability of RNNs (i.e., an RNN can approximate a broad class of nonlinear systems to any degree of accuracy), in reality, it is challenging to develop a perfect RNN model due to many reasons (e.g., insufficient number of layers and nodes). Therefore, we assume that under the optimal weight matrix $\Theta = \Theta^*$, a modeling error $\nu := F(x, u, 0) - F_{nn}(\hat{x}, u)$ exists between the RNN model of Eq. 6.4 and the nominal system of Eq. 6.1. Since we operate the nonlinear system of Eq. 6.1 in the stability region Ω_ρ only (instead of in the entire state-space), the RNN model is also developed with the goal of approximating the system dynamics for all $x \in \Omega_\rho$ and $u \in U$. As both $x(t)$ and $u(t)$ are bounded, it is straightforward to show that the modeling error $\nu(t)$ is also bounded (i.e., $|\nu(t)| \leq \nu_m$, $\nu_m > 0$). Additionally, we require the weight vector θ_i to be bounded by $|\theta_i| \leq \theta_m$, where $\theta_m > 0$, such that the weight drift issue (i.e., the RNN weights drift to infinity) is avoided during training. Following the methods in [102, 154], we develop the RNN learning algorithm to demonstrate that the state error $|e| = |\hat{x} - x|$ is ultimately bounded in the presence of a non-zero modeling error ν. Specifically, based on the RNN model of Eq. 6.4, and the modeling error that we defined earlier, the nominal system of Eq. 6.1 (i.e., $w(t) \equiv 0$) can be expressed as follows:

$$\dot{x}_i = -a_i x_i + \theta_i^{*T} y + \nu_i, \; i = 1, \ldots, n. \tag{6.7}$$

We define the optimal weight vector θ_i^* as follows:

$$\theta_i^* := \arg \min_{|\theta_i| \leq \theta_m} \{ \sum_{k=1}^{N_d} |F_i(x_k, u_k, 0) + a_i x_k - \theta_i^T y_k| \} \tag{6.8}$$

where N_d is the number of data samples in the training dataset. Using Eqs. 6.4 and 6.7, the time-derivative of the state error $e = \hat{x} - x \in \mathbf{R}^n$ is derived below

$$\dot{e}_i = \dot{\hat{x}}_i - \dot{x}_i = -a_i e_i + \zeta_i^T y - \nu_i, \; i = 1, \ldots, n \tag{6.9}$$

where ν is the modeling error, i.e., $\nu = F(x, u, 0) - Ax - \Theta^* y$, and $\zeta_i = \theta_i - \theta_i^*$ is the error between the unknown optimal weight vector θ_i^* and the current weight vector θ_i. Then, we update the weight vector θ using the following equation at the training stage:

$$\dot{\theta}_i = -\eta_i y e_i,\ i = 1, \ldots, n \tag{6.10}$$

where η is a positive definite matrix representing the learning rate. Based on the learning law of Eq. 6.10, we utilize the following theorem to demonstrate the boundedness of the state error e, and its relationship with the modeling error ν.

Theorem 6.1 (c.f. [102, Theorem 4.1]) *Consider the RNN model of Eq. 6.4 trained using the learning algorithm of Eq. 6.10. Then, the weight error ζ_i and the state error e_i are bounded, and there exist $\mu > 0$ and $\lambda \in \mathbf{R}$ such that the following inequality holds:*

$$\int_0^t |e(\tau)|^2 d\tau \leq \lambda + \mu \int_0^t |\nu(\tau)|^2 \mathrm{d}\tau. \tag{6.11}$$

Proof We first define a Lyapunov function $\tilde{V} = \frac{1}{2}\sum_{i=1}^{n}(e_i^2 + \zeta_i^T \eta_i^{-1} \zeta_i)$. Based on Eqs. 6.9, 6.10 and $\dot{\zeta}_i = \dot{\theta}_i$, we calculate the time-derivative of $\tilde{V}$ as follows:

$$\begin{aligned}\dot{\tilde{V}} &= \sum_{i=1}^{n}(e_i \dot{e}_i + \eta_i^{-1} \zeta_i \dot{\zeta}_i) \\ &= \sum_{i=1}^{n}(-a_i e_i^2 - e_i \nu_i).\end{aligned} \tag{6.12}$$

It can be seen from Eq. 6.12 that $\dot{\tilde{V}} \leq 0$ holds when there is no modeling error for the RNN model (i.e., $\nu_i = 0$). Following the proof in [102], the state error e_i and its time-derivative $\dot{e}_i$ are bounded for all times. Additionally, since $\tilde{V}$ is bounded from below and its time-derivative $\dot{\tilde{V}}$ is uniformly continuous (the uniform continuity of $\dot{\tilde{V}}$ is obtained from the fact that the second-order derivative $\ddot{\tilde{V}}$ is bounded), $\dot{\tilde{V}} \to 0$ holds as $t \to \infty$ according to Barbalat's lemma[1] [138]. This implies that the state error e_i will converge to zero ultimately if the modeling error term $-e_i \nu_i$ in Eq. 6.12 equals zero. However, in the presence of modeling error $\nu_i \neq 0$, $\dot{\tilde{V}} \leq 0$ does not hold for all times. Therefore, we derive the following equation based on Eq. 6.12:

$$\begin{aligned}\dot{\tilde{V}} &= \sum_{i-1}^{n}\left(-\frac{a_i}{2}e_i^2 - \frac{1}{2}|\zeta_i|^2\right) + \left(\frac{1}{2}|\zeta_i|^2 - \frac{a_i}{2}e_i^2 - e_i \nu_i\right) \\ &\leq -\alpha \tilde{V} + \sum_{i=1}^{n}\left(\frac{1}{2}|\zeta_i|^2 - \left(\frac{a_i}{2}e_i^2 + e_i \nu_i + \frac{1}{2a_i}\nu_i^2\right) + \frac{1}{2a_i}\nu_i^2\right) \\ &\leq -\alpha \tilde{V} + \sum_{i=1}^{n}\left(\frac{1}{2}|\zeta_i|^2 + \frac{1}{2a_i}\nu_i^2\right)\end{aligned} \tag{6.13}$$

[1] Assume f is a function of time. Barbalat's lemma says if $f(t)$ has a finite limit as $t \to \infty$, and if $\dot{f}$ is uniformly continuous, then $\dot{f}(t) \to 0$ as $t \to \infty$.

where $\alpha := \min\{a_i, 1/(\lambda_m)\}$ and λ_m represents the maximum eigenvalue of η_i^{-1}, $i = 1, \ldots, n$. Since the weight vector is bounded by $|\theta_i| \leq \theta_m$, it is derived that $\frac{1}{2}|\zeta_i|^2 \leq 2\theta_m^2$, and $\dot{\tilde{V}} \leq -\alpha\tilde{V} + \beta$ holds, where $\beta := \sum_{i=1}^{n}(2\theta_m^2 + \nu_m^2/2a_i)$. Therefore, we can show that for all $\tilde{V} \geq V_0 = \beta/\alpha$, $\dot{\tilde{V}} \leq 0$ holds, and this implies that $\tilde{V}$ is bounded. Subsequently, it is straightforward to show that e_i and ζ_i are bounded from the definition of $\tilde{V}$. Then, based on the fact that $\dot{\tilde{V}} \leq \sum_{i=1}^{n}(-\frac{a_i}{2}e_i^2 + \frac{1}{2a_i}\nu_i^2)$ (this can be derived from Eq. 6.13), the upper bound for $\tilde{V}(t)$ is obtained as follows:

$$
\begin{aligned}
\tilde{V}(t) &\leq \tilde{V}(0) + \sum_{i=1}^{n}\left(-\frac{a_i}{2}\int_0^t e_i(\tau)^2 d\tau + \frac{1}{2a_i}\int_0^t \nu_i(\tau)^2 d\tau\right) \\
&\leq \tilde{V}(0) - \frac{a_{\min}}{2}\int_0^t |e(\tau)|^2 d\tau + \frac{1}{2a_{\min}}\int_0^t |\nu(\tau)|^2 d\tau
\end{aligned}
\tag{6.14}
$$

where $a_{\min}$ is the minimum value of a_i, $i = 1, \ldots, n$. Let $\lambda = \frac{2}{a_{\min}} \sup_{t\geq 0}(\tilde{V}(0) - \tilde{V}(t))$ and $\mu = 1/a_{\min}^2$. The relationship between $|\mu|$ and $|e|$ shown in Eq. 6.11 is derived as follows:

$$
\begin{aligned}
\int_0^t |e(\tau)|^2 d\tau &\leq \frac{2}{a_{\min}}(\tilde{V}(0) - \tilde{V}(t)) + \frac{1}{a_{\min}^2}\int_0^t |\nu(\tau)|^2 d\tau \\
&\leq \lambda + \mu\int_0^t |\nu(\tau)|^2 d\tau.
\end{aligned}
\tag{6.15}
$$

Therefore, the state error $|e|$ is proportional to the modeling error $|\nu|$, and is guaranteed to be bounded. Additionally, if there exists a positive real number $C > 0$ such that $\int_0^\infty |\nu(t)|^2 dt = C < \infty$, then we can show that the state error is bounded as follows: $\int_0^\infty |e(t)|^2 dt \leq \lambda + \mu C < \infty$. Since $e(t)$ is uniformly continuous (i.e., $\dot{e}$ is bounded), the boundedness of $\int_0^\infty |e(t)|^2 dt$ implies that $e(t)$ converges to zero asymptotically.

Remark 6.3 To prevent the RNN weights from drifting to infinity in the training process, [102, 154] proposed a switching σ-modification learning algorithm to optimize the RNN weights θ_i while maintaining them within the bound for all times, i.e., $|\theta_i| \leq \theta_m$. To ensure the existence and uniqueness of solutions, the switching σ-modification approach was further improved to be continuous in an compact set in state-space. We refer the interested reader to [102, 154] for further information.

6.2.2 *Development of RNN Model*

This section presents the method for developing an RNN model from scratch for a general class of nonlinear system of Eq. 6.1. We will discuss the data generation method and the training process for building an RNN model that can well capture the process dynamics of Eq. 6.1 in a given operating region.

6.2.2.1 Data Generation

Since the nonlinear system of Eq. 6.1 is operated in the stability region Ω_ρ (i.e., a compact set in state-space), we first conduct extensive open-loop simulations for $u \in U$ and $x \in \Omega_\rho$ to generate the dataset that captures the system dynamics in the region that we considered. Specifically, the open-loop simulations of the nominal system of Eq. 6.1 are carried out with a variety of combinations of initial conditions $x_0 \in \Omega_\rho$ and inputs u, under which a large number of state trajectories (i.e., the solution of $x(t)$ for Eq. 6.1) are obtained. Ideally, to generate a dataset that fully captures the process dynamics in the operating region Ω_ρ, we should sweep over all the values that (x, u) can take in open-loop simulations. However, due to the limitation of computational resources, in practice, the targeted region in state-space may have to be discretized (see Fig. 6.2), and the range of inputs will also be discretized with sufficiently small intervals. In this study, the continuous system of Eq. 6.1 is simulated under a sequence of inputs $u \in U$ in a sample-and-hold fashion (i.e., the input is a piecewise constant function that remains constant within each sampling period Δ, i.e., $u(t) = u(t_k)$, $\forall t \in [t_k, t_{k+1})$, where $t_{k+1} := t_k + \Delta$). Then, the explicit Euler method is utilized to numerically integrate the nominal system of Eq. 6.1 with a sufficiently small integration time step $h_c < \Delta$. Subsequently, we collect time-series data of state x and input u from open-loop simulations, and separate them into a large number of time-series samples with a shorter period P_{nn}, which represents the prediction horizon of RNNs. Lastly, we partition the entire dataset into training, validation, and testing datasets.

6.2.2.2 Training Process

Next, we train RNN models using a state-of-the-art application program interface (API), i.e., Keras [48]. To ensure that the RNN model can be later utilized in the model-based controller (in sample-and-hold fashion) for predicting future state, the prediction period of RNN, P_{nn}, should be chosen as an integer multiple of the sampling period Δ. The bottom figure in Fig. 6.2 shows the data processing step, from which we split a long open-loop state trajectory into a number of short trajectories with the length of P_{nn}, and take all the states between $t = 0$ and $t = P_{nn}$ as the RNN internal states. It is noted that since we run open-loop simulation in a sample-and-hold fashion, and collect data at each integration time step, the time interval

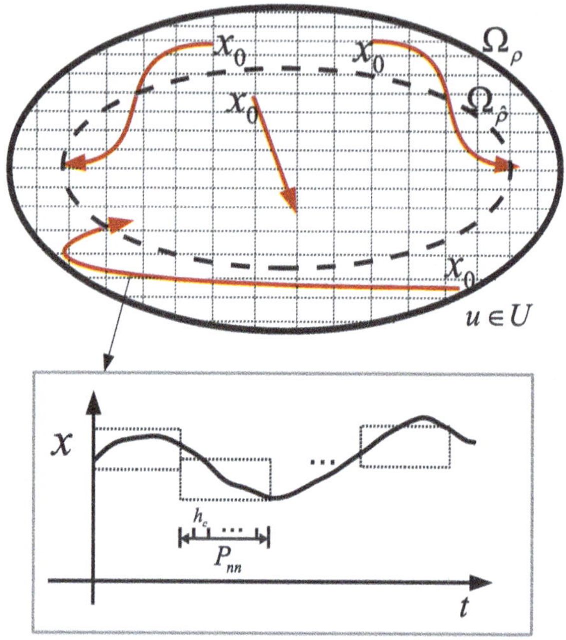

Fig. 6.2 The top figure shows the discretization of the operating region Ω_ρ for open-loop simulations with initial conditions $x_0 \in \Omega_\rho$, and the bottom figure shows the data processing step for the RNNs with a prediction horizon of P_{nn}. Ω_ρ and $\Omega_{\hat{\rho}}$ are the closed-loop stability region for the actual nonlinear system of Eq. 6.1, and the RNN model, respectively

between two consecutive internal states x_{t-1} and x_t within the prediction period P_{nn} is corresponding to one integration time step. Then, the RNN model of Eq. 6.4 is trained using the dataset from open-loop simulations to calculate the optimal weight Θ^* of Eq. 6.8 following the training algorithm that we have discussed in the previous section. Furthermore, to ensure a good approximation performance of the RNN model, the modeling error is required to be constrained by a sufficiently small bound ν_m, i.e., $|\nu| \leq \gamma|x| \leq \nu_m$, when the training process is completed.

We use adaptive moment estimation method (i.e., Adam in Keras) to solve the optimization problem of minimizing the RNN modeling error ν. The loss function is chosen to be the mean absolute percentage error (or mean squared error) between the actual states x from training data and the predicted states $\hat{x}$ from RNN models. To achieve a desired training performance and computational efficiency, we determine the optimal number of neurons and layers using a grid search. Finally, to avoid overfitting, the RNN training process is terminated once the early-stopping condition (i.e., the validation error stops decreasing) is met and the modeling error remains below the desired threshold.

Remark 6.4 In order to develop a machine learning model with a desired prediction accuracy, a high-quality dataset that can be generated from industrial process sensors, lab experiments, or extensive computer simulations is required, from which supervised machine learning models can learn the nonlinear relationship between network inputs and outputs. However, real industrial measurements often involve noise stemming from different sources, such as sensors variability and common plant variance. The training datasets consisting of noisy data or corrupt data may affect the training performance of RNNs in the following manners. On the one hand, the RNNs may capture the noisy pattern instead of the ground truth using a noisy dataset for training. On the other hand, it has been demonstrated in the literature, e.g., [33, 231], that the RNN models trained using a noisy dataset may achieve an improved generalization performance and robustness when implemented to a practical system with small perturbation in sensor measurements. Therefore, the neural network training using a noisy dataset is a critical point in machine learning modeling that needs further investigation. However, in this chapter, we perform open-loop simulations for the nominal system of Eq. 6.1 (i.e., $w(t) \equiv 0$), and thus, the RNN models are trained to approximate the dynamics of the nominal system of Eq. 6.1 based on the noise-free dataset. Additionally, it will be shown in Sect. 6.3.2.2 that the RNN models developed based on the dataset from nominal system can be applied within a model predictive controller to stabilize the disturbed system of Eq. 6.1 with sufficiently small disturbances (i.e., $|w| \leq w_m$).

Remark 6.5 The neural network modeling approach discussed in this section is a data-driven, black-box modeling approach that develops a nonlinear model of Eq. 6.4 to approximate the actual nonlinear system of Eq. 6.1 using massive amounts of process operating data. Note that in general, neural network modeling is treated as a black-box modeling approach without using any physical knowledge. However, in recent years, many researchers have also started to incorporate physical knowledge of systems into neural network formulations, trying to improve interpretability and optimality of neural network modeling. For example, it has been demonstrated in [114, 222] that the physics-based neural networks were able to improve the prediction performance than a black-box neural network. For neural networks with incorporation of process knowledge, the interest reader is referred to [25, 87, 88, 113, 114, 178, 222].

6.2.3 *Ensemble Regression Modeling*

A single RNN model may perform poorly for some states in the operating region due to many reasons, for example, inappropriate ratio between the validation and training datasets, and insufficient data in the operating region. To improve the approximation performance, ensemble learning method is proposed to combine multiple machine learning models that are trained for the same problem. Specifically, heterogeneous ensemble regression models are developed using different learning algorithms, while homogeneous models are derived following the same learning algorithm. The reasons for improved prediction performance using ensemble regression models are summarized in [127, 225], and are briefly stated as follows. First, a single RNN model may perform poorly in the region with insufficient training data; however, by training multiple RNN models and aggregating all candidate models using ensemble methods, we can reduce the risk of using one single flawed RNN model. Second, since the neural network learning algorithm is essentially an optimization problem that is NP-hard, and non-convex, the global minimum that gives the optimal weights of RNN models is not guaranteed to be found by the training process. However, ensemble learning method allows multiple RNN models to be trained with different initial weight matrices such that they will not get trapped in the same local minimum. As a result, the final prediction results from the ensemble models will be more accurate than a single RNN model. Therefore, by introducing ensemble learning method into RNN modeling of the nonlinear system of Eq. 6.1, the ensemble RNN models are expected to achieve reduced variability and improved generalization performance than a single RNN model.

A variety of ensemble-based algorithms, e.g., Bagging, Boosting, and Stacking, have been developed to improve NN performance in literature, e.g., [234]. In this chapter, we use the stacking method to develop multiple RNN models based on the same learning algorithm and obtain the final prediction results through the average of each single RNN prediction result. Note that all the RNN models are developed to approximate the nonlinear system of Eq. 6.1 using the same dataset in this study. Specifically, as shown in the dotted box in Fig. 6.3, we use k-fold cross validation method to split the dataset from extensive open-loop simulations into k subsets. Each time, we use $k - 1$ subsets as training dataset and the remaining one as validation dataset to train an RNN model following the learning algorithm in the previous section. By doing so, we will end up with k different RNN models for the same nonlinear system of Eq. 6.1. Figure 6.3 shows the training process of ensemble learning as well as the RNN structure we used in this example. Specifically, the dimension of the input layer is $m + n$, where u_i, $i = 1, \ldots, m$, and x_i, $i = 1, \ldots, n$ represent the manipulated inputs at t_k, and the real-time state measurements, respectively. We use two hidden layers for the RNN model in this example, and determine the optimal number of neurons for each layer through a grid search. The RNN outputs are the estimated states over the prediction period P_{nn}, which will be later used in the model predictive controller that will be discussed in the next section.

As shown in Fig. 6.3, based on the ensemble of k RNN models obtained from k-fold cross validation, we calculate the final predicted states by averaging all the RNN prediction results. However, note that the stacking method is not restricted to the approach of averaging all the models for the final prediction results. Remark 6.6 discusses other approaches that can be used in ensemble learning with multiple models. Additionally, it should be mentioned that in this study, we implement a normalization and a re-scaling step before and after the prediction of RNN models. Specifically, before training the RNN models, the input data consisting of manipulated inputs $u(t_k) \in \mathbf{R}^m$ and state measurements $x(t_k) \in \mathbf{R}^n$ at $t = t_k$, and the output data are first normalized such that all the process variables are in the same order of magnitude. Since we utilize the normalized data in the training process of cross-validated committee of RNNs, we will have to do the same normalization step at the prediction stage. Additionally, a re-scaling step is carried out at the end to convert the normalized output (i.e., the estimated states at $t = t_k + P_{nn}$) vector back to its normal value, such that it can be later used in MPC.

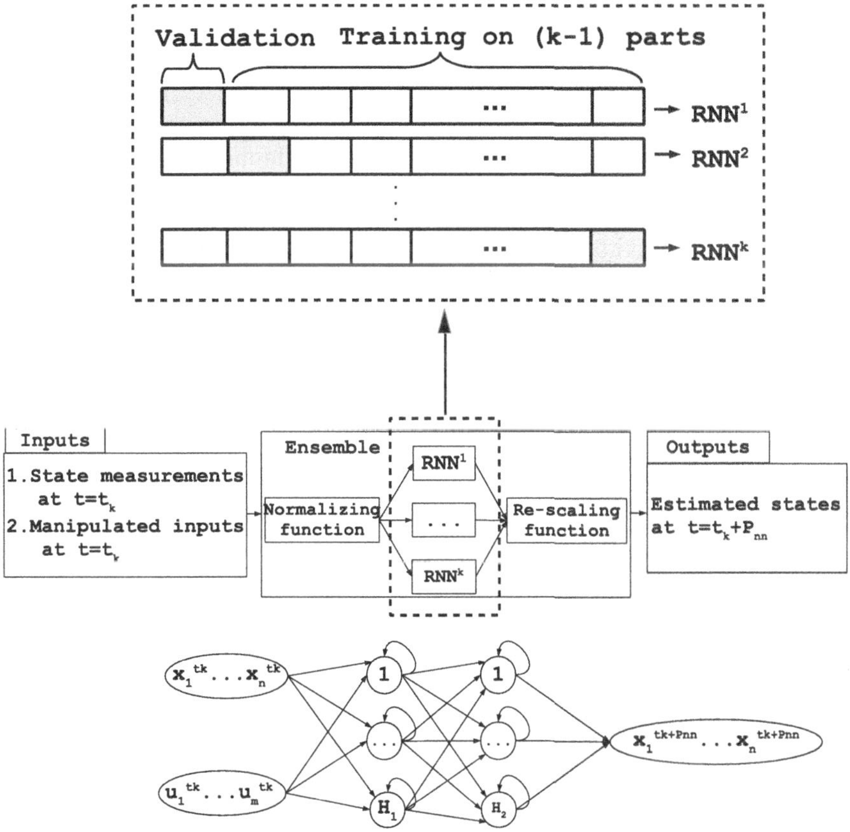

Fig. 6.3 A schematic of the implementation of ensemble learning method based on k-fold cross validation, where $u \in \mathbf{R}^m$ and $x \in \mathbf{R}^n$ are the input vector, and the state vector, respectively, and H_1, H_2 are the number of neurons in the two hidden layers

Remark 6.6 In addition to the stacking method that has been introduced in this section for combining multiple RNN models for better prediction, there are a variety of ensemble methods for improving model accuracy. For example, in the bagging method, multiple machine learning models are trained using different subsets of the training dataset. To reduce the variance error, the final results are obtained through majority voting or averaging [234]. The boosting method adds more weights to data sequences with incorrect prediction at each training iteration and finally improve the model accuracy through many training iterations. Additionally, different machine learning methods can be used to train the ensemble models, and Bayesian model averaging can be utilized to achieve further improvements through a reasonable combination of all prediction results [34, 79].

6.3 CLBF-MPC Using RNN Models

In Chap. 4, we have adopted control Lyapunov-barrier functions (CLBF) in the model predictive controllers (MPCs) to stabilize an input-constrained nonlinear system with guaranteed process operational safety simultaneously. In this section, a machine-learning-based CLBF-MPC is developed by taking advantage of the ensemble of RNN models that are developed for the nonlinear system of Eq. 6.1. Process operational safety and closed-loop stability analysis for the system of Eq. 6.1 associated with two types of unsafe regions, i.e., unbounded and bounded sets, is also provided.

To begin with, it is assumed that a set of unsafe states, i.e., $\mathscr{D} \subset \mathbf{R}^n$, exists in state-space (unsafe states could represent high temperature or pressure in a chemical plant). Also, we assume that there exists a safe stability region $\mathscr{U}$ that satisfies $\{0\} \subset \mathscr{U}$ and $\mathscr{U} \cap \mathscr{D} = \emptyset$, within which process operational safety and closed-loop stability are achieved simultaneously in the following sense:

Definition 6.1 (*Definition* 3.2) Consider the nonlinear system of Eq. 6.1 with the input constraints $u \in U$. If for any initial state $x(t_0) = x_0 \in \mathscr{U}$, there exists a control law $u = \Phi(x) \in U$ that can maintain $x(t)$ inside $\mathscr{U}$, $\forall t \geq 0$, and render the origin of the closed-loop system of Eq. 6.1 asymptotically stable, then we say that operational safety and closed-loop stability are achieved simultaneously for the nonlinear system of Eq. 6.1.

As discussed in Chaps. 3 and 4, the unsafe region can be characterized through analyses of past accidents and of past operating data, or from process first-principles knowledge. There are two types of unsafe regions that are often encountered in engineering problems: (1) unbounded sets, and (2) bounded sets. For example, in a chemical plant, there often exists an unbounded unsafe region consisting of all the operating conditions (e.g., reactor temperature or pressure) above a threshold that are considered unsafe. Bounded sets are often characterized for multivariable system where the interaction among different variables play a role in determining whether the system is safe or not. For example, in the chemical reactor example that has been discussed in Chap. 3, a bounded set of unsafe states is characterized based on the combination of temperature and concentration of reactants that reflect reaction rates. Additionally, bounded unsafe sets often occur in motion planning for self-driving cars and robots that attempt to address obstacle avoiding problems. Throughout this chapter, we will discuss both unbounded unsafe regions (denoted by $\mathscr{D}_u$) and bounded unsafe regions (denoted by $\mathscr{D}_b$), and prove closed-loop stability and process operational safety for the nonlinear system of Eq. 4.1 under CLBF-based controllers using RNN models for prediction.

6.3.1 *Stabilization and Safety via CLBF-Based Control*

The definition of the CLBF of Eq. 4.12 is restated here for convenience.

Definition 6.2 Consider the nominal system of Eq. 4.1 (i.e., $w(t) \equiv 0$) with a set of unsafe states in state-space (i.e., $\mathscr{D}$), a proper, lower-bounded and $\mathscr{C}^1$ function $W_c(x) : \mathbf{R}^n \to \mathbf{R}$ that has a minimum at the origin is a constrained CLBF if $W_c(x)$ also satisfies the following properties:

$$W_c(x) > \rho, \quad \forall x \in \mathscr{D} \subset \phi_{uc} \tag{6.16a}$$

$$L_{\hat{f}} W_c(x) < 0, \quad \forall x \in \{z \in \phi_{uc} \backslash (\mathscr{D} \cup \{0\} \cup \mathbb{X}_e) \mid L_{\hat{g}} W_c(z) = 0\} \tag{6.16b}$$

$$\mathscr{U}_\rho := \{x \in \phi_{uc} \mid W_c(x) \leq \rho\} \neq \emptyset \tag{6.16c}$$

where $\rho \in \mathbf{R}$, and $\mathbb{X}_e := \{x \in \phi_{uc} \backslash (\mathscr{D} \cup \{0\}) \mid \partial W_c(x)/\partial x = 0\}$ is a set of states where $L_{\hat{f}} W_c(x) = 0$ $(x \neq 0)$ holds for the RNN model of Eq. 6.4 due to $\partial W_c(x)/\partial x = 0$. $\hat{f}$ and $\hat{g}$ are the vector functions in the RNN model represented in the form of Eq. 6.5. Additionally, in this chapter, the safe operating region is considered to be a level set of W_c, (i.e., $\mathscr{U}_\rho$) only, and therefore, the original definition of CLBF of Eq. 4.12 can be simplified to Eq. 6.16 according to the discussion in Remark 4.1.

We assume that there exists a feedback control law $u = \Phi_{nn}(x) \in U$ that renders the origin of the RNN model of Eq. 6.4 exponentially stable for the state within an open neighborhood ϕ_{uc} around the origin. The stabilizability assumption implies that a $\mathscr{C}^1$ constrained control Lyapunov-barrier function $W_c(x)$ that has a minimum at the origin exists and satisfies the following inequalities $\forall x \in \phi_{uc}$:

$$\hat{c}_1 |x|^2 \leq W_c(x) - \rho_0 \leq \hat{c}_2 |x|^2, \tag{6.17a}$$

$$\begin{aligned} &\frac{\partial W_c(x)}{\partial x} F_{nn}(x, \Phi_{nn}(x)) \leq -\hat{c}_3 |x|^2, \forall x \in \phi_{uc} \backslash \mathscr{B}_\delta(x_e) \\ &\frac{\partial W_c(x)}{\partial x} F_{nn}(x, \Phi_{nn}(x)) \leq 0, \forall x \in \mathscr{B}_\delta(x_e), \end{aligned} \tag{6.17b}$$

$$\left| \frac{\partial W_c(x)}{\partial x} \right| \leq \hat{c}_4 |x|, \tag{6.17c}$$

where $\hat{c}_j(\cdot)$, $j = 1, 2, 3, 4$ are positive real numbers, $W_c(0) = \rho_0$ is the global minimum value of $W_c(x)$ in ϕ_{uc}, and $\mathscr{B}_\delta(x_e)$ is a small neighborhood around $x_e \in \mathbb{X}_e$. $F_{nn}(x, u)$ represents the RNN system of Eq. 6.4. It is noted that $\frac{\partial W_c(x)}{\partial x} F_{nn}(x, \Phi_{nn}(x)) \leq -\hat{c}_3 |x|^2$ does not hold for $x \in \mathscr{B}_\delta(x_e)$ since $\frac{\partial W_c(x)}{\partial x}$ is close to zero for the states in a neighborhood around the stationary point x_e, where $\frac{\partial W_c(x)}{\partial x} = 0$. The set ϕ_{uc} is characterized using numerical simulations as a set of states in the state-space where Eq. 6.17 is satisfied. Additionally, based on the continuity and the smoothness properties assumed for the functions f, g and h in the nonlinear system of Eq. 6.1, there exist positive constants L_x, L_w, M, L_x', L_w' such that the following inequalities hold for all $u \in U$, $x, x' \in \mathscr{U}_\rho$, and $w \in W$:

$$|F(x, u, w)| \leq M \tag{6.18a}$$

$$|F(x, u, w) - F(x', u, 0)| \leq L_x|x - x'| + L_w|w| \tag{6.18b}$$

$$\left|\frac{\partial W_c(x)}{\partial x}F(x, u, w) - \frac{\partial W_c(x')}{\partial x}F(x', u, 0)\right| \leq L_x^{'}|x - x'| + | \leq L_w^{'}|w_m|. \tag{6.18c}$$

The universal Sontag controller of Eq. 6.3 with $W_c(x)$ replacing the Lyapunov function $V(x)$ provides a candidate controller for $\Phi_{nn}(x)$ associated with CLBFs. Note that since the nonlinear system of Eq. 6.1 is assumed to be unknown, the set ϕ_{uc} and the CLBF of Eq. 6.16 are designed based on the RNN model of Eq. 6.4 (Eq. 6.4 can also be represented in the form of Eq. 6.5, i.e., $\dot{x} = \hat{f}(x) + \hat{g}(x)u$). To develop a constrained CLBF that meets the conditions in Eq. 6.16, we first design a CBF and a CLF separately, and combine them following the construction method in Sect. 4.3.2.

Consider the RNN model of Eq. 6.4 (also in the form of Eq. 6.5) with a constrained CLBF $W_c(x)$. Following the analysis that has been performed for the nominal system of Eq. 6.1 in Chap. 4, simultaneous process operational safety and closed-loop stability can be readily derived for the RNN model of Eq. 6.4 with both an unbounded unsafe region $\mathcal{D}_u$ and a bounded unsafe region $\mathcal{D}_b$ (see Theorems 4.2 and 4.3 for the two types of unsafe regions). Specifically, in the presence of an unbounded unsafe region, process operational safety and closed-loop stability are both guaranteed for the RNN model of Eq. 6.4 under the controller $u = \Phi_{nn}(x) \in U$. However, unlike the case of unbounded unsafe regions for which the origin is the unique stationary point in state-space, it is demonstrated in Chap. 4 that in the presence of a bounded unsafe set, stationary points (other than the origin) may exist in state-space (i.e., $\mathbb{X}_e$ in Eq. 6.16b), and thus, the origin cannot be rendered exponentially stable under a continuous controller (e.g., the Sontag control law of Eq. 6.3 with $W_c(x)$ replacing the Lyapunov function $V(x)$). To address this issue, the functional form of $W_c(x)$ should be carefully designed such that the stationary points (other than the origin) are saddle points in state-space. Subsequently, a set of discontinuous control actions that can drive the state away from the saddle points $\mathbb{X}_e$ while decreasing $W_c(x)$ at the same time will be designed ahead of time and will be implemented when the states get trapped in $\mathbb{X}_e$ in closed-loop operation. Sufficient conditions under which simultaneous process operational safety and closed-loop stability are achieved for the RNN system of Eq. 6.4 under the CLBF-based control law of Eq. 6.16 is provided by the following theorem.

Theorem 6.2 *Consider that a constrained CLBF $W_c(x)$: $\mathbf{R}^n \rightarrow \mathbf{R}$ that meets the conditions of Eq. 6.16 and has a minimum at the origin, exists for the RNN system of Eq. 6.4. The closed-loop state is guaranteed to be bounded in $\mathcal{U}_\rho$ for all times for any initial condition $x_0 \in \mathcal{U}_\rho$ under the controller $u = \Phi_{nn}(x) \in U$ that satisfies Eq. 6.17. Additionally, the controller $u = \Phi_{nn}(x) \in U$ can further render the origin exponentially stable for all $x_0 \in \mathcal{U}_\rho$, in the presence of an unbounded unsafe region $\mathcal{D}_u$; however, in the presence of a bounded unsafe region $\mathcal{D}_b$ in state-space, discontinuous control actions $u = \bar{u}(x) \in U$ that decrease $W_c(x)$ should be implemented at saddle points x_e to ensure exponential stability of the origin.*

Proof To prove the boundedness of state in the safe operating region $\mathcal{U}_\rho$, we show that there exists a controller $u = \Phi_{nn}(x) \in U$ that renders $\dot{W}_c \leq 0$ for all $x \in \mathcal{U}_\rho$. Following the proof in Theorem 4.2, it is readily shown that $\dot{W}_c \leq 0$ holds for the RNN system of Eq. 6.4 using the universal Sontag controller of Eq. 6.3, $\Phi_{nn}(x)$, that replaces the Lyapunov function $V(x)$ with the CLBF $W_c(x)$ since the RNN system of Eq. 6.4 can be represented in the same form of nonlinear system of Eq. 4.1. In the presence of an unbounded unsafe region, the origin can be rendered exponentially stable under $u = \Phi_{nn}(x) \in U$ because the operating region $\mathcal{U}_\rho$ is a level set of $W_c(x)$, within which all the states satisfy the conditions in Eq. 6.17. The discontinuous control actions $\bar{u}(x)$ (i.e., $\bar{u}(x) \neq \Phi_{nn}(x)$) are developed to handle the issue of saddle points for the case of a bounded unsafe region. Once the state leaves the saddle point under $\bar{u}(x)$, it will continue to move towards the origin with an exponential decay under the CLBF-based controller $u = \Phi_{nn}(x) \in U$. The detailed proofs for unbounded and bounded unsafe regions follow closely to those for Theorems 4.2 and 4.3 in Sect. 4.3.1.3, and are omitted here.

Remark 6.7 Note that the safe operating region $\mathcal{U}_\rho$ and the CLBF of Eq. 6.16 are characterized based on the RNN system of Eq. 6.4 since the nonlinear system of Eq. 6.1 is assumed to be unknown. Also, operational safety and closed-loop stability analyses are carried out in Theorem 6.2 for the RNN system of Eq. 6.4 with a CLBF-based controller $u = \Phi_{nn}(x) \in U$. However, as the RNN model Eq. 6.4 may not perfectly capture the process dynamics of the actual nonlinear system of Eq. 6.1 (i.e., the modeling error is non-zero), we will further demonstrate in the following section that operational safety and closed-loop stability are simultaneously achieved for the nonlinear system of Eq. 6.1 under the CLBF-based controller $u = \Phi_{nn}(x) \in U$, provided that the modeling error between the RNN system of Eq. 6.4 and the nonlinear system of Eq. 6.1 is sufficiently small.

6.3.2 CLBF-based MPC Using an Ensemble of RNN Models

In this section, we incorporate an ensemble of RNN models in MPC and formulate the optimization problem of CLBF-based MPC. Process operational safety and closed-loop stability will then be proven for the nonlinear system of Eq. 6.1 under the CLBF-based MPC using RNN models. To begin with, we demonstrate that the CLBF-based controller $u = \Phi_{nn}(x) \in U$ that is designed to stabilize the RNN system of Eq. 6.4 with guaranteed safety also guarantees stability and safety for the nominal system of Eq. 6.1 with $w(t) = 0$ (i.e., the safety and stability properties in Theorem 6.2 also hold for the nominal system of Eq. 6.1). Subsequently, to optimize process performance during dynamic operation, we develop the CLBF-MPC scheme to optimize process performance (e.g., speed of convergence and energy consumption) during the dynamic operation. As a non-zero modeling error exists between the prediction model (i.e., the ensemble of RNN models) and the actual nonlinear system of Eq. 6.1, the following proposition is developed to show the evolution of the error between the states of the nonlinear process of Eq. 6.1 and the states predicted by the RNN model of Eq. 6.4. An upper bound for the state error is also derived for the case of a bounded modeling error (i.e., $|\nu| = |F(x, u, 0) - F_{nn}(x, u)| \leq \gamma|x| \leq \nu_m$) and a bounded disturbance (i.e., $|w(t)| \leq w_m$).

Proposition 6.2 *Consider the disturbed nonlinear system $\dot{x} = F(x, u, w)$ of Eq. 6.1 with bounded disturbances, i.e., $|w(t)| \leq w_m$. Assuming that the RNN model $\dot{\hat{x}} = F_{nn}(\hat{x}, u)$ of Eq. 6.4 has the same initial condition $x_0 = \hat{x}_0 \in \mathcal{U}_\rho$ as the nonlinear system of Eq. 6.1, then, there exists a positive constant κ and a class $\mathcal{K}$ function $f_w(\cdot)$ such that the following inequalities hold $\forall x, \hat{x} \in \mathcal{U}_\rho$ and $w(t) \in W$:*

$$|x(t) - \hat{x}(t)| \leq f_w(t) := \frac{L_w w_m + \nu_m}{L_x}(e^{L_x t} - 1) \tag{6.19a}$$

$$W_c(x) \leq W_c(\hat{x}) + \frac{\hat{c}_4\sqrt{\rho - \rho_0}}{\sqrt{\hat{c}_1}}|x - \hat{x}| + \kappa|x - \hat{x}|^2. \tag{6.19b}$$

Proof Let $e(t) = x(t) - \hat{x}(t)$ represent the error vector between the solutions of the RNN model $\dot{\hat{x}} = F_{nn}(\hat{x}, u)$ and of the system $\dot{x} = F(x, u, w)$. The time-derivative of $e(t)$ can be calculated as follows:

$$\begin{aligned} |\dot{e}| &= |F(x, u, w) - F_{nn}(\hat{x}, u)| \\ &\leq |F(x, u, w) - F(\hat{x}, u, 0)| + |F(\hat{x}, u, 0) - F_{nn}(\hat{x}, u)|. \end{aligned} \tag{6.20}$$

Using Eq. 6.18b, the following inequality gives the upper bound for the first term of Eq. 6.20 for all $w(t) \in W$ and $x, \hat{x} \in \mathcal{U}_\rho$:

$$\begin{aligned} |F(x, u, w) - F(\hat{x}, u, 0)| &\leq L_x|x(t) - \hat{x}(t)| + L_w|w(t)| \\ &\leq L_x|x(t) - \hat{x}(t)| + L_w w_m. \end{aligned} \tag{6.21}$$

The second term of Eq. 6.20 represents the modeling error (i.e., $|\nu| = |F(\hat{x}, u, 0) - F_{nn}(\hat{x}, u)|$), and is bounded by $|\nu| \leq \nu_m$. Following Eq. 6.21, we can obtain the upper bound for $\dot{e}(t)$ in Eq. 6.20 as follows:

$$\begin{aligned} |\dot{e}(t)| &\leq L_x|x(t) - \hat{x}(t)| + L_w|w_m| + \nu_m \\ &\leq L_x|e(t)| + L_w|w_m| + \nu_m. \end{aligned} \tag{6.22}$$

Then, for all $|w(t)| \leq w_m$ and $x(t), \hat{x}(t) \in \mathcal{U}_\rho$, the following upper bound can be derived for $|e(t)|$ given that the initial condition equals zero (i.e., $e(0) = 0$):

$$|e(t)| = |x(t) - \hat{x}(t)| \leq f_w(t) \tag{6.23}$$

where

$$f_w(t) := \frac{L_w w_m + \nu_m}{L_x}(e^{L_x t} - 1).$$

Moreover, as $W_c(x)$ is a continuous function of x and is bounded on compact sets, the following inequality is derived for all $x, \hat{x} \in \mathcal{U}_\rho$ by implementing Taylor series expansion to $W_c(x)$ around $\hat{x}$:

$$W_c(x) \leq W_c(\hat{x}) + \frac{\partial W_c(\hat{x})}{\partial x}|x - \hat{x}| + \kappa|x - \hat{x}|^2 \tag{6.24}$$

where the term $\kappa|x - \hat{x}|^2$ (κ is a positive real number) bounds the high order terms of the Taylor series of $W_c(x)$, $\forall x, \hat{x} \in \mathcal{U}_\rho$. Using Eqs. 6.17a, 6.17c and 6.23, it follows that

$$\begin{aligned} W_c(x) &\leq W_c(\hat{x}) + \frac{\hat{c}_4\sqrt{\rho - \rho_0}}{\sqrt{\hat{c}_1}}|x - \hat{x}| + \kappa|x - \hat{x}|^2 \\ &\leq W_c(\hat{x}) + \frac{\hat{c}_4\sqrt{\rho - \rho_0}}{\sqrt{\hat{c}_1}} f_w(t) + \kappa f_w(t)^2. \end{aligned} \tag{6.25}$$

This completes the proof of Proposition 6.2.

6.3.2.1 CLBF-Based Control Using RNN Models

The following propositions are provided to demonstrate that the state of the nominal system of Eq. 6.1 can remain inside the safe operating region $\mathcal{U}_\rho$ under the controller $u = \Phi_{nn}(x) \in U$ that is designed for the RNN model of Eq. 6.4 with a sufficiently small modeling error. We first consider the nominal system of Eq. 6.1 with an unbounded unsafe region, for which we show that exponential stability is achieved by the CLBF-based controller $u = \Phi_{nn}(x) \in U$ for the closed-loop nominal system of Eq. 6.1.

Proposition 6.3 *Consider the nominal system of Eq. 6.1 (i.e., $w(t) \equiv 0$) with an unbounded unsafe region $\mathcal{D}_u$. If there exists a positive real number $\gamma < \hat{c}_3/\hat{c}_4$ such that for all $x \in \mathcal{U}_\rho$ and $u \in U$, the modeling error between the nonlinear system of Eq. 6.1 and the RNN model of Eq. 6.4 is constrained by $|\nu| = |F(x, u, 0) - F_{nn}(x, u)| \leq \gamma|x|$, then the stability and safety properties in Theorem 6.2 also hold for the nominal closed-loop system of Eq. 6.1 under the CLBF-based controller $u = \Phi_{nn}(x) \in U$ that satisfies Eq. 6.17.*

Proof To show that the CLBF-based controller $u = \Phi_{nn}(x) \in U$ can render the origin of the nominal system of Eq. 6.1 (i.e., $w(t) \equiv 0$) exponentially stable, we prove that there exists a positive real number $\tilde{c}_3$ such that $\frac{\partial W_c(x)}{\partial x} F(x, \Phi_{nn}(x), 0) \leq -\tilde{c}_3|x|^2$ holds for all $x \in \mathcal{U}_\rho$. Since the origin is the unique stationary point in state-space for the case of an unbounded unsafe region, we derive the following time-derivative of W_c using Eqs. 6.17b and 6.17c:

$$
\begin{aligned}
\dot{W}_c &= \frac{\partial W_c(x)}{\partial x} F(x, \Phi_{nn}(x), 0) \\
&= \frac{\partial W_c(x)}{\partial x} (F_{nn}(x, \Phi_{nn}(x)) + F(x, \Phi_{nn}(x), 0) - F_{nn}(x, \Phi_{nn}(x))) \\
&\leq -\hat{c}_3|x|^2 + \hat{c}_4|x|(F(x, \Phi_{nn}(x), 0) - F_{nn}(x, \Phi_{nn}(x))) \\
&\leq -\hat{c}_3|x|^2 + \hat{c}_4\gamma|x|^2
\end{aligned}
\tag{6.26}
$$

Let $\tilde{c}_3 = -\hat{c}_3 + \hat{c}_4\gamma$. It follows that $\dot{W}_c \leq -\tilde{c}_3|x|^2 \leq 0$ if γ is chosen to satisfy $\gamma < \hat{c}_3/\hat{c}_4$. Therefore, following the proof of operational safety and closed-loop stability for the RNN system of Eq. 6.4 in Theorem 6.2, it is straightforward to show that the controller $u = \Phi_{nn}(x) \in U$ can drive the state of the nominal system of Eq. 6.1 to the origin and avoid the unbounded unsafe region $\mathscr{D}_u$ at all times. This completes the proof of simultaneous operational safety and closed-loop stability for the nominal system of Eq. 6.1 with any initial condition x_0 in the safe operating region $\mathscr{U}_\rho$.

The following proposition provides sufficient conditions under which process operational safety and closed-loop stability are guaranteed for the nominal system of Eq. 6.1 when a bounded unsafe region $\mathscr{U}_b$ exists in state-space. The existence of saddle points x_e in the safe operating region $\mathscr{U}_\rho$ will also be taken into account for the case of a bounded unsafe region.

Proposition 6.4 *Consider the nominal system of Eq. 6.1 with a bounded unsafe region $\mathscr{D}_b$. If there exists a positive real number $\gamma < \hat{c}_3/\hat{c}_4$ such that the modeling error is constrained by $|\nu| = |F(x, u, 0) - F_{nn}(x, u)| \leq \gamma|x|$ for all $x \in \mathscr{U}_\rho$ and $u \in U$, and there exist discontinuous control actions $u = \bar{u}(x) \in U$ such that Eq. 6.27 is satisfied when $x(t_k) = \hat{x}(t_k) \in \mathscr{B}_\delta(x_e)$,*

$$
W_c(\hat{x}(t)) < W_c(\hat{x}(t_k)) - f_e(t - t_k), \ \forall t > t_k \tag{6.27}
$$

where

$$
f_e(t - t_k) := \frac{\hat{c}_4\sqrt{\rho - \rho_0}}{\sqrt{\hat{c}_1}} f_w(t - t_k) - \kappa f_w(t - t_k)^2
$$

and $f_w(t)$ is defined in Eq. 6.23, then the safety and stability properties in Theorem 6.2 also hold for the nominal system of Eq. 6.1 under the controllers $u = \bar{u}(x) \in U$ for $x(t_k) \in \mathscr{B}_\delta(x_e)$ and $u = \Phi_{nn}(x) \in U$ for $x(t_k) \in \mathscr{U}_\rho \backslash \mathscr{B}_\delta(x_e)$, where $\Phi_{nn}(x)$ is the CLBF-based controller that satisfies Eq. 6.17.

Proof Note that the continuous controller $u = \Phi_{nn}(x) \in U$ cannot render the origin of the nominal system of Eq. 6.1 (i.e., $w(t) \equiv 0$) exponentially stable when a bounded unsafe region exists in state-space due to the existence of saddle points x_e ($x_e \neq 0$) in the safe operating region $\mathscr{U}_\rho$. To prevent the states from converging to the stationary points x_e, we design another set of control actions $\bar{u}$ that can drive the state away from saddle points when the state enters a neighborhood around x_e. Specifically, it is readily shown that Eq. 6.26 still holds for all $x \in \mathcal{U}_\rho \backslash \mathcal{B}_\delta(x_e)$ since $\frac{\partial W_c(x)}{\partial x} F_{nn}(x, \Phi_{nn}(x)) \leq -\hat{c}_3|x|^2$ is satisfied in $\mathcal{U}_\rho \backslash \mathcal{B}_\delta(x_e)$ for the case of a bounded unsafe region. This implies that the controller $u = \Phi_{nn}(x) \in U$ that is designed to ensure safety and stability for the RNN model of Eq. 6.4 is also able to maintain the state of the closed-loop system of Eq. 6.1 within $\mathcal{U}_\rho$ for all times in the presence of a bounded unsafe region.

Subsequently, we prove that the state of the nonlinear system of Eq. 6.1 can escape from the saddle points using the discontinuous control actions $u = \bar{u}(x) \in U$ that are designed for the RNN model of Eq. 6.4. Proposition 6.2 has shown the boundedness of state error between the nonlinear system of Eq. 6.1 and the RNN system of Eq. 6.4 under the same initial condition and the same control actions. Also, the evolution of $W_c(x)$ for the state of the disturbed system of Eq. 6.1 is shown to be bounded by Eq. 6.25 accounting for bounded disturbances and sufficiently small modeling error. Now we assume that at time $t = t_k$, the state enters a neighborhood around the saddle points (i.e., $\hat{x}(t_k) = x(t_k) \in \mathcal{B}_\delta(x_e)$). The following inequality can be derived from Eq. 6.25 if Eq. 6.4 is met by the discontinuous control actions $\bar{u}(\hat{x})$ that are developed for the RNN model of Eq. 6.4 for all $x \in \mathcal{B}_\delta(x_e)$.

$$\begin{aligned} W_c(x(t)) &\leq W_c(\hat{x}(t)) + \frac{\hat{c}_4\sqrt{\rho - \rho_0}}{\sqrt{\hat{c}_1}} f_w(t - t_k) + \kappa f_w(t - t_k)^2, \\ &< W_c(\hat{x}(t_k)). \end{aligned} \tag{6.28}$$

The above inequality shows that the value of $W_c(x)$ for the state of the actual nonlinear system of Eq. 6.1 is decreasing $\forall t > t_k$, which implies that the state of the nonlinear system of Eq. 6.1 can escape from saddle points under the discontinuous control actions. This completes the proof that the closed-loop state of the nonlinear system of Eq. 6.1 can converge to the origin and avoid the bounded unsafe region $\mathcal{D}_b$ in state-space under the controllers $u = \bar{u}(x) \in U$ and $u = \Phi_{nn}(x) \in U$, for any initial condition $x_0 \in \mathcal{U}_\rho$.

Remark 6.8 Propositions 6.3 and 6.4 demonstrate that the controller $u = \Phi_{nn}(x) \in U$ designed for the RNN system of Eq. 6.4 (i.e., $\dot{\hat{x}} = F_{nn}(\hat{x}, u)$) also guarantees operational safety and closed-loop stability for the nominal system of Eq. 6.1 (i.e., $w(t) \equiv 0$). Specifically, in the presence of an unbounded unsafe region, the origin is rendered exponentially stable and the state is bounded in the safe operating region at all times using the CLBF-based controller $u = \Phi_{nn}(x) \in U$. However, for the case of an unbounded unsafe region, in addition to the CLBF-based controller $u = \Phi_{nn}(x) \in U$, a set of discontinuous control actions $u = \bar{u}(x) \in U$ satisfying Eq. 6.27 should be implemented when the state enters a neighborhood around saddle points (i.e., $\mathcal{B}_\delta(x_e)$) to achieve the stability and safety properties for the nominal system of Eq. 6.1.

6.3.2.2 Sample-and-Hold Implementation of CLBF-Based Controller

This section presents the stability properties for the sample-and-hold implementation of the CLBF-based controllers $u = \Phi_{nn}(x) \in U$ and of the discontinuous control actions $u = \bar{u}(x) \in U$ (for a bounded unsafe region) for the disturbed system of Eq. 6.1 with bounded disturbances (i.e., $|w(t)| \leq w_m$). To proceed, the following proposition is developed to show that the closed-loop state $x(t)$ of the nonlinear system of Eq. 6.1 remains inside the safe operating region $\mathcal{U}_\rho$ for all times, and is ultimately bounded in a small region $\mathcal{U}_{\rho_{\min}}$ around the origin, under the controllers $u = \bar{u}(x) \in U$ and $u = \Phi_{nn}(x) \in U$ implemented in a sample-and-hold fashion, i.e., $u(t) = u(t_k), \forall t \in [t_k, t_{k+1})$, where Δ is the sampling period and $t_{k+1} := t_k + \Delta$.

Proposition 6.5 *Consider the nonlinear system of Eq. 6.1 under the sample-and-hold implementation of the CLBF-based controller $u = \Phi_{nn}(x) \in U$ satisfying the conditions of Eq. 6.17. If the controller $u = \bar{u}(x) \in U$ (in a sample-and-hold fashion) meets Eq. 6.27 for all $x \in \mathcal{B}_\delta(x_e)$, and there exist $\varepsilon_w > 0$, $\Delta > 0$ and $\rho > \rho_{\min} > \rho_{nn} > \rho_s$ that satisfy*

$$-\frac{\tilde{c}_3}{\hat{c}_2}(\rho_s - \rho_0) + L_x^{'} M\Delta + L_w^{'} w_m \leq -\varepsilon_w \tag{6.29}$$

and

$$\rho_{nn} := \max\{W_c(\hat{x}(t+\Delta)) \mid \hat{x}(t) \in \mathcal{U}_{\rho_s}, u \in U\} \tag{6.30a}$$

$$\rho_{\min} \geq \rho_{nn} + f_e(\Delta) \tag{6.30b}$$

where $f_e(t)$ is defined in Eq. 6.27, then for any $x(t_k) \in \mathcal{U}_\rho \backslash \mathcal{U}_{\rho_s}$, we can show that the value of $W_c(x(t))$ associated with the state of the nonlinear system of Eq. 6.1 is decreasing within each sampling period. As a result, the state is bounded in the safe operating region $\mathcal{U}_\rho$ for all times and is ultimately bounded in $\mathcal{U}_{\rho_{\min}}$.

Proof When $x(t_k) = \hat{x}(t_k) \in \mathcal{U}_\rho \backslash \mathcal{U}_{\rho_s}$, the time-derivative of $W_c(x)$ for the disturbed system of Eq. 6.1 (i.e., $|w| \leq w_m$) is derived as follows:

$$\begin{aligned}\dot{W}_c(x(t)) =& \frac{\partial W_c(x(t))}{\partial x} F(x(t), \Phi_{nn}(x(t_k)), w) \\ =& \frac{\partial W_c(x(t_k))}{\partial x} F(x(t_k), \Phi_{nn}(x(t_k)), 0) + \frac{\partial W_c(x(t))}{\partial x} F(x(t), \Phi_{nn}(x(t_k)), w) \\ &- \frac{\partial W_c(x(t_k))}{\partial x} F(x(t_k), \Phi_{nn}(x(t_k)), 0).\end{aligned} \tag{6.31}$$

Based on Eqs. 6.17b, 6.26 and the Lipschitz condition in Eq. 6.18, when $x(t_k) \in \mathcal{U}_\rho \backslash (\mathcal{U}_{\rho_s} \cup \mathcal{B}_\delta(x_e))$, the upper bound for $\dot{W}_c(x(t))$ over $t \in [t_k, t_{k+1})$ is derived by the following inequality:

$$
\begin{aligned}
\dot{W}_c(x(t)) \leq & -\frac{\tilde{c}_3}{\hat{c}_2}(\rho_s - \rho_0) + \frac{\partial W_c(x(t))}{\partial x} F(x(t), \Phi_{nn}(x(t_k)), w) \\
& - \frac{\partial W_c(x(t_k))}{\partial x} F(x(t_k), \Phi_{nn}(x(t_k)), 0) \\
\leq & -\frac{\tilde{c}_3}{\hat{c}_2}(\rho_s - \rho_0) + L_x^{'}|x(t) - x(t_k)| + L_w^{'}|w| \\
\leq & -\frac{\tilde{c}_3}{\hat{c}_2}(\rho_s - \rho_0) + L_x^{'} M \Delta + L_w^{'} w_m .
\end{aligned} \tag{6.32}
$$

It is noted that Eq. 6.32 does not hold for the state close to the saddle points, i.e., $x \in \mathscr{B}_\delta(x_e)$ since Eq. 6.26 may not hold for those states where $\frac{\partial W_c(x)}{\partial x}$ is close to zero. Based on Eq. 6.32, the following inequality is derived for all $x(t_k) \in \mathscr{U}_\rho \backslash \mathscr{U}_{\rho_s}$ and $t \in [t_k, t_{k+1})$ if Eq. 6.29 is met:

$$
\dot{W}_c(x(t)) \leq -\varepsilon_w . \tag{6.33}
$$

Since the level set of W_c is an invariant set, from Eq. 6.33, it is straightforward to show that for any initial condition $x_0 \in \mathscr{U}_\rho$, the state is guaranteed to be bounded in the safe operating region $\mathscr{U}_\rho$ for all times under the CLBF-based controller $u = \Phi_{nn}(x) \in U$ in sample-and-hold fashion.

Additionally, to ensure that the state of the nonlinear system of Eq. 6.1 does not get trapped in any saddle points, and can be ultimately driven into a small neighborhood $\mathscr{U}_{\rho_s}$ around the origin, we implement the controller $u = \bar{u}(x(t_{k+i})) \in U$, $\forall t \in [t_{k+i}, t_{k+i+1}), i = 0, 1, 2, \ldots$ when $x(t_k) = \hat{x}(t_k) \in \mathscr{B}_\delta(x_e)$. Using Eq. 6.28 in Proposition 6.4, it follows that $W_c(x(t)) < W_c(x(t_k))$, $\forall t > t_k$ holds for the nonlinear system of Eq. 6.1 if Eq. 6.27 is met by the sample-and-hold implementation of $u = \bar{u}(\hat{x}) \in U$. This implies that the value of $W_c(x)$ will continuously decrease until the state moves away from the saddle points, and the CLBF-based controller $u = \Phi_{nn}(x) \in U$ can again drive the state towards the origin.

Next, we show that once the state enters a small region around the origin, i.e., $x(t_k) = \hat{x}(t_k) \in \mathscr{U}_{\rho_s}$, the state will be bounded in $\mathscr{U}_{\rho_{\min}}$, where $\mathscr{U}_{\rho_{\min}}$ is a level set of $W_c(x)$ that is slightly larger than $\mathscr{U}_{\rho_s}$, for the remaining time $t \geq t_k$. Based on the definition of $\mathscr{U}_{\rho_{nn}}$ in Eq. 6.30a, $\mathscr{U}_{\rho_{nn}}$ is characterized as the largest level set of $W_c(\hat{x})$ that the state of the RNN system of Eq. 6.4 can reach within one sampling period if starting from $\mathscr{U}_{\rho_s}$. Correspondingly, $\mathscr{U}_{\rho_{\min}}$ of Eq. 6.30b is the largest level set of $W_c(x)$ associated with the state of the nonlinear system of Eq. 6.1 when the state $\hat{x}$ of the RNN system of Eq. 6.4 is inside $\mathscr{U}_{\rho_{nn}}$. Since the states in $\mathscr{U}_{\rho_s}$ are very close to the origin, $\dot{W}_c \leq -\varepsilon_w$ may not hold under the sample-and-hold implementation of $u = \Phi_{nn}(x) \in U$. Therefore, we characterize the sets $\mathscr{U}_{\rho_{\min}}$ and $\mathscr{U}_{\rho_{nn}}$ to ensure the boundedness of the state for the nonlinear system of Eq. 6.1 and for the RNN system of Eq. 6.4, respectively. In practice, we can determine the size of $\mathscr{U}_{\rho_{nn}}$ from extensive open-loop simulations with various $u \in U$ and $x \in \mathscr{U}_{\rho_s}$. Subsequently, $\mathscr{U}_{\rho_{\min}}$ of Eq. 6.30b can be characterized from the open-loop simulations that account for the bounded disturbances and modeling error. This completes the proof of the boundedness of state in the safe operating region $\mathcal{U}_\rho$, and of the convergence of state into a small neighborhood around the origin $\mathcal{U}_{\rho_{\min}}$ under the sample-and-hold implementation of $u = \bar{u}(x) \in U$ and $u = \Phi_{nn}(x) \in U$.

6.3.2.3 Formulation of CLBF-MPC

The CLBF-MPC design is represented by the following optimization problem [219]:

$$\min_{u\in S(\Delta)} \int_{t_k}^{t_{k+N}} l_t(\tilde{x}(t), u(t))dt \tag{6.34a}$$

$$\text{s.t.} \quad \dot{\tilde{x}}(t) = \frac{1}{N_e}\sum_{j=1}^{N_e} F_{nn}^j(\tilde{x}(t), u(t)) \tag{6.34b}$$

$$\tilde{x}(t_k) = x(t_k) \tag{6.34c}$$

$$u(t) \in U, \ \forall\, t \in [t_k, t_{k+N}) \tag{6.34d}$$

$$\dot{W}_c(x(t_k), u(t_k)) \le \dot{W}_c(x(t_k), \Phi_{nn}(t_k)), \ \text{if } W_c(x(t_k)) > \rho_{nn} \text{ and } x(t_k) \notin \mathcal{B}_\delta(x_e) \tag{6.34e}$$

$$W_c(\tilde{x}(t)) \le \rho_{nn}, \ \forall\, t \in [t_k, t_{k+N}), \ \text{if } W_c(x(t_k)) \le \rho_{nn} \tag{6.34f}$$

$$W_c(\tilde{x}(t)) < W_c(x(t_k)) - f_e(t - t_k), \ \forall\, t \in (t_k, t_{k+N}), \ \text{if } x(t_k) \in \mathcal{B}_\delta(x_e) \tag{6.34g}$$

where Δ is the sampling period, $S(\Delta)$ is the set of piecewise constant functions with time interval Δ, $\tilde{x}(t)$ is the predicted state trajectory, and N is the number of sampling steps in the prediction horizon. We use $\dot{W}_c(x, u)$ to represent $\frac{\partial W_c(x)}{\partial x} F_{nn}(x, u)$. The optimization problem of Eq. 6.34 is to minimize the object function of Eq. 6.34a subject to the constraints of Eqs. 6.34b–6.34g. Specifically, the objective function is the integral of $l_t(\tilde{x}(t), u(t))$ over the prediction horizon, in which $l_t(\tilde{x}(t), u(t))$ is developed to satisfy $l_t(\tilde{x}(t), u(t)) > 0, \forall(\tilde{x}(t), u(t)) \neq (0, 0)$ and $l_t(0, 0) = 0$ such that it attains the minimum value at the steady-state of the nonlinear system of Eq. 6.1. The ensemble of RNN models F_{nn}^j, $j = 1, \ldots, N_e$ are used to calculate the predicted states $\tilde{x}(t)$, $t \in [t_k, t_{k+N})$ in Eq. 6.34b, where N_e is the number of RNN models in the ensemble. The final prediction results are obtained through the average of the state trajectories predicted by the ensemble of RNN models. The input constraints that will be applied over the entire prediction horizon are defined by Eq. 6.34d. The initial condition for the prediction model of Eq. 6.34b is the current state measurement defined by Eq. 6.34c. The CLBF-based constraints of Eqs. 6.34e–6.34g are utilized to ensure process operational safety and closed-loop stability. Specifically, the constraint of Eq. 6.34e drives the closed-loop state into a smaller level set of $W_c(x)$ by decreasing the value of $W_c(\tilde{x})$ along the predicted state trajectory at least at the rate under the CLBF-based controller $u = \Phi_{nn}(x) \in U$ when $W_c(x(t_k)) > \rho_{nn}$ and $x(t_k) \notin \mathcal{B}_\delta(x_e)$. When $W_c(x(t_k)) \le \rho_{nn}$, the constraint of Eq. 6.34f is activated to maintain the predicted state of the RNN system within

$\mathcal{U}_{\rho_{nn}}$ such that the closed-loop state of the nonlinear system of Eq. 6.1 is bounded $\mathcal{U}_{\rho_{\min}}$. Additionally, if $x(t_k) \in \mathcal{B}_\delta(x_e)$, the constraint of Eq. 6.34g drives the state in the direction of decreasing $W_c(x)$ over the prediction horizon such that within finite sampling steps, the state can escape from saddle points x_e . The CLBF-MPC optimization problem of Eq. 6.34 is solved with state measurements available at each sampling time. After the optimal solution $u^*(t)$ is computed from CLBF-MPC, we implement only the first control action of $u^*(t)$ to the actual nonlinear system for the next sampling period. Then, at the next sampling time $t_{k+1} := t_k + \Delta$, the optimization problem is solved again with a new state measurement. Additionally, in the CLBF-MPC formulation of Eq. 6.34, we use x instead of $\hat{x}$ to represent the RNN state in CLBF-MPC to simplify the notations as the MPC predictions are based on the RNN models only.

The following theorem is established to show guaranteed process operational safety and closed-loop stability for the nonlinear system of Eq. 6.1 under the CLBF-MPC of Eq. 6.34.

Theorem 6.3 *Consider the system of Eq. 6.1 with a constrained CLBF W_c that has a minimum at the origin and satisfies Eq. 6.16. The CLBF-MPC optimization problem of Eq. 6.34 is guaranteed to be solved with recursive feasibility for all times for any initial state $x_0 \in \mathcal{U}_\rho$. Additionally, it is guaranteed that the state initiating from $\mathcal{U}_\rho$ is bounded in $\mathcal{U}_\rho$, $\forall\, t \geq 0$, and can be ultimately bounded in $\mathcal{U}_{\rho_{\min}}$ as $t \to \infty$ under the sample-and-hold implementation of the CLBF-MPC that uses an ensemble of RNN models satisfying the conditions in Proposition 6.5 and the constraint on the modeling error, i.e., $|\nu| = |F(x, u, 0) - F_{nn}(x, u)| \leq \gamma|x| \leq \nu_m$.*

Proof The proof consists of two parts. We first show that recursive feasibility of CLBF-MPC is guaranteed for all states $x(t) \in \mathcal{U}_\rho$. Then, we show the simultaneous process operational safety and closed-loop stability for the nonlinear system of Eq. 6.1 under the CLBF-MPC using ensemble of RNN models of Eq. 6.4 for prediction.

Part 1: We prove that the CLBF-based controller $u = \Phi_{nn}(x) \in U, \forall x \in \mathcal{U}_\rho \backslash \mathcal{B}_\delta(x_e)$ and the discontinuous controller $u = \bar{u}(x) \in U, \forall x \in \mathcal{B}_\delta(x_e)$ implemented in a sample-and-hold fashion are feasible solutions to the CLBF-MPC optimization problem of Eq. 6.34, (i.e., they satisfy the CLBF-MPC constraints of Eqs. 6.34d–6.34g). Specifically, both $u = \Phi_{nn}(x)$ and $u = \bar{u}(x)$ meet the input constraint of Eq. 6.34d as they are both saturated when the values of control actions exceed the upper or lower bound. Next, it is readily shown that the constraint of Eq. 6.34e is met by letting $u(t_k) = \Phi_{nn}(x(t_k))$ for $x(t_k) \in \mathcal{U}_\rho \backslash (\mathcal{B}_\delta(x_e) \cup \mathcal{U}_{\rho_{nn}})$. When the state enters $\mathcal{U}_{\rho_{nn}}$, we can show that the CLBF-based controller $u(t) = \Phi_{nn}(x(t_{k+i})) \in U$, $\forall t \in [t_{k+i}, t_{k+i+1})$ with $i = 0, \ldots, N-1$ is again a feasible solution that satisfies the constraint of Eq. 6.34f because it is shown in Proposition 6.5 that the state remains inside $\mathcal{U}_{\rho_{nn}}$ after it enters this region. Lastly, when the state enters a neighborhood of saddle points, i.e., $x(t_k) \in \mathcal{B}_\delta(x_e)$, the sample-and-hold implementation of $u(t) = \bar{u}(x(t_{k+i})) \in U, \forall t \in [t_{k+i}, t_{k+i+1})$ with $i = 0, \ldots, N-1$ meets the constraint of Eq. 6.34g because the controller $u = \bar{u}(x) \in U$ is developed to satisfy Eq. 6.27. Therefore, recursive feasibility is guaranteed for the CLBF-MPC optimization problem of Eq. 6.34.

Part 2: Consider the nonlinear system of Eq. 6.1 with an unbounded unsafe region $\mathscr{D}_u$. The last constraint of Eq. 6.34g in the CLBF-MPC optimization problem remains inactive in this case as the origin is the unique stationary point (i.e., $\mathbb{X}_e = \emptyset$) in state-space for an unbounded unsafe region $\mathscr{D}_u$. Therefore, starting from any initial condition $x_0 \in \mathscr{U}_\rho \backslash \mathscr{U}_{\rho_{nn}}$, the closed-loop state is forced to move towards the origin and enter $\mathscr{U}_{\rho_{nn}}$ within finite time under the constraint of Eq. 6.34e. Once the state enters $\mathscr{U}_{\rho_{nn}}$, the constraint of Eq. 6.34f ensures boundedness of the state in $\mathscr{U}_{\rho_{nn}}$ for the remaining time. Correspondingly, as shown in Proposition 6.5, the state of the nonlinear system of Eq. 6.1 is guaranteed to be bounded in $\mathscr{U}_{\rho_{\min}}$ (a small neighborhood around the origin that contains $\mathscr{U}_{\rho_{nn}}$ inside). Therefore, the nonlinear system of Eq. 6.1 is considered practically stable as the state is ultimately bounded in a compact set close to the origin. Additionally, process operational safety is guaranteed for the system of Eq. 6.1 under CLBF-MPC because the boundedness of the state in $\mathscr{U}_\rho$ also implies the avoidance of the unsafe region $\mathscr{D}_u$ in state-space (i.e., $\mathscr{U}_\rho$ does not intersect with $\mathscr{D}_u$).

The proof of process operational safety and closed-loop stability for a bounded unsafe region $\mathscr{D}_b$ follows closely to the above analysis; however, in the presence of a bounded unsafe region, we need to show that the state can converge to $\mathscr{U}_{\rho_{nn}}$ instead of getting trapped in saddle points. Similarly, for any initial condition $x_0 \in \mathscr{U}_\rho \backslash \mathscr{U}_{\rho_{nn}}$, the state is driven towards the origin under the constraint of Eq. 6.34e. However, along its trajectory towards the origin, the state may settle at saddle points (local minima of the CLBF) if no further action is taken by CLBF-MPC. To address this issue, we activate the constraint of Eq. 6.34g when $x(t_k) \in \mathscr{B}_\delta(x_e)$ to continuously decrease the value of $W_c(x)$ such that the state can move away from the saddle points. Once the state leaves $\mathscr{B}_\delta(x_e)$ (i.e., the neighborhood around the saddle points), the constraints of Eqs. 6.34e–6.34f are activated again to ensure process operation safety and closed-loop stability by maintaining the state in the safe operating region $\mathscr{U}_\rho$ for all times, and ultimately bounding the state in $\mathscr{U}_{\rho_{nn}}$. This completes the proof of operational safety and closed-loop stability for the system of Eq. 6.1 with both a bounded unsafe region and an unbounded unsafe region.

6.3.3 Parallel Computing and Ensemble of RNN Models

An ensemble of RNN models have been utilized in CLBF-MPC to provide more accurate prediction of future states through the average of multiple RNN prediction results. As a result, an improvement of the closed-loop performance is achieved in the sense that the closed-loop state of the system of Eq. 6.1 is able to converge to the origin quickly and smoothly. While ensemble learning improves model prediction accuracy by using multiple RNN models, computation time for running CLBF-MPC is inevitably increased. Therefore, considering the significant increase of computation time arising from the use of multiple RNN models, we introduce parallel computing, which is a type of computation that carries out the execution of multiple processes simultaneously [13], in this section to reduce real-time computation time. Generally, a complex, computationally-heavy problem is first broken into discrete parts (i.e., subproblems) that can be solved concurrently. Then, we take advantage of many computers connected by a network or multiple compute resources (e.g., a single computer with multiple cores/processors) to solve all the subproblems simultaneously. Based on whether a communication exists between networked computers/processors, there are two types of parallel computing: (1) In parallel computing without communication, multiple tasks are executed in multiple processors, and the results are generated independently. (2) In parallel computing with communication (we sometimes call it distributed computing), multiple processors or networked computers use message passing interface (MPI) to communicate and coordinate the computation task to obtain final results. In this example, we implement the first type and the second type of parallel computing to the off-line training process of multiple RNN models and the prediction of future states in CLBF-MPC, respectively, to improve computational efficiency in both cases.

6.3.3.1 Training Multiple RNNs in Parallel

k-fold cross validation method in Sect. 6.2.3 is used in this study to construct multiple RNN models. Since we need to train k RNN models, the computation time for the training processes in series is approximately k times longer than that for a single RNN model. However, the increase of computation time is unnecessary as k RNN models can be trained independently using their own datasets. To that end, we distribute k training processes to multiple processors and utilize parallel computing to execute all the training processes simultaneously. Specifically, the implementation strategy of training k RNN models in parallel is as follows: (1) we first reserve k processors with sufficient memory for each processor. (2) According to the k-fold cross validation method, we partition the entire dataset into k subsets of the same size and distribute them to all reserved processors. (3) In the kth processor, we train the RNN model using the training dataset consisting of all the subsets except the $k-1$ subset. The remaining kth subset is used as the validation dataset. (4) To further improve computational efficiency, we create a bash script to run all k concurrently. Ideally, the computation time of k RNN models using parallel computing should be the same as that for a single RNN model. However, as the training processes may not be terminated at the same time due to different training datasets and settings, the total computation time is actually determined by the slowest training process.

6.3.3.2 Parallel Operation of CLBF-MPC Using an Ensemble of RNNs

We use an ensemble of RNN models in the CLBF-MPC of Eq. 6.34 to improve the prediction accuracy while closed-loop stability remains valid for the nonlinear system of Eq. 6.1. Since the optimal solution $u^*(t)$ is now computed through the average of all the state trajectories predicted by multiple RNN models, the computation time for the calculation of N_e RNN models increases rapidly under serial computation of

Eq. 6.34b, where N_e is the number of RNN models used in CLBF-MPC. In fact, the computation time for N_e RNN prediction in series is expected to be at least N_e times larger than the computation time for a single RNN model prediction in CLBF-MPC. From a practical perspective, the computation burden greatly limits the real-time implementation of CLBF-MPC that uses an ensemble of machine learning models for prediction. Therefore, parallel computing is utilized in this subsection to reduce the computation time of running multiple RNN predictions of Eq. 6.34b.

Specifically, we notice that the state prediction of Eq. 6.34b in CLBF-MPC can be broken apart into N_e sub-tasks that can be carried out independently and simultaneously. Consider the CLBF-MPC with an ensemble of N_e ($N_e \leq k$) RNN models for predicting future states. The implementation strategy of running parallel computing for the calculation of Eq. 6.34b is stated as follows: (1) As shown in Fig. 6.4, we first reserve $N_e + 1$ nodes with one as host node (e.g., node 0) and the rest as worker nodes. The host node communicates with other programs, for example, the dynamic simulation of the nonlinear system of Eq. 6.1, and receive/send the information (e.g., state measurements) from/to the worker nodes. The computation tasks are mainly carried out in worker nodes. (2) In this study, the optimization problem of CLBF-MPC is carried out on the host node while multiple RNN predictions are assigned to the worker nodes. For example, after we receive the state measurement at the sampling time t_k and send it to CLBF-MPC, the host node will broadcast the new state measurement $x(t_k)$ to all the worker nodes. Note that all the worker nodes share the same guess of control actions $u(t)$ and the same initial condition $x(t_k)$ at $t = t_k$. (3) Each worker node is assigned with an RNN model for prediction. At the end of parallel computation, the host node gathers the prediction results from worker nodes and obtain the final result following the ensemble learning algorithm. (4) We send the optimal control action $u^*(t_k)$ from CLBF-MPC to the nonlinear system for the next sampling period through the host node. Then, at the next sampling time, the above process (steps 1–4) is repeated to parallelize the computation of Eq. 6.34b with a new state measurement at t_{k+1}.

Remark 6.9 Computational efficiency of the CLBF-MPC optimization problem of Eq. 6.34 is significantly improved by introducing parallel computing into the calculation of N_e independent RNN models. However, it should be mentioned that due to the waiting and communication time between the worker nodes and the host node, the overall computation time of parallel computing may not be reduced exactly by N_e times compared to that under serial computing. Additionally, as shown in Fig. 6.4, before the host node calculates the final prediction results, a synchronization operation is needed to ensure that all the computation tasks in workers nodes are completed, and the results have been sent to the host node.

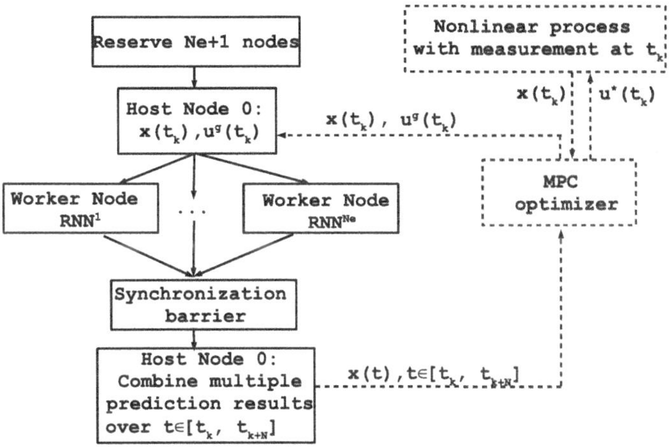

Fig. 6.4 Parallel computation of the ensemble of RNN models in CLBF-MPC, where $u^g(t_k)$ represents the guess of control action sent to the RNN models

6.3.4 Online Learning of RNN Models

Now we consider the nonlinear system of Eq. 6.1 subject to bounded time-varying disturbances (i.e., $|w(t)| \leq w_M$, where w_M is greater than the sufficiently small bound w_m in Eq. 6.1) that cannot be fully eliminated by the CLBF-MPC using the RNN models that are developed for the nominal system of Eq. 6.1 (i.e., $w(t) \equiv 0$). In this case, it is readily shown that the closed-loop system of Eq. 6.1 may be rendered unstable under the CLBF-based predictive controllers using the nominal RNN model (i.e., the RNN model obtained from open-loop simulations of the nominal system of Eq. 6.1 with $w(t) \equiv 0$) for all times since the modeling error between the uncertain system of Eq. 6.1 and the nominal RNN model no longer satisfies the constraint $|\nu| = |F(x, u, w) - F_{nn}(x, u)| \leq \gamma |x| \leq \nu_m$.

To account for the impact of disturbances in the predictions of the CLBF-MPC of Eq. 6.34, the RNN models of Eq. 6.34b need to be updated via online learning using the most recent process data to capture the nonlinear dynamics of the system of Eq. 6.1 subject to the time-varying disturbances $w(t)$. Error-triggered and event-triggered mechanisms can be utilized to implement online learning of RNN models, e.g., [223, 224]. Specifically, the event-triggered mechanism updates the RNN model if the following inequality is violated for any $x \in \mathscr{U}_\rho \backslash \mathscr{U}_{\rho_w}$:

$$W_c(x(t)) \leq W_c(x(t_k)) - \varepsilon_w (t - t_k), \; t \in [t_k, t_{k+1}) \tag{6.35}$$

where $\varepsilon_w > 0$. $\mathscr{U}_{\rho_w}$ with $\rho_w < \rho$ is characterized to be the largest level set of $W_c(x)$ within $\mathscr{U}_\rho$ such that if the current state is in $\mathscr{U}_{\rho_w}$, the value of $W_c(x)$ does not increase under the stabilizing controller $u = \Phi_{nn}(x)$ within one sampling period in the presence of bounded disturbances $|w(t)| \leq w_M$, i.e., $W_c(x(t)) < W_c(x(t_k))$,

$\forall t \in [t_k, t_{k+1})$. Additionally, it also ensures that the closed-loop state is bounded in $\mathcal{U}_\rho$ and ultimately enters $\mathcal{U}_{\rho_w}$ for any initial state in $\mathcal{U}_\rho$. From Eq. 6.35, it is shown that the event-triggered mechanism activates the online learning of RNN models if the decreasing rate of CLBF $W_c(x)$ is not satisfied within one sampling period. As a result, the RNN prediction accuracy is improved once the online learning is activated using the most recent process data and the closed-loop state can be driven into $\mathcal{U}_{\rho_w}$ at a faster rate.

In addition to the event-triggered mechanism, the following moving horizon error metric $E_{rnn}(t_k)$ is developed to indicate the RNN model prediction accuracy at $t = t_k$:

$$E_{rnn}(t_k) = \sum_{i=0}^{N_b} \frac{|x_p(t_{k-i}) - x(t_{k-i})|}{|x(t_{k-i})| + \delta} \tag{6.36}$$

where $x(t_{k-i})$ and $x_p(t_{k-i})$, $i = 0, \ldots, N_b$ are the past state measurements from the actual nonlinear system of Eq. 6.1, and the predictions of the past states using RNN models under the same control actions, respectively. We design a moving window with the length N_b (i.e., $N_b$is the number of sampling periods in the window) to account for the prediction errors before the current time step. Also, we introduce a small positive real number, δ, in the denominator of Eq. 6.36 to avoid the division by small numbers when the state $x(t_{k-i})$ is close to the origin. The RNN model of Eq. 6.34b is updated if the error $E_{rnn}(t_k)$ of Eq. 6.36 exceeds the threshold E_T:

$$E_{rnn}(t_k) > E_T \tag{6.37}$$

where the threshold E_T is determined in advance via extensive closed-loop simulations. It should be noted that when an online learning of RNN models is activated, all the data points since the last model update will be used as the training and validation data for the new RNN model. As the number of available data points has a great impact on the RNN model accuracy, N_b and E_T need to be carefully chosen to achieve a desired training performance. Specifically, the moving horizon length N_b is first determined via extensive closed-loop simulations to ensure that there are enough data points that can be utilized in the online update of RNN models, and meanwhile, will not cause data-storage burden. Subsequently, the threshold E_T is determined via simulations off-line to trigger an online learning of RNN models when the state error has reached an undesired level while accounting for common measurement noise, which is sufficiently small and should not trigger an update of RNN models in most times. Additionally, when the state approaches the unsafe region, the threshold E_T should be adjusted to update online learning more frequently such that the new RNN models are able to capture the most recent dynamics subject to disturbances in a timely manner, and therefore, provide a sufficiently accurate prediction for the CLBF-MPC of Eq. 6.34 to avoid the unsafe region. Lastly, after we update the RNN model at a certain sampling step $t = t_k$, all the errors before $t = t_k$ are reset to zero.

Remark 6.10 The event-triggered and the error-triggered mechanisms are developed to be activated when the conditions of Eq. 6.35 is violated, and when the prediction error of Eq. 6.37 exceeds its threshold, respectively, at a time instant $t = r_k$ that is within one sampling period, i.e., $r_k \in [t_k, t_{k+1})$. However, since the CLBF-MPC of Eq. 6.34 are implemented in a sample-and-hold fashion where the control actions remain the same for each sampling period Δ, i.e., $u = u(t_k)$, $\forall t \in [t_k, t_{k+1})$, the control actions will not be immediately updated after the update of RNN models within one sampling period. In other words, if the online update of RNN models is triggered at $t = r_k \in [t_k, t_{k+1})$, the control actions will still be calculated at the next sampling time, i.e., $t = t_{k+1}$, using the updated RNN models. The asynchronization between the online learning of RNN models and the calculation of control actions using the new RNN models ensures that the sample-and-hold implementations of the CLBF-MPC of Eq. 6.34 remain unchanged, and also leaves enough computation time for RNN models to be updated using the most recent process data.

Remark 6.11 The main objective of triggered model update is to improve the prediction accuracy of RNN models such that they are able to capture the most recent process dynamics subject to time-varying disturbances. Since the event-triggered mechanism updates RNN models only if the condition is violated, the event-triggered mechanism is demonstrated to update RNN models less frequently, and therefore, achieves better approximation performance due to more data available than the regular model update that is triggered every sampling period. Additionally, the frequency of online update depends on the threshold E_T. As a result, we determine the optimal value of E_T through extensive closed-loop simulations to achieve the desired closed-loop performance under disturbances.

6.3.4.1 Implementation Strategy For Online RNN Learning within CLBF-MPC

Based on the event-triggered and the error-triggered schemes, we implement the online learning of RNN models within the machine-learning-based CLBF-MPC of Eq. 6.34 through the following steps:

Step 1: We first derive an initial RNN model for the nominal system of Eq. 6.1 following the construction method in Sect. 6.2.2, and use it in the closed-loop simulation under the CLBF-MPC of Eq. 6.34.

Step 2: As shown in Fig. 6.5, the closed-loop state initiating from an initial condition $x_0 \in \mathcal{U}_\rho$ approaches the origin under the sample-and-hold implementation of CLBF-MPC. The process states are continuously monitored and collected at each sampling time. The online learning of RNN models is triggered once the decreasing rate of CLBF $W_c(x)$ of Eq. 6.35 is violated for any $x(t) \in \mathcal{U}_\rho \backslash \mathcal{U}_{\rho_w}$, or the moving horizon error detector of Eq. 6.36 exceeds its threshold E_T for any $x \in \mathcal{U}_{\rho_w} \backslash \mathcal{U}_{\rho_{\min}}$. At the next sampling time, the new RNN model will replace the old model in the CLBF-MPC of Eq. 6.34 to solve for the optimal control actions $u^*(t)$ for the next sampling period.

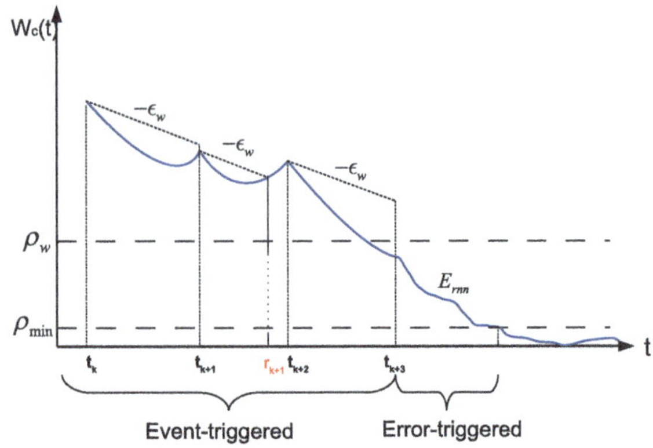

Fig. 6.5 Evolution of CLBF $W_c(x)$ (blue trajectory) under the CLBF-MPC of Eq. 6.34 with error-triggered condition of Eq. 6.37 and event-triggered condition of Eq. 6.35, where the threshold lines in Eq. 6.35 are represented by the dashed lines with the slope $-\varepsilon_W$

Step 3: When the closed-loop state finally enters $\mathscr{U}_{\rho_{\min}}$ (i.e., the small neighborhood around the origin). The closed-loop system is considered to be practically stable, and the error-triggering mechanism is taken off-line until the state leaves $\mathscr{U}_{\rho_{\min}}$ again due to time-varying disturbances.

Remark 6.12 It is noted that the online learning of RNN models is performed using the most recent process data only by loading the old RNN models with the previous RNN structure and weight matrices as initialization. Therefore, the new RNN models that are trained using new data points inherit some important features of the nominal process from the old RNN models and also capture the recent dynamics subject to time-varying disturbances from new data points. Additionally, instead of training a new RNN model from scratch, the training process based on the most recent data and the previous RNN model is more computationally tractable, and thus, can be readily incorporated in the real-time implementation of CLBF-MPC.

Remark 6.13 To ensure that there are enough data points for the online training of RNN models, an additional constraint for the number of collected data points can be employed with the event-triggered and the error-triggered mechanisms without affecting closed-loop stability or safety. Specifically, based on the definition of $\mathscr{U}_{\rho_W}$ in Eq. 6.35, it is guaranteed that the closed-loop state moves towards the origin every sampling period (maybe slowly) even if not updating the RNN models. Therefore, it allows us to collect enough data points from multiple sampling periods to achieve a better training performance while maintaining the state in the closed-loop stability region. Additionally, in the error-triggered mechanism of Eq. 6.37, the moving horizon window length N_b for the prediction error of Eq. 6.36 needs to be carefully chosen to obtain a sufficient number of data points that will be utilized in the online update of RNN models.

6.3.5 Computational Implementation Issues of RNN Models

In this section, we address computational implementation issues for the RNN models obtained following the training algorithm in Sect. 6.2. Specifically, the implementation of RNN models for long prediction horizon is first discussed. Then, numerical methods are employed to evaluate modeling error and approximate the CLBF-based constraints in CLBF-MPC, respectively.

6.3.5.1 Long Prediction Horizon

Although the ensemble of RNN models developed in Sect. 6.2 is to predict future states over $t \in [t_k, t_k + P_{nn}]$ given the states and inputs at $t = t_k$, where P_{NN} is an integer multiple of the sampling period Δ, it is noted that ensemble regression models can be applied to predict states for longer period of time (i.e., $t \in [t_k, t_k + N P_{nn}]$, $N > 1$) in practical applications, e.g., model predictive control. Specifically, the obtained RNN models will be utilized successively at every prediction step $t = t_k + i P_{nn}, i = 0, 1, \ldots, N-1$, to predict all the states within the entire prediction horizon $t \in [t_k, t_k + N P_{nn}]$, in which the prediction results (i.e., the output vector $x(t_k + i P_{nn})$) from the previous RNN models will be used as the initial states for the current prediction to predict states over $[t_k + i P_{nn}, t_k + (i+1) P_{nn}]$, $i = 0, 1, \ldots, N-1$. Additionally, since the means and the standard deviations for normalizing inputs and re-scaling outputs could be slightly different, intermediate re-scaling and normalizing steps should be performed between two successive ensemble prediction steps during the entire prediction horizon.

Before we apply the obtained RNN models within LMPC, the testing dataset that has not been used in the training process is utilized to test the prediction performance of RNNs. In this case, the normalizing and re-scaling functions before and after the ensemble of RNN models (Fig. 6.3) should be updated with the statistics of the testing dataset. Specifically, the normalizing and re-scaling functions during the training process are constructed based on the statistics of the training dataset only instead of the entire dataset due to the following reasons. First, the training and testing datasets may not be equally representative of the operating region considered, and thus, the training and testing datasets should be normalized separately. Second, data leakage that introduces information from outside, e.g., testing dataset, into RNN model should be prevented during the training process to avoid creating an overly optimistic but potentially invalid predictive model. Therefore, based on the normalizing and re-scaling functions designed for the testing dataset, the prediction performance of RNN models is evaluated by the mean absolute percentage error between the predicted states of the RNN models and the actual states derived from the nominal nonlinear system $\dot{x} = f(x) + g(x)u$.

Remark 6.14 While the use of a longer prediction horizon by recursively performing RNN predictions in CLBF-MPC can improve the closed-loop performance, a short horizon may be computationally advantageous for real-time application. Also, it should be noted that closed-loop stability and safety properties derived in the previous sections hold for any prediction horizon size. Therefore, the length of the prediction horizon should be determined via closed-loop simulations to balance optimality of the CLBF-MPC solutions and its computational complexity.

6.3.5.2 Approximation Via Numerical Methods

Since we mainly discuss the continuous RNN models in Sect. 6.2, while in practice, the datasets for training RNN models are mostly generated by a sample-data collection from industrial processes, lab experiments or numerical simulation, necessary approximations should be performed to incorporate the RNN model trained on sample data within LMPC. Specifically, numerical methods are utilized to compute modeling error, characterize the closed-loop stability and safety region $\mathcal{U}_\rho$ for the RNN model and calculate $\dot{W}_c(x(t_k), u(t_k))$ in the CLBF-MPC constraint of Eq. 6.34e, respectively.

(a) Approximation of modeling error

Since the RNN is trained to predict future states over $t \in [t_k, t_k + P_{nn})$, in which the RNN output is the state at $t_k + P_{nn}$ and the time interval between internal states is chosen as the integration time step h_c, the modeling error $\nu = \dot{x}(t_k) - \dot{\hat{x}}(t_k)$ at the state $x(t_k) = \hat{x}(t_k)$ is approximated using a forward finite difference method during the training process as follows:

$$
\begin{aligned}
|\nu| &= \left| \frac{x(t_k + h_c) - x(t_k)}{h_c} - \frac{\hat{x}(t_k + h_c) - \hat{x}(t_k)}{h_c} \right| \\
&= \left| \frac{x(t_k + h_c) - \hat{x}(t_k + h_c)}{h_c} \right|
\end{aligned}
\tag{6.38}
$$

where h_c is a sufficiently small time interval. $x(t_k + h_c)$ is obtained via explicit Euler method with an integration time step h_c, and $\hat{x}(t_k + h_c)$ is the first internal state of the RNN model. Then, the constraint $|\nu| \leq \gamma |x|$ is satisfied if the following equation holds:

$$
\left| \frac{x(t_k + h_c) - \hat{x}(t_k + h_c)}{x(t_k + h_c)} \right| \leq \gamma h_c. \tag{6.39}
$$

According to Eq. 6.39, the mean absolute percentage error between predicted states $\hat{x}$ and targeted states x in training data can be utilized as a metric to indicate the modeling error of RNNs.

(b) Characterization of the closed-loop operating region

The stabilizing controller $u = \Phi_{nn}(x) \in U$ is initially utilized to characterize the set ϕ_{uc} and the closed-loop stability and safety region $\mathcal{U}_\rho$ based on the RNN model written in the form of $\dot{\hat{x}} = \hat{f}(\hat{x}) + \hat{g}(\hat{x})u$. However, since it is difficult to derive the explicit forms of $\hat{f}(\cdot)$ and $\hat{g}(\cdot)$ for an RNN with a complex structure, numerical methods are utilized to approximate $\hat{f}(\cdot)$ and $\hat{g}(\cdot)$. For example, $\hat{f}(\cdot)$ can be approximated by the predicted $\dot{\hat{x}}$ with $u = 0$, where $\dot{\hat{x}}$ is obtained using the forward finite difference method as shown in the previous section. Then, $\hat{g}(\cdot)$ is approximated by $\hat{g}(\hat{x}) = (\dot{\hat{x}} - \hat{f}(\hat{x}))/u$ with a non-zero u. Since the minimum prediction step in RNNs is the sufficiently small integration time step h_c, the approximation results via numerical methods can be regarded as a good representation of the actual $\hat{f}(\cdot)$ and $\hat{g}(\cdot)$ of an RNN model. After $\hat{f}(\cdot)$ and $\hat{g}(\cdot)$ are obtained, a simulation with a full sweep over the entire state-space based on the stabilizing controller $u = \Phi_{nn}(x) \in U$ is performed to characterize the region ϕ_{uc} in which Eq. 6.17 is satisfied and $\dot{W}_c(x) = \frac{\partial W_c(x)}{\partial x} F_{nn}(x, u)$ is approximated via forward finite difference method. Subsequently, the closed-loop stability region $\mathcal{U}_\rho$ is characterized as a level set of $W_c(x)$ in ϕ_{uc}.

(c) Approximation of the CLBF-based constraints

Additionally, $\dot{W}_c(x(t_k), u(t_k))$ in the CLBF-based constraint of Eq. 6.34e is approximated via the same numerical method (i.e., forward finite difference method). It is noted that the approximation of $\dot{W}_c(x(t_k), u(t_k))$ does not affect closed-loop stability of the actual nonlinear system (i.e., $\dot{x} = F(x, u, w) := f(x) + g(x)u + h(x)w$) under the constraint of Eq. 6.34e since the same numerical method is used to approximate both $\dot{W}_c(x(t_k), u(t_k))$ and $\dot{W}_c(x(t_k), \Phi_{nn}(x))$. Specifically, it has been shown that the controller $u = \Phi_{nn}(x) \in U$ is able to stabilize the actual nonlinear system at the origin with safety guarantees for all x in $\mathcal{U}_\rho$ since Eq. 6.17 is satisfied in $\mathcal{U}_\rho \subset \phi_{uc}$ that is characterized via the numerical computation of $\dot{W}_c(x(t_k), \Phi_{nn}(x))$. Therefore, closed-loop stability and operational safety hold for the nonlinear system under CLBF-EMPC when the same numerical method is utilized.

6.3.6 *Application to a Chemical Process Example*

In this section, a chemical process example is utilized to illustrate the application of the proposed machine-learning-based CLBF-MPC scheme to nonlinear systems with a bounded/unbounded unsafe region. We consider a well-mixed, non-isothermal continuous stirred tank reactor (CSTR) where an irreversible second-order exothermic reaction takes place. The reaction converts the reactant A to the product B via the chemical reaction $A \rightarrow B$. A heating jacket that supplies or removes heat from the reactor is used. The CSTR dynamic model derived from material and energy balances is given below

$$\frac{\mathrm{d}C_A}{\mathrm{d}t} = \frac{F}{V}(C_{A0} - C_A) - k_0 e^{\frac{-E}{RT}} C_A^2 \tag{6.40a}$$

$$\frac{\mathrm{d}T}{\mathrm{d}t} = \frac{F}{V}(T_0 - T) + \frac{-\Delta H}{\rho_L C_p} k_0 e^{\frac{-E}{RT}} C_A^2 + \frac{Q}{\rho_L C_p V} \tag{6.40b}$$

Table 6.1 Parameter values of the CSTR system

$T_0 = 300$ K	$F = 5$ m^3/h
$V = 1$ m^3	$E = 5 \times 10^4$ kJ/kmol
$k_0 = 8.46 \times 10^6$ m^3/kmol h	$\Delta H = -1.15 \times 10^4$ kJ/kmol
$C_p = 0.231$ kJ/kg K	$R = 8.314$ kJ/kmol K
$\rho_L = 1000$ kg/m^3	$C_{A0_s} = 4$ kmol/m^3
$Q_s = 0.0$ kJ/h	

where C_A is the concentration of reactant A in the reactor, V is the volume of the reacting liquid in the reactor, T is the temperature of the reactor, and Q denotes the heat input rate. The concentration of reactant A in the feed is C_{A0}. The feed temperature and volumetric flow rate are T_0 and F, respectively. The reacting liquid has a constant density of ρ_L and a heat capacity of C_p. ΔH, k_0, E, and R represent the enthalpy of reaction, pre-exponential constant, activation energy, and ideal gas constant, respectively. Process parameter values are listed in Table 6.1.

The CSTR is initially operated at the unstable steady-state $(C_{As}, \ T_s) = (1.95 \text{ kmol/m}^3, \ 402 \text{ K})$, and $(C_{A0_s} \ Q_s) = (4 \text{ kmol/m}^3, \ 0 \text{ kJ/h})$. The manipulated inputs are the inlet concentration of species A and the heat input rate, which are represented by the deviation variables $\Delta C_{A0} = C_{A0} - C_{A0_s}$, $\Delta Q = Q - Q_s$, respectively. The manipulated inputs are bounded as follows: $|\Delta C_{A0}| \leq 3.5 \text{ kmol/m}^3$ and $|\Delta Q| \leq 5 \times 10^5$ kJ/h. The states and the inputs of the closed-loop system are $x^T = [C_A - C_{As} \ T - T_s]$ and $u^T = [\Delta C_{A0} \ \Delta Q]$, respectively, such that the equilibrium point of the system is at the origin of the state-space, (i.e., $(x_s^*, u_s^*) = (0, 0)$). The explicit Euler method with an integration time step of $h_c = 2 \times 10^{-5}$ h is applied to numerically simulate the dynamic model of Eq. 6.40. The nonlinear optimization problem of the CLBF-MPC of Eq. 6.34 is solved using the python module of the IPOPT software package [201], named PyIpopt with the sampling period $\Delta = 2 \times 10^{-3}$ h.

6.3.6.1 Development of RNN Models

We run extensive open-loop simulations with various inputs $u \in U$ and initial states in state-space for finite sampling steps to generate the dataset for training an ensemble of RNN models. The sampled data points including inputs u and states x are collected every integration time step h_c. We construct an RNN model with two hidden layers consisting of 96 and 64 recurrent units, respectively. The RNN inputs are the control actions $u(t_k)$ and the state measurement $x(t_k)$ at $t = t_k$, $k = 0, 1, \ldots$, and the RNN outputs are the predicted state trajectory over the next sampling period (i.e., $t \in [t_k, t_{k+1}]$), where all the data points within one sampling period (i.e., the data collected at every integration time step) are used as the internal states for RNN models. We use the sigmoid function as the activation function in RNN hidden layers, and introduce

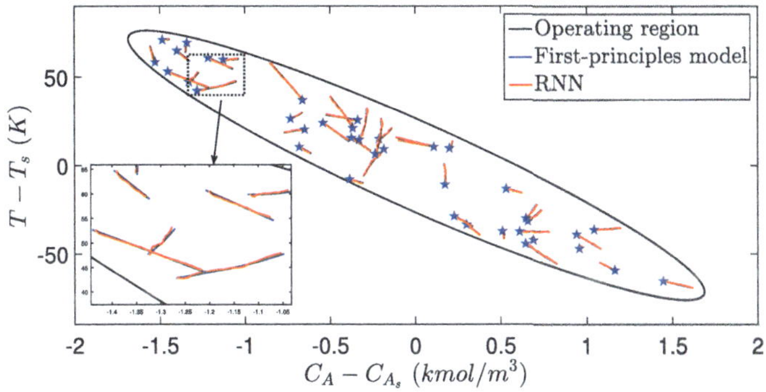

Fig. 6.6 The state-space profiles for the open-loop simulation using the first-principles model of Eq. 6.40 and the RNN model, respectively, for various sets of inputs and initial conditions (marked as blue stars) x_0 in the operating region

early stopping into the training process to avoid over-fitting. Additionally, a 10-fold cross validation is utilized to train an ensemble of 10 RNN models for the CLBF-MPC of Eq. 6.34. However, note that not all the RNN models are needed in real-time implementation of CLBF-MPC. In general, we determine the optimal number of RNN models in MPC based on the size of datasets and the complexity of process dynamics. In this example, we determine the optimal number through closed-loop simulations. Specifically, after we obtain k RNN models using a k-fold cross-validation, we start with a single RNN model and keep increasing the number of models used in MPC until no further improvement of closed-loop performance is noticed with the increase of RNN models being used. We first carry out open-loop simulation using the RNN model and the first-principles model of the CSTR system of Eq. 6.40, respectively. It should be noted that the machine learning approach is used when only data are available. The first-principles model in this study substitutes for the role of the experimental/industrial process. In other words, the simulation using first-principles model only serves as a benchmark to determine the best performance that any data-driven modeling method can achieve. In Fig. 6.6, it is demonstrated that starting from the same initial condition $x_0 \in \Omega_{\hat{\rho}}$ with the same input sequences, the state trajectories for a fixed finite interval of time under the RNN model are close to those under the first-principles model of the nonlinear CSTR of Eq. 6.40. This implies that the well-trained RNN model can be regarded as a good representation for the CSTR first-principles model of Eq. 6.40 within the operating region.

6.3.6.2 Closed-Loop Simulation Results

The control objective of CLBF-MPC is to operate the CSTR at the unstable equilibrium point $(C_{As},\ T_s)$ and maintain the state in the safe operating region for all times. The inlet concentration ΔC_{A0} and the heat input rate ΔQ are the two manipulated inputs. Both the unbounded and bounded unsafe regions in state-space will be studied

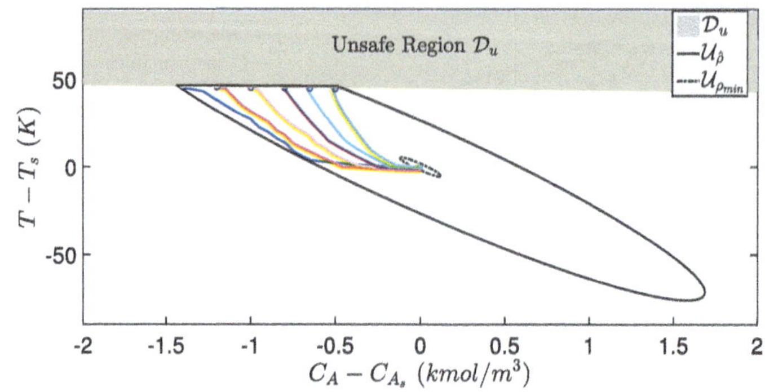

Fig. 6.7 State trajectories for the closed-loop CSTR of Eq. 6.40 under the CLBF-MPC using an ensemble of RNN models. The gray area on the top represents the set of unbounded unsafe states $\mathscr{D}_u$, and the circles represent the initial conditions

in this section. We first carry out the closed-loop simulation for the system under the proposed CLBF-MPC control scheme when an unbounded unsafe region $\mathscr{D}_u$ exists in the state-space. The unbounded unsafe region is a set of states with high temperature and concentration for the CSTR of Eq. 6.40: $\mathscr{D}_u := \{x \in \mathbf{R}^2 \mid F(x) = x_1 + x_2 > 47\}$. It is noted that with the form of $F(x) = x_1 + x_2$, the temperature in the reactor is considered to be the dominant factor in characterizing the unsafe region $\mathscr{D}_u$; however, note that we also account for the reactant concentration because of its impact on the reaction rate $r = k_0 e^{-E/RT} C_A^2$. Following the construction method of a CLBF in Sect. 4.3.2, we first design a control Lyapunov function with the standard quadratic form $V(x) = x^T P x$, where P is a positive definite matrix as follows:

$$P = \begin{bmatrix} 1060 & 22 \\ 22 & 0.52 \end{bmatrix}. \tag{6.41}$$

Then, we characterize a set H that contains $\mathscr{D}_u$: $H := \{x \in \mathbf{R}^2 \mid F(x) > 45\}$, and design the control barrier function $B(x)$ as follows:

$$B(x) = \begin{cases} e^{F(x)-47} - 2 \times e^{-2}, & \text{if } x \in H \\ -e^{-2}, & \text{if } x \notin H \end{cases}. \tag{6.42}$$

The control Lyapunov-barrier function $W_c(x) = V(x) + \mu B(x) + \nu$ is constructed with the following parameters: $\hat{\rho} = 0$, $c_1 = 0.1$, $c_2 = 1061$, $c_3 = 5808$, $c_4 = 2259$, $\nu = \hat{\rho} - c_1 c_4 = -225.9$, and $\mu = 4.6 \times 10^7$. It is shown in Fig. 6.7 that under the CLBF-MPC of Eq. 6.34, all the trajectories starting from initial states in $\mathscr{U}_{\hat{\rho}}$ (a subset of the safe operating region $\mathscr{U}_\rho$ in state-space) avoid the unbounded unsafe region $\mathscr{D}_u$ on the top and converge to $\mathscr{U}_{\rho_{\min}}$.

The second example is used to demonstrate that the state of the closed-loop system of Eq. 6.40 can avoid a bounded unsafe region $\mathscr{D}_b$ in state-space and converge to a small neighborhood around the origin under the CLBF-MPC of Eq. 6.34. To

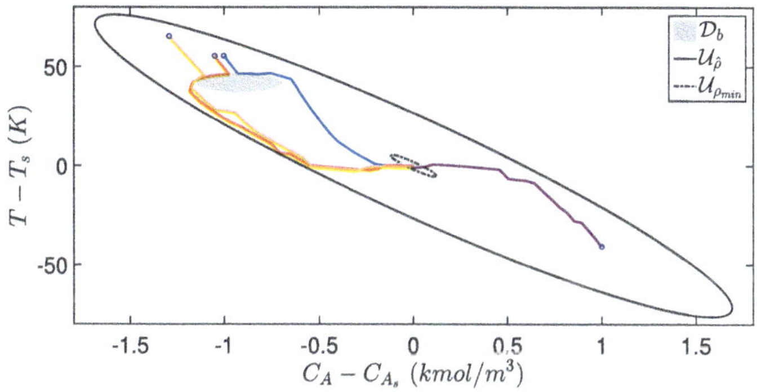

Fig. 6.8 State trajectories for the closed-loop system of Eq. 6.40 under the CLBF-MPC using an ensemble of RNN models. The gray area embedded within $\mathcal{U}_{\hat{\rho}}$ represents the set of bounded unsafe states, and the circles represent the initial conditions

demonstrate that the state is able to pass around the unsafe region along the trajectory towards the origin, we design a bounded unsafe region $\mathcal{D}_b$ embedded within the safe operating region as shown in the above example. Specifically, the unsafe region is defined as an ellipse: $\mathcal{D}_b := \{x \in \mathbf{R}^2 \mid F(x) = \frac{(x_1+0.92)^2}{1} + \frac{(x_2-42)^2}{500} < 0.06\}$. H is defined as $H := \{x \in \mathbf{R}^2 \mid F(x) < 0.07\}$. The control barrier function $B(x)$ is defined as follows:

$$B(x) = \begin{cases} e^{\frac{F(x)}{F(x)-0.07}} - e^{-6}, & \text{if } x \in H \\ -e^{-6}, & \text{if } x \notin H \end{cases}. \tag{6.43}$$

Using the same control Lyapunov function $V(x)$ as in the first example, the control Lyapunov-barrier function $W_c(x) = V(x) + \mu B(x) + \nu$ is constructed with the following parameters: $\rho_c = 0$, $c_1 = 0.1$, $c_2 = 1061$, $c_3 = \max_{x\in\partial\mathcal{H}} |x|^2 = 2295$, $c_4 = \min_{x\in\partial\mathcal{D}} |x|^2 = 1370$, $\nu = \rho_c - c_1c_4 = -160$. Hence, μ is chosen to be 1×10^9 to satisfy the construction rules in Sect. 4.3.2 and $\mathcal{U}_{\hat{\rho}}$ with $\hat{\rho} = -2.47 \times 10^6$ is the stability and safety region in the simulation. Additionally, since the unsafe region is a bounded set in state-space, we calculate all the stationary points of $W_c(x)$ in state-space. It is shown that by letting $\frac{\partial W_c(x)}{\partial x} = 0$, $x_e = (-1.004,\ 47.48)$ is a stationary point (other than the origin). Furthermore, we verify that it is a saddle point in state-space via partial derivative test (i.e., if the determinant of the Hessian matrix of $W_c(x)$ at x_e is negative, then x_e is a saddle point).

In Fig. 6.8, it is demonstrated that all the closed-loop trajectories initiating from initial states x_0 in $\mathcal{U}_{\hat{\rho}}$ (marked by circles) avoid the bounded unsafe region $\mathcal{D}_b$ that is embedded within $\mathcal{U}_{\hat{\rho}}$, and ultimately converges to $\mathcal{U}_{\rho_{\min}}$ under CLBF-MPC.

6.3.6.3 Comparison with A Linear State-Space Model

Additionally, a linear state-space model is developed and compared with the ensemble of RNN models in the context of MPC to demonstrate the merits of the machine-learning-based CLBF-MPC in terms of guaranteed process operational safety and desired prediction accuracy. Specifically, we develop the following linear state-space model using the same dataset for the ensemble of RNN models to approximate the process dynamics in the operating region under consideration:

$$\dot{x} = A_s x + B_s u \tag{6.44}$$

where u and x are the manipulated input vector and the state vector. The space system identification algorithm is utilized following the system identification method in [93] to obtain the coefficient matrices A_s and B_s as follows:

$$A_s = 100 \times \begin{bmatrix} -0.154 & -0.003 \\ 5.19 & 0.138 \end{bmatrix}, \quad B_s = \begin{bmatrix} 4.03 & 0 \\ 1.23 & 0.004 \end{bmatrix}. \tag{6.45}$$

The eigenvalues of matrix A_s are calculated to be $\lambda_1 = -5$ and $\lambda_2 = 3.14$, which is consistent with the fact that the steady-state $(C_{As}, T_s) = (1.95 \text{ kmol/m}^3, 402 \text{ K})$ is an unstable equilibrium point of CSTR. Figure 6.9 shows three state trajectories under the MPC using a linear model. It is seen that all the trajectories avoid the unsafe region in the path towards the steady-state However, it is demonstrated in Fig. 6.10 that the state trajectories initiating from some other initial conditions (with dashed line) enter the unsafe region in the simulation. Since the optimization problem of MPC is solved with a feasible solution all the time, the undesired closed-loop performance is resulting from a considerable model mismatch of the linear state-space model. In general, it is sufficient to use a simple linear state-space model in a model predictive controller to stabilize the nonlinear system in a neighborhood around the origin, provided that the modeling error between the actual nonlinear system and the linear model is small in this neighborhood. However, the linear state-space model performs poorly in this example because in addition to closed-loop stability, process operational safety is required for which, a sufficiently small modeling error is necessary for the state to avoid the unsafe region. This motivates us to improve the prediction model in MPC by using an ensemble of RNN models with a desire model accuracy for approximating the nonlinear process dynamics in the operating region.

From the above two case studies, it is demonstrated that the ensemble of RNN models that are developed from extensive open-loop simulations provide a sufficiently accurate state prediction for MPC. Additionally, for both the cases of bounded and unbounded unsafe regions in state-space, we show that the state of the closed-loop system of Eq. 6.40 is bounded within the safe operating region at all times, and is ultimately bounded within a small neighborhood around the origin ($\mathcal{U}_{\rho_{\min}}$) for any initial condition in the safe operating region. Thus, process operational safety

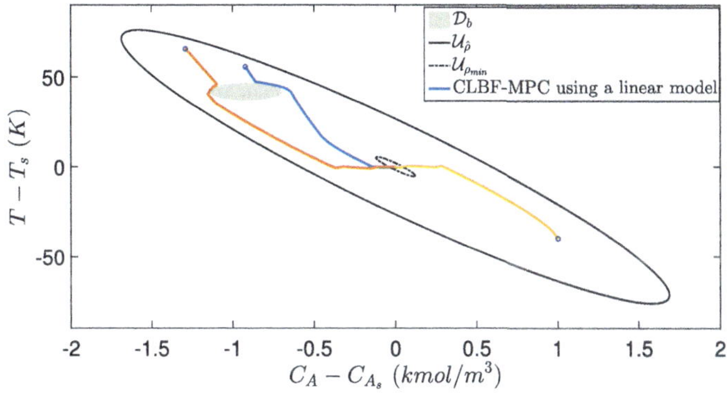

Fig. 6.9 State trajectories for the closed-loop CSTR system the CLBF-MPC using a linear state-space model. The gray ellipse in state-space represents the set of bounded unsafe states $\mathscr{D}_b$, and the circles represent the initial conditions

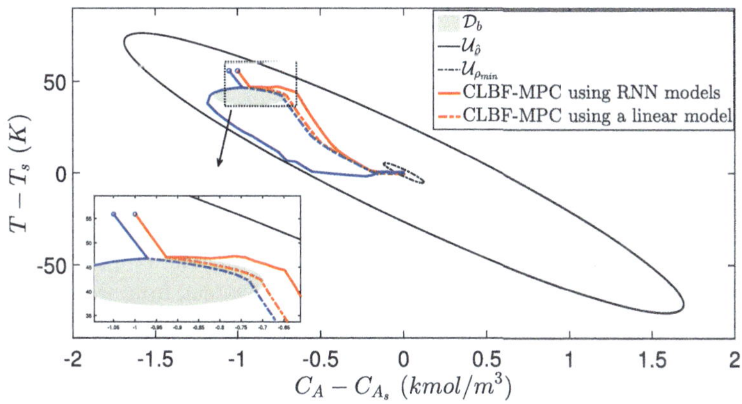

Fig. 6.10 Closed-loop state trajectories under the CLBF-MPC using an ensemble of RNN models (solid trajectory) and a linear state-space model (dashed trajectory), respectively. The gray ellipse in state-space represents the set of bounded unsafe states $\mathscr{D}_b$, and the circles represent the initial conditions

and closed-loop stability are guaranteed under the CLBF-MPC of Eq. 6.34 using an ensemble of RNN models.

6.3.6.4 Real-Time CLBF-MPC with Online Learning of RNN Models

Under the CLBF-MPC of Eq. 6.34, we consider the model variations due to the following disturbances: (1) the feed flow rate F is changing from 5 m^3/h to 7 m^3/h at $t = 0$ h, and (2) the actual value of the pre-exponential constant k_0 used in the process model is reduced by half to represent a change in the reaction rate at the simulation time $t = 0$ hr. The closed-loop simulation results for the CSTR of Eq. 6.40 under

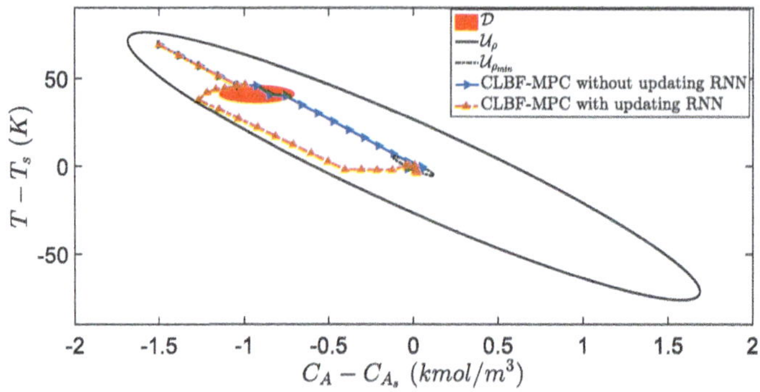

Fig. 6.11 The state-space profiles for the closed-loop CSTR subject to time-varying disturbances under the CLBF-MPC of Eq. 6.34 with (red trajectory) and without online RNN update (blue trajectory), respectively, for an initial condition (−1.5,70)

the CLBF-MPC with and without online learning of RNN models, respectively, are shown in Figs. 6.11, 6.12 and 6.13. Specifically, in Fig. 6.11, it is demonstrated that in the presence of disturbances, the closed-loop state trajectory under the CLBF-MPC using online update of RNN models is able to avoid the unsafe region and converge to a small neighborhood around the origin, while the one under the CLBF-MPC without online RNN update crosses the red unsafe region $\mathscr{D}$ due to a considerable model mismatch between the initial RNN model for the nominal process of Eq. 6.40 and the actual process subject to disturbances. Figure 6.12 shows the input profiles under the CLBF-MPC with and without online RNN update, from which recursive feasibility and satisfaction of input constraints are demonstrated for both optimization problems. Additionally, it is observed in Fig. 6.12 that since RNN models are updated in a timely manner under the CLBF-MPC with online learning, the oscillation of u_1 is reduced near the end of the operation period compared to that without online update.

In the closed-loop simulation, it is demonstrated that the event-triggered mechanism of Eq. 6.35 is not activated as the closed-loop state moves towards the origin quickly. Therefore, the value of the accumulated prediction errors $E_{rnn}(t)$ of Eq. 6.36 is plotted in Fig. 6.13 for CLBF-MPCs with and without online RNN update, respectively, to show the real-time prediction accuracy of the RNN models. Fig. 6.13 demonstrates that without online learning, the error (blue lines) exceeds the threshold (left y-axis) quickly and increases to an undesired level during the operation, which implies the failure of the initial RNN model in capturing the actual CSTR dynamics in the presence of disturbances. However, under the CLBF-MPC with online RNN learning, it is demonstrated that the RNN model update is triggered six times during the entire operation period (i.e., from t = 0 h to t = 0.06 h) to maintain the error (red lines) below its threshold (right y-axis) for most of the time. Therefore, by using online learning, the RNN models in CLBF-MPC always capture the latest process dynamics subject to disturbances, and lead to a desired closed-loop performance for the CSTR of Eq. 6.40 in terms of simultaneous closed-loop stability and operational safety.

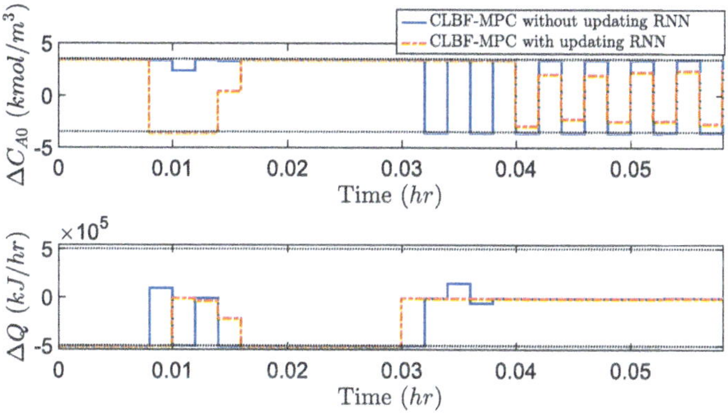

Fig. 6.12 Manipulated input profiles ($u_1 = \Delta C_{A0}, u_2 = \Delta Q$) for the closed-loop CSTR subject to time-varying disturbances under the CLBF-MPC of Eq. 6.34 with (red profile) and without online RNN update (blue profile), respectively, for an initial condition (−1.5,70)

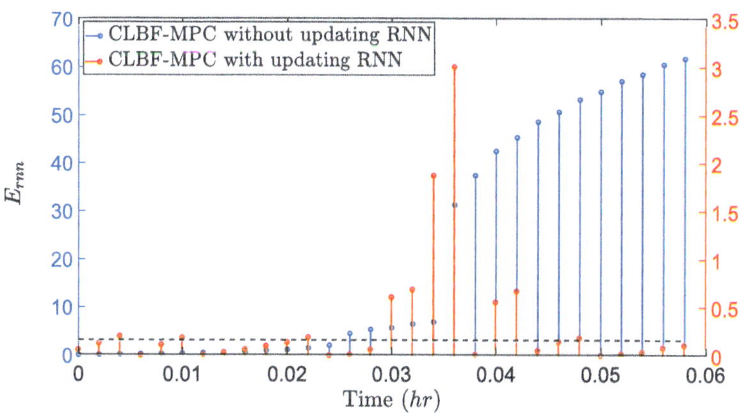

Fig. 6.13 Value of $E_{rnn}(t)$ at each sampling time for the closed-loop CSTR subject to time-varying disturbances under the CLBF-MPC of Eq. 6.34 with (red, right y-axis) and without online RNN update (blue, left y-axis), respectively, where the threshold E_T is set to 0.15 (dashed horizontal line corresponding to the right y-axis)

6.4 CLBF-EMPC Using RNN Models

To achieve higher economic profitability than the steady-state operation of the nonlinear system of Eq. 6.1 (i.e., the system is operated at steady-state for all times), economic model predictive control scheme (EMPC) that is formulated with an economic objective function to operate the system in a time-varying fashion is utilized in this section. See, also, Sects. 3.3 and 4.5 for designs of EMPC accounting for operational safety. Specifically, based on the RNN-based CLBF-MPC of Eq. 6.34, an RNN-based economic model predictive controller with CLBF-based constraints (i.e., CLBF-EMPC) is developed in this section. Similarly, the ensemble learning of multiple RNN models is used to improve the overall prediction performance. k-fold cross-validation is used to train k distinct RNN models for the same process, (i.e., the nonlinear system of Eq. 6.1), and the final prediction results are obtained by taking average of k RNN predictions. Based on the ensemble of RNN models, the CLBF-EMPC scheme using RNN models is represented by the following optimization problem [218]:

$$\max_{u \in S(\Delta)} \int_{t_k}^{t_{k+N}} l_e(\tilde{x}(t), u(t))\mathrm{d}t \tag{6.46a}$$

$$\text{s.t.} \quad \dot{\tilde{x}}(t) = \frac{1}{N_e} \sum_{j=1}^{N_e} F_{nn}^j(\tilde{x}(t), u(t)) \tag{6.46b}$$

$$\tilde{x}(t_k) = x(t_k) \tag{6.46c}$$

$$u(t) \in U, \ \forall\, t \in [t_k, t_{k+N}) \tag{6.46d}$$

$$W_c(\tilde{x}(t)) \le \rho_e, \ \forall\, t \in [t_k, t_{k+N}), \ \text{ if } W_c(x(t_k)) \le \rho_e \tag{6.46e}$$

$$\dot{W}_c(x(t_k), u(t_k)) \le \dot{W}_c(x(t_k), \Phi_{nn}(t_k)), \ \text{ if } W_c(x(t_k)) > \rho_e \tag{6.46f}$$

where the notation follows that in Eq. 6.34 and the CLBF-EMPC is implemented in a sample-and-hold fashion. Unlike the CLBF-MPC objective function $l_t(x, u)$ of Eq. 6.34a that has its minimum value at the steady-state, the objective function $l_e(x, u)$ of Eq. 6.46a represents the process economic performance and will be maximized over the prediction horizon. To ensure boundedness of the state in the safe operating region $\mathcal{U}_\rho$, two CLBF-based constraints are incorporated in the design of CLBF-EMPC. Specifically, when the state $x(t_k)$ is in $\mathcal{U}_{\rho_e}$, where $\rho_e < \rho$, we activate the constraint of Eq. 6.46e. However, if the state leaves $\mathcal{U}_{\rho_e}$ due to model mismatch or disturbances (which will be discussed in the following section), the constraint of Eq. 6.46f is implemented to drive the state towards the origin, and thus into $\mathcal{U}_{\rho_e}$ within finite time. The state measurements of the closed-loop system of Eq. 6.1 are assumed to be available at each sampling time. An optimal input sequence $u^*(t)$, $\forall t \in [t_k, t_{k+N})$ is calculated by the CLBF-EMPC optimization problem of Eq. 6.46 at each sampling time, from which only the first control action of $u^*(t)$ will be applied to the nonlinear system for the next sampling period.

6.4.1 Stability and Safety Under CLBF-EMPC

In this section, process operational safety and closed-loop stability for the nonlinear system of Eq. 6.1 will be proven for the CLBF-EMPC of Eq. 6.46. Note that unlike the CLBF-MPC that requires the state to be operated at the steady-state, the system is considered stable and safe under EMPC, if the state can be bounded in a safe stability region for all times for any initial condition inside of this region. In this way, we will show that economic performance is much improved under time-varying operation of EMPC than the steady-state operation. We first develop the following proposition to demonstrate that the CLBF-based controller $u = \Phi_{nn}(x) \in U$ that maintains the state of the RNN model of Eq. 6.4 in the safe operating region $\mathcal{U}_\rho$ also guarantees the boundedness of the state of the nonlinear system of Eq. 6.1 within $\mathcal{U}_\rho$ accounting for sample-and-hold implementation of control actions, bounded modeling error (i.e., $|\nu| = |F(x, u, 0) - F_{nn}(x, u)| \leq \gamma|x| \leq \nu_m$) and bounded disturbances (i.e., $|w(t)| \leq w_m$).

Proposition 6.6 *Consider the system of Eq. 6.1 under the sample-and-hold implementation of the controller $u = \Phi_{nn}(x) \in U$ that meets the conditions of Eq. 6.17. If there exists a positive real number $\gamma < \hat{c}_3/\hat{c}_4$ such that for all $x \in \mathcal{U}_\rho$ and $u \in U$, the modeling error between the RNN model of Eq. 6.4 and the nonlinear system of Eq. 6.1 is constrained by $|\nu| = |F(x, u, 0) - F_{nn}(x, u)| \leq \gamma|x|$, and there exist $\varepsilon_w > 0$, $\Delta > 0$ and $\rho > \rho_e$ that satisfy*

$$-\frac{\tilde{c}_3}{\hat{c}_2}(\rho_e - \rho_0) + L_x^{'} M\Delta + L_w^{'} w_m \leq -\varepsilon_w \tag{6.47a}$$

$$\rho_e \leq \rho - f_e(f_w(\Delta)) \tag{6.47b}$$

$$\mathbb{X}_e \subset \mathcal{U}_{\rho_e} \tag{6.47c}$$

where $f_w(t)$ and $f_e(t)$ are given by Eqs. 6.19a and 6.27, respectively, then for any $x(t_k) \in \mathcal{U}_\rho$, the state of the nonlinear system of Eq. 6.1 is guaranteed to be bounded in $\mathcal{U}_\rho$ for all times.

Proof We first prove that $\dot{W}_c(x)$ based on the state of the nonlinear system of Eq. 6.1 can be rendered negative under continuous implementation of $u = \Phi_{nn}(x) \in U$ for any $x \in \mathcal{U}_\rho \backslash \mathcal{U}_{\rho_e}$. The time-derivative of $W_c(x)$, $\forall x \in \mathcal{U}_\rho \backslash \mathcal{U}_{\rho_e}$ is derived as follows using Eqs. 6.17b and 6.17c:

$$\begin{aligned}
\dot{W}_c =& \frac{\partial W_c(x)}{\partial x} F(x, \Phi_{nn}(x), 0) \\
=& \frac{\partial W_c(x)}{\partial x} (F_{nn}(x, \Phi_{nn}(x)) + F(x, \Phi_{nn}(x), 0) - F_{nn}(x, \Phi_{nn}(x))) \\
\leq& -\hat{c}_3|x|^2 + \hat{c}_4|x|(F(x, \Phi_{nn}(x), 0) - F_{nn}(x, \Phi_{nn}(x))) \\
\leq& -\hat{c}_3|x|^2 + \hat{c}_4\gamma|x|^2.
\end{aligned} \tag{6.48}$$

Therefore, if γ is constrained by $\gamma < \hat{c}_3/\hat{c}_4$, it holds that $\dot{W}_c \leq -\tilde{c}_3|x|^2 < 0$, $\forall x \in \mathcal{U}_\rho \backslash \mathcal{U}_{\rho_e}$ by letting $\tilde{c}_3 = -\hat{c}_3 + \hat{c}_4\gamma$. Next, we consider the impacts of bounded disturbances and of the sample-and-hold implementation of control actions (i.e., $u(t) = u(t_k)$, $\forall t \in [t_k, t_{k+1})$, where $t_{k+1} := t_k + \Delta$ and Δ is the sampling period) on closed-loop stability of the nonlinear system of Eq. 6.1. Assuming $x(t_k) = \hat{x}(t_k) \in$

$\mathcal{U}_\rho \backslash \mathcal{U}_{\rho_s}$, the time-derivative of $W_c(x)$ in Eq. 6.48 for the nonlinear system of Eq. 6.1 subject to bounded disturbances (i.e., $|w| \leq w_m$) can be derived as follows:

$$\begin{aligned}\dot{W}_c(x(t)) =& \frac{\partial W_c(x(t))}{\partial x} F(x(t), \Phi_{nn}(x(t_k)), w) \\ =& \frac{\partial W_c(x(t_k))}{\partial x} F(x(t_k), \Phi_{nn}(x(t_k)), 0) + \frac{\partial W_c(x(t))}{\partial x} F(x(t), \Phi_{nn}(x(t_k)), w) \\ &- \frac{\partial W_c(x(t_k))}{\partial x} F(x(t_k), \Phi_{nn}(x(t_k)), 0).\end{aligned} \tag{6.49}$$

Using Eq. 6.17b, Eq. 6.48 and the Lipschitz condition in Eq. 6.18, $\dot{W}_c(x(t))$ is bounded by the the following inequality for all $t \in [t_k, t_{k+1})$ and $x(t_k) \in \mathcal{U}_\rho \backslash \mathcal{U}_{\rho_e}$:

$$\begin{aligned}\dot{W}_c(x(t)) \leq & -\frac{\tilde{c}_3}{\hat{c}_2}(\rho_e - \rho_0) + \frac{\partial W_c(x(t))}{\partial x} F(x(t), \Phi_{nn}(x(t_k)), w) \\ & - \frac{\partial W_c(x(t_k))}{\partial x} F(x(t_k), \Phi_{nn}(x(t_k)), 0) \\ \leq & -\frac{\tilde{c}_3}{\hat{c}_2}(\rho_e - \rho_0) + L_x^{'}|x(t) - x(t_k)| + L_w^{'}|w| \\ \leq & -\frac{\tilde{c}_3}{\hat{c}_2}(\rho_e - \rho_0) + L_x^{'} M \Delta + L_w^{'} w_m.\end{aligned} \tag{6.50}$$

From Eq. 6.50, it is obtained that $\dot{W}_c(x(t)) \leq -\varepsilon_w$ holds for all $x(t_k) \in \mathcal{U}_\rho \backslash \mathcal{U}_{\rho_e}$ and $t \in [t_k, t_{k+1})$ if Eq. 6.47a is satisfied.

So far we have demonstrated that for any state $x(t_k) \in \mathcal{U}_\rho \backslash \mathcal{U}_{\rho_e}$, the state does not leave $\mathcal{U}_\rho$ under the the sample-and-hold implementation of $u = \Phi_{nn}(x) \in U$. It remains to show that for $x(t_k) \in \mathcal{U}_{\rho_e}$, the state of the nonlinear system of Eq. 6.1 will not leave $\mathcal{U}_\rho$ within one sampling period if the state predicted by the RNN system of Eq. 6.4 is bounded in $\mathcal{U}_{\rho_e}$. Specifically, for any $x(t_k) = \hat{x}(t_k) \in \mathcal{U}_{\rho_e}$, the following inequality is derived based on Eqs. 6.19 and 6.27 for $t \in [t_k, t_{k+1})$:

$$\begin{aligned}W_c(x(t)) &\leq W_c(\hat{x}(t)) + f_e(|x(t) - \hat{x}(t)|) \\ &\leq W_c(\hat{x}(t)) + f_e(f_w(t - t_k)) \\ &\leq \rho_e + f_e(f_w(\Delta)).\end{aligned} \tag{6.51}$$

Therefore, if $\mathcal{U}_{\rho_e}$ satisfies Eq. 6.47b, it is straightforward to show that $W_c(x(t)) \leq \rho$ holds, which implies the boundedness of the state within $\mathcal{U}_\rho$ during one sampling period. This completes the proof that for any initial condition x_0 in the safe operating region $\mathcal{U}_\rho$, the closed-loop state of the disturbed nonlinear system of Eq. 6.1 (i.e., $|w(t)| \leq w_m$) is guaranteed to be bounded in $\mathcal{U}_\rho$ under the sample-and-hold implementation of $u = \Phi_{nn}(x) \in U$.

Remark 6.15 The convergence of state to the stationary points (for $x \neq 0$) in the presence of a bounded unsafe region $\mathcal{D}_b$ is no longer an issue in EMPC because the set $\mathcal{U}_{\rho_e}$ is designed to include the set of stationary points inside (i.e., Eq. 6.47c).

Specifically, since the EMPC does not require the nonlinear system of Eq. 6.1 to be operated at the origin, the state will not move towards the origin (or any stationary points) within $\mathscr{U}_{\rho_e}$ under the constraint of Eq. 6.46e. As a result, the state will not get trapped in a stationary point unless it is exactly the state where the objective function of CLBF-EMPC of Eq. 6.46 attains its maximum value. Therefore, the sample-and-hold implementation of $u = \Phi_{nn}(x) \in U$ guarantees the boundedness of the state in $\mathscr{U}_\rho$ for the nonlinear system of Eq. 6.1 with both unbounded and bounded unsafe regions. However, if the system is required to be operated at the steady-state under a tracking MPC, e.g., CLBF-MPC, it has been demonstrated in Theorem 6.3 that we need to design a set of discontinuous control actions for handling the stationary points such that the state can escape from them and ultimately converge to the origin.

Based on Propositions 6.2 and 6.6, we utilize the following theorem to demonstrate closed-loop stability and process operational safety guarantees for the nonlinear system of Eq. 6.1 under the CLBF-EMPC of Eq. 6.46.

Theorem 6.4 *Consider the system of Eq. 6.1 with a CLBF W_c that satisfies Eq. 6.16. If there exist $\rho > \rho_e$ and $\gamma < \hat{c}_3/\hat{c}_4$ that satisfy the conditions in Propositions 6.2 and 6.6, then given any initial state $x_0 \in \mathscr{U}_\rho$, recursive feasibility of the CLBF-EMPC optimization problem of Eq. 6.46 and boundedness of the state in the safe stability region $\mathscr{U}_\rho$ are guaranteed for all times.*

Proof We first prove the existence of feasible solutions for the CLBF-EMPC optimization problem of Eq. 6.46 for all states $x(t) \in \mathscr{U}_\rho$. We show that the input trajectories $u(t) = \Phi_{nn}(x(t_{k+i})) \in U, \forall t \in [t_{k+i}, t_{k+i+1})$ with $i = 0, \ldots, N-1$ are the feasible solutions that satisfy the constraints of the CLBF-EMPC optimization problem of Eq. 6.46. We will mainly discuss the constraints of Eqs. 6.46e–6.46f as the satisfaction of the input constraint $u \in U$ of Eq. 6.46d is readily shown for the controller $u = \Phi_{nn}(x) \in U$. Specifically, when $x(t_k) \in \mathscr{U}_{\rho_e}$, the sample-and-hold implementation of $u = \Phi_{nn}(x) \in U$ satisfies the constraint of Eq. 6.46e because the state of the RNN system of Eq. 6.4 will be steered towards the origin (or the stationary points in the case of a bounded unsafe region). In any case, the state is bounded within $\mathscr{U}_{\rho_e}$ under $u = \Phi_{nn}(x) \in U$. On the other hand, when $x(t_k) \in \mathscr{U}_\rho \backslash \mathscr{U}_{\rho_e}$, the set of control actions $u(t) = \Phi_{nn}(x(t_{k+i})) \in U, i = 0, \ldots, N-1$ is again a feasible solution that meets the constraints of Eq. 6.46f (i.e., the inequality constraint of Eq. 6.46f becomes an active constraint). This completes the proof of recursive feasibility for the CLBF-EMPC of Eq. 6.46.

The proof of boundedness of the state in $\mathscr{U}_\rho$ follows the conclusions in Propositions 6.2 and 6.6. We first consider the case where $x(t_k) \in \mathscr{U}_{\rho_e}$. As it is required by the constraint of Eq. 6.46e that the state $x(t)$, $\forall t \in [t_k, t_{k+1})$ predicted by the ensemble of RNN models of Eq. 6.46b be bounded in $\mathscr{U}_{\rho_e}$, it follows from Eq. 6.51 that the state of the nonlinear system of Eq. 6.1 does not leave $\mathscr{U}_\rho$ within one sampling period. At the next sampling period, if $x(t_{k+1})$ remains in $\mathscr{U}_{\rho_e}$, it is again bounded in $\mathscr{U}_\rho$ following the above analysis. However, if $x(t_{k+1})$ enters $\mathscr{U}_\rho \backslash \mathscr{U}_{\rho_e}$, the constraint of Eq. 6.46e is activated to drive the state of the RNN model of Eq. 6.4 towards the origin. Since it is proven in Proposition 6.6 that $\dot{W}_c$ based on the state of the nonlinear system of Eq. 6.1 can be rendered negative accounting for bounded disturbances and modeling error within one sampling period under $u = \Phi_{nn}(x) \in U$ (a feasible solution to the CLBF-EMPC optimization problem), the state of the nonlinear system of Eq. 6.1 is also able to move towards the origin and ultimately enters $\mathcal{U}_{\rho_e}$ within finite sampling periods. This completes the proof of closed-loop stability of the nonlinear system of Eq. 6.1 under CLBF-EMPC.

Additionally, since the safe stability region $\mathcal{U}_\rho$ does not intersect with the (bounded and unbounded) unsafe region (i.e., $\mathcal{U}_\rho \cap \mathcal{D} = \emptyset$) according to the definition of the constrained CLBF of Eq. 6.16, the state trajectory under the time-varying operation of the nonlinear system of Eq. 6.1 does not enter the unsafe region for all times. Therefore, process operational safety in terms of avoidance of the unsafe region is also guaranteed under CLBF-EMPC.

6.4.1.1 Online Learning of RNN Models

Operational safety and closed-loop stability properties in Theorem 6.4 hold for the system of Eq. 6.1 subject to bounded disturbances (i.e., $|w(t)| \leq w_m$) since the sample-and-hold implementation of the control actions already account for the effects of disturbances (i.e., the sampling period Δ and the disturbances $w(t)$ are required to be sufficiently small to meet Eq. 6.50). However, in the presence of time-varying disturbances that are not sufficiently small, e.g., $|w(t)| \leq w_M$ where $w_M > w_m$, the nonlinear system of Eq. 6.1 may lose operational safety and closed-loop stability due to a considerable model mismatch between the RNN models that are developed for the nominal system of Eq. 6.1 with $w(t) \equiv 0$ and the actual nonlinear process under disturbances. In this case, we develop real-time adaptive machine-learning-based predictive control to mitigate the impact of disturbances by using the most recent process data to update RNN models online. Following the discussion of online learning of RNN models in Sect. 6.3.4, the implementation strategy of online update of RNN models within CLBF-EMPC is given as follows:

Step 1: An initial ensemble of RNN models for the nominal system of Eq. 6.1 (i.e., $w(t) \equiv 0$) are developed to approximate the nonlinear dynamics in the operating region $\mathcal{U}_\rho$.

Step 2: The nonlinear system of Eq. 6.1 is operated under CLBF-EMPC in a sample-and-hold fashion and the states are continuously monitored and collected. For any $x(t) \in \mathcal{U}_\rho \backslash \mathcal{U}_{\rho_e}$, where $\mathcal{U}_{\rho_e} \subset \mathcal{U}_{\rho_w}$, the online learning of RNN models is activated following the event-triggered mechanism of Eq. 6.35. For any $x(t) \in \mathcal{U}_{\rho_e}$, the error-triggered mechanism of Eq. 6.37 is utilized to adapt the RNN models to the time-varying disturbances using the most recent process data. Similarly, at the next sampling time, the CLBF-EMPC of Eq. 6.46 will use the updated RNN model to calculate the optimal control actions $u^*(t)$ for the next sampling period.

6.4.2 *Application to a Chemical Process Example*

We consider the same chemical process example as in Sect. 6.3.6 to illustrate the application of CLBF-EMPC using an ensemble of RNN models. The dynamic process model of the continuous stirred tank reactor (CSTR) and the parameter values are given in Eq. 6.40 and Table 6.1, respectively, and are omitted here. The CSTR is initially operated at the unstable steady-state $(C_{As}, T_s) = (1.95 \text{ kmol/m}^3, 402 \text{ K})$, and $(C_{A0_s} Q_s) = (4 \text{ kmol/m}^3, 0 \text{ kJ/h})$. The states x and the manipulated inputs u of the closed-loop CSTR system are represented in deviation forms, i.e., $x^T = [C_A - C_{As} \; T - T_s]$ and $u^T = [\Delta C_{A0} \; \Delta Q]$, respectively. Additionally, the manipulated inputs are bounded as follows: $|\Delta C_{A0}| \leq 3.5$ kmol/m^3 and $|\Delta Q| \leq 5 \times 10^5$ kJ/h. The explicit Euler method with an integration time step of $h_c = 2 \times 10^{-5}$ h is applied to numerically simulate the dynamic model of Eq. 6.40. Additionally, the ensemble of RNN models is developed following the same approach as performed in Sect. 6.3.6.1.

6.4.2.1 Closed-Loop Simulation Results

The control objective of CLBF-EMPC is to maximize the profit of the CSTR process of Eq. 6.40 while maintaining the closed-loop state trajectories in the safe stability region $\mathscr{U}_\rho$ for all times. The inlet concentration ΔC_{A0} and the heat input rate ΔQ are the two manipulated inputs. The objective function of the CLBF-EMPC is of the following form:

$$l_e(\tilde{x}, u) = k_0 e^{-E/RT} C_A^2. \tag{6.52}$$

Additionally, we introduce a material constraint into the CLBF-EMPC of Eq. 6.46 to make the averaged reactant material available within the entire operating period t_p to be its steady-state value, C_{A0s}. The material constraint is formulated as follows:

$$\frac{1}{t_p} \int_0^{t_p} u_1(\tau) \mathrm{d}\tau = 0 \text{ kmol/m}^3 \tag{6.53}$$

where the averaged reactant material in deviation form, u_1, is equal to 0. Similar to the application of CLBF-MPC scheme in Sect. 6.3.6, we also demonstrate the application of the RNN-based CLBF-EMPC control scheme to both bounded and unbounded unsafe regions in state-space.

We first consider the case of an unbounded unsafe region in state-space, where the unsafe region, CLBF $W_c(x)$, and the parameter values are the same as those in Sect. 6.3.6. The Lyapunov-based EMPC (LEMPC) that accounts for closed-loop stability only is also used here for comparison. Specifically, based on the formulation of the standard LEMPC of Eq. 2.35, the LEMPC using RNN models is presented as follows [216]:

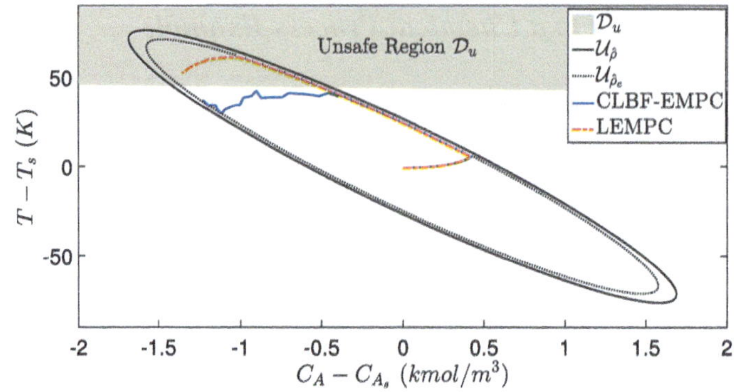

Fig. 6.14 State trajectories for the closed-loop system of Eq. 6.40 within one operating period under LEMPC and CLBF-EMPC, respectively, where the gray area on the top of $\mathscr{U}_\rho$ represents the unbounded set of unsafe states $\mathscr{D}_u$, and the initial condition is (0, 0)

$$\max_{u \in S(\Delta)} \int_{t_k}^{t_{k+N}} l_e(\tilde{x}(\tau), u(\tau))\, \mathrm{d}\tau \tag{6.54a}$$

$$\text{s.t.} \quad \dot{\tilde{x}}(t) = \frac{1}{N_e} \sum_{j=1}^{N_e} F_{nn}^{j}(\tilde{x}(t), u(t)) \tag{6.54b}$$

$$\tilde{x}(t_k) = x(t_k) \tag{6.54c}$$

$$u(t) \in U, \ \forall\, t \in [t_k, t_{k+N}) \tag{6.54d}$$

$$V(\tilde{x}(t)) \le \rho_e, \ \forall\, t \in [t_k, t_{k+N}), \ \text{if } x(t_k) \in \Omega_{\rho_e} \tag{6.54e}$$

$$\dot{V}(x(t_k), u(t_k)) \le \dot{V}(x(t_k), \Phi_{nn}(x(t_k))), \ \text{if } x(t_k) \in \Omega_\rho \backslash \Omega_{\rho_e} \tag{6.54f}$$

where the notations follow those in Eqs. 2.35 and 6.46. As the objective of EMPC is to maximize the production rate $r = k_0 e^{-E/RT} C_A^2$ by dynamically operating the CSTR process of Eq. 6.40, it is observed in Fig. 6.14 that the closed-loop state enters the top of the operating region with a much higher temperature value than the steady-state value, to increase economic profits under the LEMPC of Eq. 6.54 that does not account for safety concerns. Additionally, we design the unbounded unsafe region in the form of $F(x) = x_1 + x_2$ such that the temperature in the reactor plays a dominant role in characterizing the unsafe region $\mathscr{D}_u$. This is also consistent with the operation of an exothermic reaction in CSTR, where rapid increases in temperature might lead to potential safety problems. However, note that we still account for reactant concentration in the characterization of the unbounded unsafe region $\mathscr{D}_u$ since the reaction rate also depends on the reactant concentration.

Figures 6.14, 6.15, and 6.16 show the closed-loop simulation results for the system of Eq. 6.40 under the LEMPC of Eq. 6.54, and the RNN-based CLBF-EMPC of Eq. 6.46. Specifically, Fig. 6.14 compares the state trajectories under CLBF-EMPC and LEMPC, respectively. It is demonstrated that starting from the initial condition

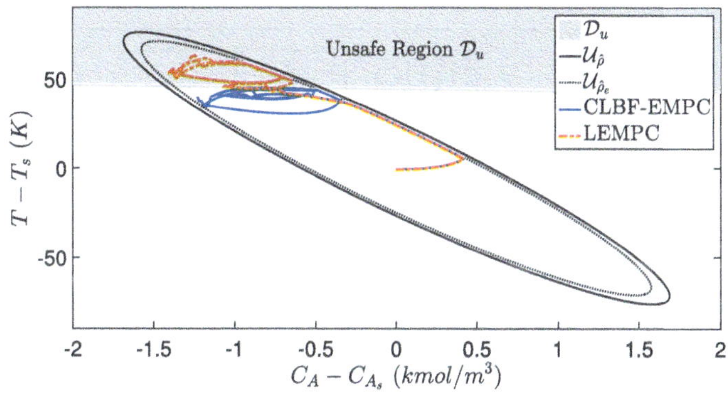

Fig. 6.15 Closed-loop state trajectories for the system of Eq. 6.40 within four operating periods under CLBF-EMPC and LEMPC, respectively, where the initial condition is (0, 0) and the unbounded set of unsafe states $\mathscr{D}_u$ is the gray area on the top of $\mathscr{U}_\rho$

(0, 0), the state trajectory for one simulation period $t_p = 0.128$ h under CLBF-EMPC is maintained below the unbounded unsafe region $\mathscr{D}_u$ for all times, while the one under LEMPC exceeds the threshold and enters $\mathscr{D}_u$ near the end of simulation. Additionally, we run the closed-loop simulation for successive four operating period, where each operating period is $t_p = 0.128$ h. The material constraint is imposed in each operating period such that the averaged reactant material (in deviation form) within each operating period equals zero. It is demonstrated in Fig. 6.15 that the state trajectory under the CLBF-EMPC of Eq. 6.46 remains in the safe stability region $\mathscr{U}_\rho$ within four operating periods, while the one under LEMPC enters the unsafe region during the first operating period and stays there for the remainder of the process operation. Both state trajectories progress in a circular manner in the stability region (the solid ellipse) because the material constraint forces the decrease of the reactant concentration near the end of each operating period. This can also be observed in the input profiles for the closed-loop system of Eq. 6.40 within four operating periods shown in Fig. 6.16, where CLBF-EMPC consumes the maximum allowable ΔC_{A0} at the beginning of each operating period and lowers the consumption near the end.

We also calculate the total economic profits over four operating periods, i.e., $L_E = \int_{t=0}^{4t_p} k_0 e^{-E/RT} C_A^2 dt$, for the closed-loop system of Eq. 6.40 under the different controllers. It was obtained that the L_E values are 8.42, 8.01 and 5.24 for the closed-loop CSTR under LEMPC, CLBF-EMPC, and steady-state operation, respectively, from which it is demonstrated that economic profits are significantly improved (around 52%) under EMPC compared to the steady-state operation. The reason for a slightly larger L_E under LEMPC than CLBF-EMPC is that the state under LEMPC enters the unsafe region during the simulation where increased production rate is obtained due to higher temperature (Fig. 6.15).

The second example is to demonstrate the effectiveness of the CLBF-EMPC of Eq. 6.46 for the CSTR system with a bounded unsafe region $\mathscr{D}_b$ in state-space. The bounded unsafe region $\mathscr{D}_b$ is designed to be a set embedded within the stability region

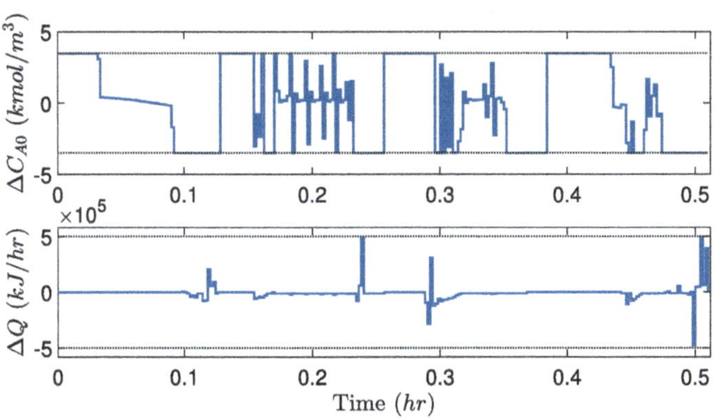

Fig. 6.16 Input profiles for the closed-loop system of Eq. 6.40 within four operating periods under CLBF-EMPC, where the unsafe region is the gray area on the top of $\mathscr{U}_\rho$

as shown in the above example to demonstrate that the CLBF-EMPC of Eq. 6.46 is able to achieve economic optimality while maintaining the state out of $\mathscr{D}_b$ for all times. The bounded unsafe region as well as the CLBF and its parameters are the same as those in Sect. 6.3.6.

The simulation results for the closed-loop system of Eq. 6.40 under CLBF-EMPC are shown in Figs. 6.17, 6.18, and 6.18. Specifically, in Fig. 6.17, it is demonstrated that the state trajectory under CLBF-EMPC is maintained in the safe stability region $\mathscr{U}_\rho$ for all times (i.e., four successive operating periods with $t_p = 0.128$ h). However, the state trajectory under LEMPC enters the bounded unsafe region $\mathscr{D}_b$ since the design of the LEMPC of Eq. 6.54 does not account for any safety constraints. Similarly, Fig. 6.18 shows the input profiles for the closed-loop system of Eq. 6.40 within four operating periods under the CLBF-EMPC of Eq. 6.46, where ΔC_{A0} shows variation due to the material constraint of Eq. 6.53 applied in each operating period. Additionally, the accumulated economic profits are calculated for the closed-loop system of Eq. 6.40 in the presence of a bounded unsafe region. It was found that the L_E values are 8.42, 8.47 and 5.24 for LEMPC, CLBF-EMPC, and steady-state operation, respectively. This again demonstrates that process economics is optimized under EMPC while closed-loop stability and process operational safety are both guaranteed. It is noted that the total economic profits under LEMPC and under CLBF-EMPC are very close since the two state trajectories both stay in a region above the unsafe set for most of the simulation time (Fig. 6.17). The only difference is that the state trajectory under CLBF-EMPC avoids the bounded unsafe region for all times, while the one under LEMPC does not.

Additionally, it is noted that the RNN-based MPC is computationally more demanding than the first-principles-model-based MPC because the RNN model is essentially a complicated nonlinear function which requires more computation time for prediction. In our example, the computation time for running RNN-based MPC is around 2.3 s, which is less than one sampling period (i.e., 2×10^{-3} h = 7.2 s)

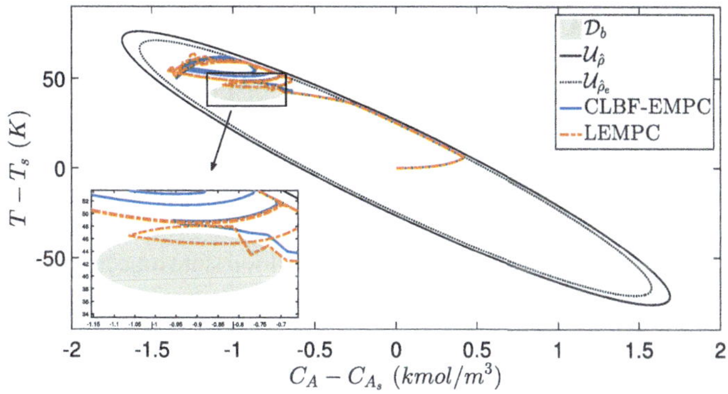

Fig. 6.17 Closed-loop state trajectories for the system of Eq. 6.40 within four operating periods under CLBF-EMPC and LEMPC, respectively, where the initial condition is (0, 0) and the bounded set of unsafe states $\mathscr{D}_b$ is embedded within $\mathscr{U}_\rho$

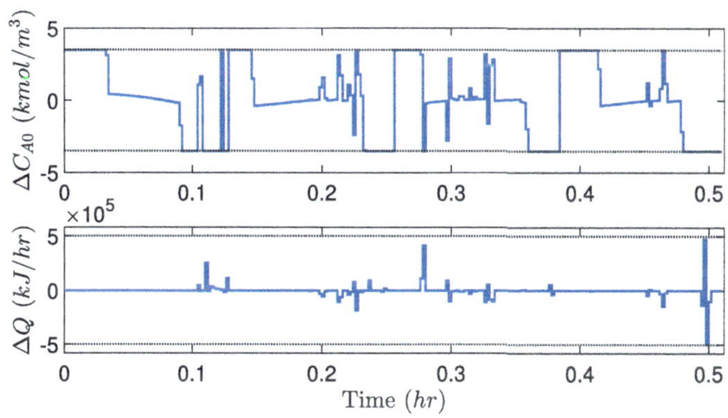

Fig. 6.18 Input profiles for the closed-loop system of Eq. 6.40 within four operating periods under CLBF-EMPC, where the bounded set of unsafe states $\mathscr{D}_b$ is embedded within $\mathscr{U}_\rho$

such that it can implemented in real-time optimization and control. The above case studies demonstrate that the CLBF-EMPC of Eq. 6.46 based on an ensemble of RNN models achieved desired model prediction results for the nonlinear system of Eq. 6.40, and thus, is able to optimize control actions that maintain the closed-loop state within the safe stability region $\mathscr{U}_\rho$ for all times. Additionally, we demonstrate the applicability of the CLBF-EMPC of Eq. 6.46 to both bounded and unbounded unsafe regions in a CSTR example. The economic profits over multiple operating periods are calculated and compared under LEMPC, CLBF-EMPC and steady-state operation, respectively, from which it can be concluded that significant improvement of economic benefits can be achieved under EMPC.

6.4.2.2 Real-Time CLBF-EMPC with Online Learning of RNN Models

The closed-loop simulation results for the CSTR of Eq. 6.40 under the machine-learning-based CLBF-EMPC of Eq. 6.46 with and without online learning of RNN models, respectively, are shown in this subsection. The disturbance on the feed flow rate F which varies from 5 m^3/h to 10 m^3/h at $t = 0.11$ h is introduced into the closed-loop system. The simulation results are shown in Figs. 6.19, 6.20, 6.21, and 6.22. In Fig. 6.19, it is demonstrated that the closed-loop state trajectory under CLBF-EMPC with updating RNN models avoids the unsafe region while the one under the CLBF-EMPC using the initial RNN model for all times enters the unsafe region $\mathscr{D}$ near the end of operating period due to the disturbed feed flow rate F and reaction rate.

Moreover, the closed-loop simulation of the CSTR system under CLBF-EMPC with multiple operating periods is performed with the following disturbances: (1) the feed flow rate F is changing from 5 m^3/h to 11.5 m^3/h at $t = 0.1$ h during the first operating period from $t = 0$ h to $t = 0.128$ h, and (2) the actual value of the pre-exponential constant k_0 used in the process model is reduced by 20% to represent a change in the reaction rate at $t = 0.148$ h during the second operating period from $t = 0.128$ h to $t = 0.256$ h. Figures 6.20 and 6.21 show the closed-loop simulation results under the above settings. Specifically, Fig. 6.20 demonstrates that with online learning of RNN models, the closed-loop state trajectory under CLBF-EMPC is able to avoid the unsafe region for all times within two consecutive EMPC operating periods. Figure 6.21 shows the corresponding input profiles under CLBF-EMPC, from which it is observed that the inlet concentration ΔC_{A0} consumes its maximum allowable value at the beginning of each operating period, and thus, decreases to its lower bound near the end of each operating period to meet the material constraint of Eq. 6.53. Additionally, the accumulated prediction error diagram under CLBF-EMPC with and without online learning of RNN models is shown in Fig. 6.22. It is

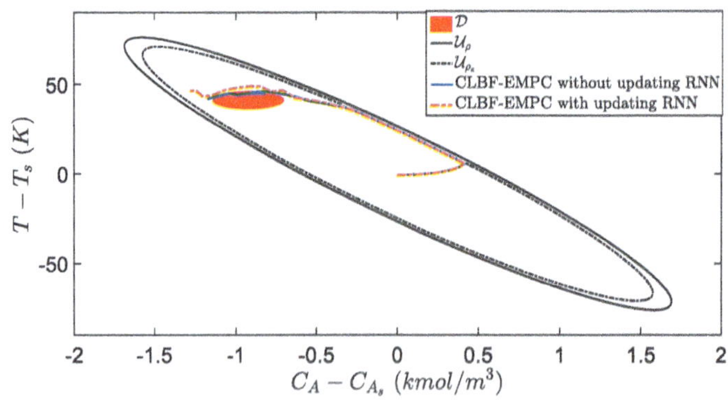

Fig. 6.19 The state-space profiles for the closed-loop CSTR subject to time-varying disturbances under CLBF-EMPC with (red trajectory) and without online RNN update (blue trajectory), respectively, for an initial condition (0,0)

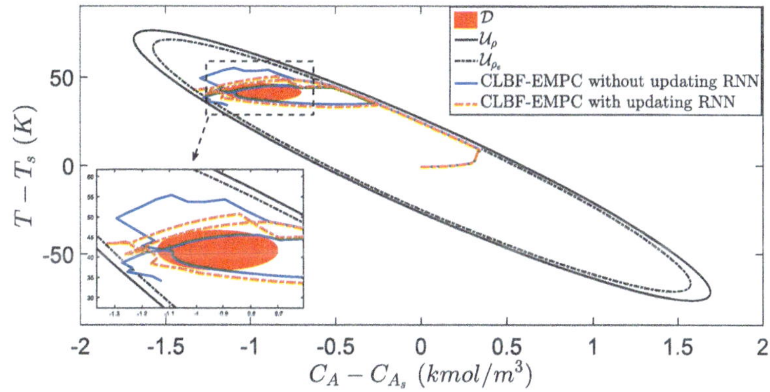

Fig. 6.20 The state-space profiles for the closed-loop CSTR subject to time-varying disturbances under CLBF-EMPC with (red trajectory) and without online RNN update (blue trajectory), respectively, for two consecutive operating periods with an initial condition (0,0)

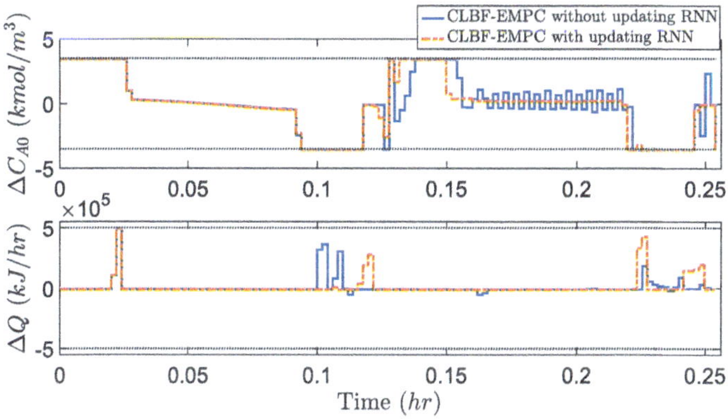

Fig. 6.21 Manipulated input profiles ($u_1 = \Delta C_{A0}$, $u_2 = \Delta Q$) for the closed-loop CSTR subject to time-varying disturbances under CLBF-EMPC with (red trajectory) and without online RNN update (blue trajectory), respectively, for two consecutive operating periods with an initial condition (0,0)

demonstrated that the prediction error (red lines) for the CLBF-EMPC with updating RNN models is maintained at a very low level during the two consecutive EMPC operating periods. However, the prediction error (blue lines) derived from the CLBF-EMPC without updating RNN models indicates a large model mismatch between the initial RNN model for the nominal CSTR of Eq. 6.40 and the actual disturbed system.

Lastly, to demonstrate the improved process economic benefits under the time-varying operation of EMPC, accumulated economic profits over the entire operating period, i.e., $L_E = \int_0^{t=0.256\text{ h}} l_e(x, u)d\tau$ is compared for the CLBF-EMPC and the steady-state operation (i.e., the CSTR of Eq. 6.40 is operated at the steady-state for all times). It is obtained that $L_E = 4.93$ for the closed-loop system under CLBF-EMPC with online update of RNN models and $L_E = 2.61$ for the steady-state operation

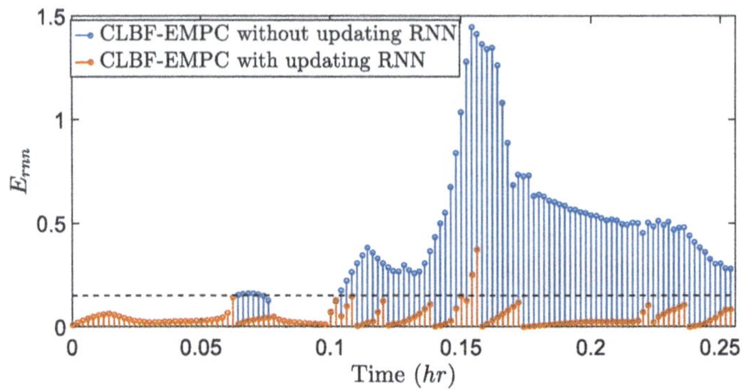

Fig. 6.22 Value of $E_{rnn}(t)$ at each sampling time for the closed-loop CSTR subject to time-varying disturbances under CLBF-EMPC with and without online RNN update, respectively, where the threshold E_T is set to 0.15

within 0.256 h. Therefore, it is concluded that closed-loop stability, process operational safety, and economic optimality are achieved simultaneously for the disturbed CSTR process of Eq. 6.40 under the CLBF-EMPC of Eq. 6.46 with online learning of RNN models.

6.5 Conclusions

In this chapter, CLBF-MPC and CLBF-EMPC methods formulated with machine learning models were developed for nonlinear process systems. RNN models were first trained using extensive open-loop simulation data to capture process dynamics in a certain operating region such that the modeling error between the recurrent neural network model and the actual nonlinear process model was sufficiently small. Then, the well-fitting RNN models were incorporated in the formulation of CLBF-MPC/EMPC to predict process dynamics, for which ensemble learning was employed to improve prediction accuracy and parallel computing was used to reduce computation time of multiple RNN models.

The stability analysis of the closed-loop system under the CLBF-MPC/EMPC schemes using RNN models established the boundedness of the closed-loop state in the safety and stability region for MPC and EMPC and demonstrated the ultimate convergence to a small neighborhood around the origin for MPC. Additionally, event-triggered and error-triggered mechanisms were designed for the real-time implementation of CLBF-MPC and CLBF-EMPC schemes to update the RNN models online using the most recent process data that account for nonlinear dynamics in the presence of time-varying disturbances. The application of the machine-learning-based control schemes to a chemical reactor demonstrated the applicability and effectiveness of the schemes in stabilizing nonlinear systems with simultaneous stability and safety guarantees, and the ability to deal with time-varying disturbances using online learning of machine learning models.

Chapter 7
Process Cybersecurity Via Machine Learning Detection

7.1 Introduction

Chemical plants are cyber-physical systems (CPSs) that integrate physical process components, computation, and communication networks to ensure automated real-time operation in a seamless manner. To operate cyber-physical systems stably and securely, accurate information is required via reliable communication technologies. As more communication networks are complemented or replaced by wireless networks in addition to point-to-point communications [2, 49], cybersecurity has become increasingly important in the operation of chemical process networks. While these new developments and communication technologies improve operation performance and efficiency, they also increase the risk of the chemical plant getting attacked by cyber-attacks. As more components are included, there is a higher probability that an accurate and continuous feedback measurement is unavailable due to bursts of network transmission errors, which poses a challenge for closed-loop control systems that rely on accurate feedback measurements. Malicious cyber-attacks could target any communication channels or device in the control network with a variety of attacking strategies, for example, modifying control actions, and/or compromising sensor measurements that could affect process stability, integrity, operational cost, and other safety considerations. Being aware of the technical details of the control system, the targeted cyber-attacks pose severe threats to the control system as they are intelligently designed to compromise fundamental process safety beyond disrupting process operation. Therefore, plant-wide risk assessments should be carefully developed to incorporate safety measures addressing cybersecurity needs.

On the other hand, with the increase in computing power and digital connectivity, the massive amount of archived plant data could provide a potential solution to handling cyber-attacks beyond the use in day-to-day process monitoring and operations. Since physical and cyber components are closely interacted in a chemical plant, operational cybersecurity of control systems would mandate a cyber-attack

Z. Wu and P. D. Christofides, *Process Operational Safety and Cybersecurity*, Advances in Industrial Control, https://doi.org/10.1007/978-3-030-71183-2_7

mitigation strategy that is different from the traditional information technology (IT) approaches—one that combines an advanced detection scheme with robust control strategies using the process data at hand [164]. Additionally, due to the accessibility to control system information, the cyber-attacks are intended to disrupt the closed-loop operation while remaining undetectable by control engineers or by conventional detection methods. Situations where intelligent cyber-attacks cannot be efficiently detected by conventional detection methods (e.g., model-based detection schemes) can be potentially tackled by data-based detection methods [43]. Machine learning, a method of data analysis that can help engineers learn from data, identify patterns and make decisions with minimal human intervention, has attracted an increasing attention and has demonstrated promising potential for use in detection of cyber-attacks. In recent years, we have witnessed an increasing number of the applications of machine learning methods in traditional engineering fields, and more specifically in the field of systems engineering, e.g., [154, 166, 196, 226]. Machine learning techniques such as support vector machines and artificial neural networks have demonstrated success in detecting machine and plant anomalies, e.g., [34, 40, 45, 46, 78, 136, 147, 192, 208, 212], and can be readily adopted in the context of cyber-physical security for industrial control systems.

In this chapter, machine-learning-based detection systems and resilient control schemes are developed to detect and mitigate the impact of stealthy, intelligent cyber-attacks. In the first section, the concept of stealthy cyber-attacks is presented, followed by several common cyber-attacks discussed in the literature. The second section presents the construction of data-based machine learning detection algorithms which can effectively detect multiple classes of intelligent cyber-attacks. Subsequently, we design several resilient control strategies to promptly contain and eliminate the impact of cyber-attacks upon detection. The application to a benchmark multivariable nonlinear process example is presented to evaluate the ability of the proposed cyber-attack detection and mitigation schemes.

7.1.1 Class of Nonlinear Systems

We consider the class of continuous-time nonlinear systems described by the following state-space form:

$$\dot{x}(t) = f(x(t), u(t), w(t)) \tag{7.1a}$$

$$\bar{x}(t) = h(x(t)) \tag{7.1b}$$

where $x(t) \in D \subset \mathbf{R}^n$, $u(t) \in \mathbf{R}^m$, and $w \in W$ are the state vector, the manipulated input vector, and the noise vector, respectively. The control action is constrained by $u \in U := \{u_i^{\min} \leq u_i \leq u_i^{\max}, i = 1, \ldots, m\} \subset \mathbf{R}^m$, where $u_i^{\max}$ and $u_i^{\min}$ are the upper and lower bounds for the input vector. The noise is assumed to be bounded within the set $W := \{w \in \mathbf{R}^l : \mid |w| \leq \theta,\ \theta \geq 0\}$. $\bar{x}(t) \in \mathbf{R}^n$ represents the sensor measurements of process states, which may be compromised by cyber-attacks (i.e., $\bar{x}(t) = x(t)$, when no cyber-attacks are present in the system). The vector functions $f(\cdot, \cdot, \cdot)$ and $h(\cdot)$ are assumed to be sufficiently smooth with $f(0, 0, 0) = 0$ and $h(0) = 0$. Thus, the origin is a steady-state of the nominal system of Eq. 7.1 (i.e., with $w(t) \equiv 0$) under. Additionally, the initial time t_0 is taken to be zero ($t_0 = 0$).

Throughout this chapter, we assume that there exists a stabilizing feedback controller $u = \Phi(x) \in U$ for the nominal system of Eq. 7.1 (i.e., $w(t) \equiv 0$) that renders the origin of the nominal closed-loop system asymptotically stable. According to converse Lyapunov theorems, the stabilizability assumption implies the existence of a $\mathscr{C}^1$ Lyapunov function $V : D \rightarrow \mathbf{R}_+$ that satisfies the following conditions for all x in an open neighborhood D around the origin:

$$\alpha_1(|x|) \leq V(x) \leq \alpha_2(|x|), \tag{7.2a}$$

$$\frac{\partial V(x)}{\partial x} f(x, \Phi(x), 0) \leq -\alpha_3(|x|), \tag{7.2b}$$

$$\left| \frac{\partial V(x)}{\partial x} \right| \leq \alpha_4(|x|), \tag{7.2c}$$

where $\alpha_i(\cdot)$, $i = 1, 2, 3, 4$, are class $\mathscr{K}$ functions. The stability region Ω_ρ of the closed-loop system of Eq. 7.1 is characterized as a level set of $V(x)$ inside D in which Eq. 7.2 is satisfied under $u = \Phi(x) \in U$, i.e., $\Omega_\rho := \{x \in D \mid V(x) \leq \rho, \rho > 0\}$. Therefore, when the sensor measurements received by the controller are reliable and secure (i.e., $\bar{x}(t) = x(t)$), the controller $u = \Phi(x) \in U$ guarantees closed-loop stability in the sense that the closed-loop state of the system of Eq. 7.1 is bounded in Ω_ρ for all times, and asymptotically converges to the origin for any initial conditions $x_0 \in \Omega_\rho$. Additionally, based on the smoothness property assumed for the vector function $f(x, u, w)$ and the boundedness of $u \in U$, there exist positive constants $L_x^{'}$, L_x, and M such that the following inequalities hold for all $x, x' \in D$:

$$|f(x, u, 0)| \leq M, \tag{7.3a}$$

$$|f(x, u, 0) - f(x', u, 0)| \leq L_x |x - x'| \tag{7.3b}$$

$$\left| \frac{\partial V(x)}{\partial x} f(x, u, 0) - \frac{\partial V(x')}{\partial x} f(x', u, 0) \right| \leq L_x^{'} |x - x'|. \tag{7.3c}$$

7.1.2 Lyapunov-Based MPC and EMPC

Cyber-attack detection systems and resilient control schemes in this chapter are developed in the context of model predictive control, and more specifically, Lyapunov-based model predictive control (LMPC) and Lyapunov-based economic model predictive control (LEMPC). Therefore, the LMPC and LEMPC formulations in Chap. 2

are presented here again for convenience. Specifically, the LMPC optimization problem is formulated as follows:

$$\min_{u \in S(\Delta)} \int_{t_k}^{t_{k+N}} l_t(\tilde{x}(\tau), u(\tau))\, \mathrm{d}\tau \tag{7.4a}$$

$$\text{s.t.} \quad \dot{\tilde{x}}(t) = f(\tilde{x}(t), u(t), 0) \tag{7.4b}$$

$$\tilde{x}(t_k) = x(t_k) \tag{7.4c}$$

$$u(t) \in U, \ \forall\, t \in [t_k, t_{k+N}) \tag{7.4d}$$

$$\frac{\partial V(x(t_k))}{\partial x} f(x(t_k), u(t_k), 0) \le \frac{\partial V(x(t_k))}{\partial x} f(x(t_k), \Phi(x(t_k)), 0),$$
$$\text{if } x(t_k) \in \Omega_\rho \backslash \Omega_{\rho_{\min}} \tag{7.4e}$$

$$V(\tilde{x}(t)) \le \rho_{\min}, \ \forall\, t \in [t_k, t_{k+N}), \ \text{if } x(t_k) \in \Omega_{\rho_{\min}} \tag{7.4f}$$

where the notations follow those in Eq. 2.23. The LEMPC is represented by the following optimization problem:

$$\max_{u \in S(\Delta)} \int_{t_k}^{t_{k+N}} l_e(\tilde{x}(\tau), u(\tau))\, \mathrm{d}\tau \tag{7.5a}$$

$$\text{s.t.} \quad \dot{\tilde{x}}(t) = f(\tilde{x}(t), u(t), 0) \tag{7.5b}$$

$$\tilde{x}(t_k) = x(t_k) \tag{7.5c}$$

$$u(t) \in U, \ \forall\, t \in [t_k, t_{k+N}) \tag{7.5d}$$

$$V(\tilde{x}(t)) \le \rho_e, \ \forall\, t \in [t_k, t_{k+N}), \ \text{if } x(t_k) \in \Omega_{\rho_e} \tag{7.5e}$$

$$\frac{\partial V(x(t_k))}{\partial x} f(x(t_k), u(t_k), 0) \le \frac{\partial V(x(t_k))}{\partial x} f(x(t_k), \Phi(x(t_k)), 0),$$
$$\text{if } x(t_k) \in \Omega_\rho \backslash \Omega_{\rho_e} \tag{7.5f}$$

where the notations follow those in Eq. 2.35. For EMPC, it is common that chemical processes are subject to periodic feedstock constraints, which are specified as a limitation on the quantity of feed reactant materials within a fixed period of time t_{N_p}, and is included in the input constraint set U. The total feed reactant material within one operating period is constrained to a constant value C as follows:

$$\frac{1}{t_{N_p}} \int_{t_0}^{t_{N_p}} u_m(\tau)\mathrm{d}\tau = C \tag{7.6}$$

where the feed material used within each sampling period is denoted by u_m. The EMPC renews the material consumption constraint every t_{N_p} sampling steps. In other words, the total consumption limit is renewed at the start of a new material constraint period, as new feed materials become available for the next constraint period. Therefore, if the total operation time is longer than one material constraint period, the material constraint of Eq. 7.6 will lead to a cyclic operation of the plant, under which, the state-space trajectories and the manipulated input profiles will show a periodic behavior.

It is demonstrated in Chap. 2 that when a secure state measurement x is available at every sampling time, closed-loop stability is ensured for the nonlinear system of Eq. 7.1 in the sense that for any initial condition $x_0 \in \Omega_\rho$, the closed-loop state is guaranteed to be bounded in Ω_ρ for all times under LMPC/LEMPC, and can be ultimately driven to a small neighborhood $\Omega_{\rho_{min}}$ around the origin under LMPC. However, under cyber-attacks that compromise sensor measurements or communication networks between sensors and controllers, closed-loop stability under LMPC/LEMPC is no longer guaranteed because the evolution of the true state will be different from the MPC predicted state trajectory based on falsified state measurements.

7.2 Intelligent Cyber-Attacks

Stealthy, intelligent cyber-attack diagnosis and defense span a much broader scope than classical fault diagnosis problems because the sensor, the actuator, and the control implementation based on process and control system information can all be modified by intelligent adversaries. Being aware of the control strategy and the plant model, cyber-attacks are strategically programmed with the goal of disruption, and are fundamentally different from ordinary actuator and sensor faults. Specifically, among sensor cyber-attacks, deception attacks, replay attacks and Denial-of-Service attacks (e.g., Surge, Geometric, Min-Max) are some of the most common attacks that are easily implementable by attackers [180]. They are designed to intentionally disrupt the control objectives of the system, degrading control performance, and disrupting system safety and stability. Moreover, the impact of these cyber-attacks may be only observed in changes of the closed-loop dynamic behavior. Therefore, the detection schemes that use hardware performance counters to track code modifications are not feasible for detecting intelligent cyber-attacks [94].

In this section, we consider the cyber-attacks on sensor measurements. Under normal operation, in order to ensure closed-loop stability, accurate sensor feedback measurements of the true state of the process should be sent to the controller; falsified measurements may lead to undesired control actions that reduce process economic benefits and drive the true process states outside of the stability region. Some standard types of cyber-attacks are discussed in literature [180]. For example, min-max cyber-attacks will compromise sensor measurements to disrupt process operation within shortest amount of time. Similar to the min-max cyber-attack, surge attacks maximize the disruptive impact for an initial "surge" period. Then, the attacked value is changed to a reduced value for the remainder of the attack duration such that it will not be detected by conventional detection methods such as Cumulative Sum that uses a certain threshold to trigger alarms [43, 132]. Unlike min-max and surge cyber-attacks that achieve maximum disruptive impact at the beginning, geometric attacks geometrically increase the deviation of the attacked value from its true value and stops increasing when the alarming threshold is about to be triggered. Being controller and process behavior aware, the intelligent cyber-attacks have access to information on the existing alarms configured on the process state variables, and the operating region of the closed-loop system under MPC/EMPC. Specifically, when attacks (e.g., min-max or surge attacks) intend to induce maximum disruption, the attacked value will be set to the minimum or maximum value beyond which the process safety systems (e.g., alarms system) that monitor the current state measurement will be triggered immediately. These intelligent cyber-attacks are designed with the falsified state measurement remaining inside the alarm window or the operating stability region such that no alarms will be sounded. Additionally, by maintaining the falsified state measurement within the operating region, feasible control actions are still available, but will deviate from the optimal solutions based on the true state, and might have large enough variations such that closed-loop stability and economic optimality will be lost.

7.2.1 Types of Intelligent Cyber-Attacks

Consider the system of Eq. 7.1 under LMPC/LEMPC within the operating region Ω_ρ. The cyber-attacks on the sensor measurements are designed to have a falsified measurement within the operating region Ω_ρ and to avoid triggering any immediate alarms that monitor state measurements. The following subsections present the mathematical formulations of min-max, geometric, surge, and replay attacks.

7.2.1.1 Min-Max Cyber-Attack

Min-max attacks aim to maximize destabilizing impact within a short time period while avoiding triggering any alarms. Therefore, the falsified state measurements are designed to take values that are inside the operating region Ω_ρ, and furthest from the steady-state (maximum or minimum). The min-max attack is formulated as follows:

$$\bar{x}(t_i) = \min_{x \in \mathbf{R}^n} / \max_{x \in \mathbf{R}^n} \{x \mid V(x(t_i)) = \rho\}, \quad \forall\, i \in [i_0, i_0 + L_a] \tag{7.7}$$

where ρ is the size of the operating region represented by the level set of the Lyapunov function $V(x)$ (i.e., $\Omega_\rho := \{x \in \mathbf{R}^n \mid V(x) \le \rho, \rho > 0\}$) for the nonlinear system of Eq. 7.1 under LMPC/LEMPC. i_0 is the time instant that the min-max attack is introduced, $\bar{x}$ is the compromised sensor measurement, and L_a is the total duration of the cyber-attack.

7.2.1.2 Geometric Cyber-Attack

Closed-loop stability under geometric cyber-attacks deteriorates at a geometric speed until the cyber-attack reaches the minimum or maximum allowable value as characterized by the operating region. Initially, the geometric cyber-attack adds a small constant $\beta \in \mathbf{R}$ to the true measured output $x(t_{i_0})$ at $t = t_{i_0}$, where $x(t_{i_0}) + \beta$ is maintained below the alarm threshold. Subsequently, β is multiplied by a factor $(1 + \alpha)$ at each subsequent time step, where $\alpha \in (0, 1)$, until $\bar{x}$ reaches the maximum allowable value bounded by Ω_ρ. Thus, the two parameters β and α will be chosen by attackers based on the the attack duration and the size of Ω_ρ. The geometric attack is formulated as follows:

$$\bar{x}(t_i) = x(t_i) + \beta \times (1 + \alpha)^{i-i_0}, \quad \forall\, i \in [i_0, i_0 + L_a] \tag{7.8}$$

where the parameters α and β define the speed and magnitude of the geometric cyber-attack.

7.2.1.3 Replay Cyber-Attack

Replay cyber-attacks have access to all previous system outputs corresponding to secure nominal operating conditions without any cyber-attacks. The attacker extracts segments of these previous state measurements and injects some of them into the current measurement readings. Since the replay attack takes past secure sensor measurements as the replayed values, which are supposedly inside the operating region, classical detectors will not be able to recognize any abnormalities. Replay attacks are represented by the following equations:

$$\bar{x}(t_i) = x(t_k), \ \forall\, k \in [k_0, k_0 + L_a], \quad \forall\, i \in [i_0, i_0 + L_a] \tag{7.9}$$

where $x(t_k)$, $\bar{x}$, L_a are the true plant measurement, the series of replay attacks, and the length of the attack in terms of sampling periods (which is also the length of the replay segment), respectively. Note that, the replay attack is added at time t_{i_0} using segments of the previous state measurements starting from time t_{k_0}. For example, the duration of the attack could be exactly the length of one material constraint period. In this case, the tampered state trajectory would look identical to the nominal state trajectory within one operating period from a different initial condition.

7.2.1.4 Surge Cyber-Attack

Surge cyber-attack is a stealthy cyber-attack that cannot be detected by conventional detection methods such as cumulative sum (CUSUM). Specifically, based on the process model of Eq. 7.1, CUSUM statistic detection method [43] is developed to minimize the detection time when a cyber-attack occurs. The CUSUM statistic method calculates the cumulative sum of the deviation between measured and expected/predicted states to detect cyber-attacks:

$$S(k) = (S(k-1) + z(k))^{+}, \ S(0) = 0 \tag{7.10a}$$

$$D(S(k)) = \begin{cases} 1, & \text{if } S(k) > S_{TH} \\ 0, & \text{otherwise} \end{cases} \tag{7.10b}$$

where $S(k)$ and S_{TH} are the nonparametric CUSUM statistic and the threshold for detection of cyber-attacks, respectively. $(S)^{+} = S$, if $S \geq 0$ and $(S)^{+} = 0$ otherwise. We develop a binary detection indicator D, where $D = 0$ and $D = 1$ indicate no attack, and under attack, respectively. $z(k)$ is the error between measured states $x(t_k)$ and expected states $\tilde{x}(t_k)$ at time $t = t_k$: $z(k) := |\tilde{x}(t_k) - x(t_k)| - b$ where b is a small positive constant that eliminates the impact of common disturbances that should not be considered as cyber-attacks. Note that if the process model is available, $\tilde{x}(t_k)$ can be derived based on the state measurement at $t = t_{k-1}$, and the control action for all $t \in [t_{k-1}, t_k)$. With a carefully selected S_{TH}, the model-based detection method is able to efficiently detect a variety of sensor attacks. However, the above model-based method may become ineffective in detecting intelligent cyber-attacks that have access to process knowledge (e.g., the system model and the principles of the detection method). For example, surge attacks can avoid detection by maximizing the disruptive impact for an initial short period of time, and then remaining at a lower attack value (this value can be defined based on the operating region Ω_ρ) for the remaining time such that $S(k)$ is maintained below S_{TH} for all times. The reduced value after the surge and the length of the initial surge period can be designed in many ways as long as the cumulative error from t_{i_0} to $t_{i_0+L_a}$ between the predicted true values of the states and their compromised measurements does not exceed the CUSUM threshold S_{TH}. In this study, we set the reduced value after the surge to be a sufficiently small bounded noise imposed on the attacked sensor. The surge attack is formulated as follows:

$$\begin{aligned} \bar{x}(t_i) &= \min_{x \in \mathbf{R}^n} / \max_{x \in \mathbf{R}^n} \{x \mid V(x(t_i)) = \rho\}, \ \text{if } i_0 \leq i \leq i_0 + L_s \\ \bar{x}(t_i) &= x(t_i) + \eta(t_i), \ \text{if } i_0 + L_s < i \leq i_0 + L_a \end{aligned} \tag{7.11}$$

where i_0 and L_s are the start time of the attack, and the duration of the initial surge, respectively. $\eta_l \leq \eta(t_u) \leq \eta_u$ is the bounded noise imposed on the sensor measurement after the initial surge period, where η_u and η_l are the upper and lower bounds of the noise, respectively.

Remark 7.1 The four cyber-attacks introduced in this section are among the most common deception cyber-attacks in literature. By feeding falsified measurement values into process control systems, they will drive the closed-loop states away from their expected values and finally ruin the stability of the closed-loop system. To collect the process operational data that can be later used in the development of data-based detectors and data-based state reconstructor, extensive computational simulations of the system under attacks need to be carried out. In the following subsection, we will discuss the design procedure for the open-loop and closed-loop simulations under sensor attacks.

7.2.1.5 Simulation Design Guide

To capture the process dynamics in the presence of cyber-attacks, we will run computer simulations and introduce the aforementioned cyber-attacks into the measurements of process variables during operation. Specifically, to mimic the execution of digital controllers in industrial control systems, the continuous-time process model of Eq. 7.1 is simulated using explicit Euler method with a sufficiently small integration time step, and with control actions u implemented in sample-and-hold (i.e., the input is a piecewise constant function that remains constant within each sampling period Δ, i.e., $u(t) = u(t_k)$, $\forall t \in [t_k, t_{k+1})$, where $t_{k+1} := t_k + \Delta$). In simulation studies, it is important to note that only the sensor measurements of Eq. 7.1b are tampered at each integration time step (or at each sampling time, depending on the attacking scenario), while the process model of Eq. 7.1a is integrated using the true state values x all the time. This is consistent with the fact that the sensor attacks cannot directly disrupt system operation, but will compromise sensor measurements to mislead the controller to compute incorrect control actions. As a result, one simulation run will generate two state profiles: (falsified) measured state profile and true state profile, and one control action profile. The simulation results provide a visualization of attacking pattern for low-dimensional systems that can help control engineer recognize abnormal behavior in process monitoring; for high-dimensional systems, they also provide a better understanding of intelligent cyber-attacks through appropriate data analysis techniques. In this chapter, we will utilize the simulation data to develop machine-learning-based state reconstructor and cyber-attack detector based on open-loop and closed-loop simulation results, respectively. The simulation design guide for the generation of open-loop and closed-loop data using computer simulations is presented as follows.

In open-loop simulations, we need to determine the initial condition x_0 and control actions u that will be implemented during the entire simulation period. Some common data generation methods in machine learning and in classical system identification field (e.g., subspace algorithms) include extensive open-loop simulations with combinations of different initial conditions x_0 and control actions u, and a single continuous trajectory under a pseudorandom binary input sequence (PRBS). While data generation through PRBS is easy to implement in real industrial processes, extensive open-loop simulations may better capture the process dynamics in the operating region by sweeping over all the possible initial conditions and control actions in simulations. Also, with the use of parallel computing, the computation time of extensive open-loop simulations with a short period of time can be significantly reduced compared to the simulation of a single, long state trajectory. Therefore, in this study, we use extensive open-loop simulations following the procedure below.

Consider the nonlinear system of Eq. 7.1 operated in the stability region Ω_ρ (i.e., a compact set in state-space). The open-loop simulations of the nominal system of Eq. 7.1 are carried out with a variety of combinations of initial conditions $x_0 \in \Omega_\rho$ and inputs $u \in U$, under which a large number of state trajectories (i.e., the solution of $x(t)$ for Eq. 7.1) are obtained. As it is impractical to sweep over all the values that (x, u) can take in state-space due to the limitation of computational resources, we will discretize the range of inputs and of the initial conditions with sufficiently small intervals (see Fig. 6.2). Each simulation run will be executed with finite sampling periods, and the control actions are varying at each sampling step. The simulation period should be chosen based on the use of these data. For example, if they will be used to develop machine-learning-based state reconstructors, then it should cover at least the number of state measurements that the state reconstructor requires. If we use the data to build process models that will be used in model predictive controllers (MPCs) (see Chap. 6), then the simulation period should be no shorter than the sampling period used in MPC. Also, for the system operated around an unstable equilibrium point, the simulation period needs to be carefully chosen to make sure that the process state does not diverge and exits the stability region during operation. An alternative way to generate data for unstable systems is running closed-loop simulations, which will be introduced in the next paragraph. After we collect time-series data of measured states $\bar{x}$, true states x and control actions u, we will do normalization and partition the entire dataset into training, validation and testing datasets.

Closed-loop simulations follow a similar procedure, whereas the control actions are no longer pre-determined, but will be computed by the controller. Similarly, the continuous-time process model of Eq. 7.1 is simulated using explicit Euler method, and the sensor measurements of Eq. 7.1b are tampered by cyber-attacks at each integration time step or sampling step. As the sensor measurements are compromised, the controller will compute control actions based on the falsified state measurements and send them to the actual system to be applied over the next sampling period. Note that in closed-loop simulations, the control actions will be applied to the actual nonlinear system of Eq. 7.1a based on true state values since sensor attacks only mask the readings of sensor measurements and do not directly affect the process dynamics. With knowledge on the control formulation and the plant model, intelligent cyber-attacks—which are strategically programmed with the goal of disruption—can quickly drive the system to instability while avoiding conventional detection triggers; this fundamentally distinguishes them from process faults, which might also disrupt process operation, but would not lead to unsafe process operation or catastrophic consequences in general. Additionally, the attacking mode of sensor cyber-attacks may vary depending on the formulation of control systems. For example, in a centralized control system that computes control actions using full state measurements, cyber-attacks can destabilize the process and degrade control performance by compromising the measurements of safety-critical process variables. However, in a distributed control systems with inter-controller communication between each distributed local controller, cyber-attacks can also target communication channels to affect the calculation of control actions. As a result, it is more challenging to detect cyber-attacks in a distributed control system since additional communication channels create a greater exposure surface, and erroneous calculations of control actions may propagate as the local controllers exchange incorrect control actions through communication. Therefore, the closed-loop simulations of the system of Eq. 7.1 under cyber-attacks should be carried out accounting for the control schemes used in practice, such that the data-based detector using these simulation data can detect attacks for the system using the same control scheme when implementing online.

Although both open-loop and closed-loop simulations can be used to generate data, they are typically implemented with different purposes. Open-loop simulations are often used in modeling processes to capture the underlying process dynamics under nominal operation (e.g., machine learning modeling of nonlinear processes in Chap. 6), or under cyber-attacks (i.e., state reconstructor in this chapter). There are many advantages of generating data through open-loop simulations. For example, extensive open-loop simulations can be done with low computational burden as no control algorithms are needed for computing control actions. Also, they are not restricted to any specific control schemes used in a closed-loop system as the data generation does not rely on any control algorithms. In other words, the machine learning model developed using open-loop simulation data can be readily used in a new control system under a different control scheme without having to regenerate data or retrain the model. However, closed-loop simulation data is preferred in cyber-attack detection because the data captures the closed-loop behavior accounting for the interplay between the compromised sensor measurements and erroneous control actions (i.e., a falsified state measurement will lead to an erroneous control action, while this control action will then drive the state away from the desired reference trajectory/set-point). Therefore, although the computation time for generating closed-loop simulation data is longer than that in open-loop simulation, they show great potential in machine-learning-based cyber-attack detector that identifies the occurrence of cyber-attacks based on real-time process data, and can significantly reduce the detection time since the training data is consistent with the closed-loop data in online implementation. The detailed development of machine-learning-based detector is introduced in the next section.

7.3 Detection of Cyber-Attacks Targeting MPC Systems

While conventional detection methods have demonstrated their effectiveness in detecting suspicious process variable deviations, most of these methods are model-based—either dependent on physical process models or on network and computer system models. Certain classes of intelligent cyber-attacks either render traditional detection methods ineffective, or remain undetected until a significant deviation occurs to drive the system an undesirable operating point, which triggers the existing alarm systems. To handle the potential cyber-attacks in process control systems, a robust cyber-attack detector should be developed to identify attacks from subtle

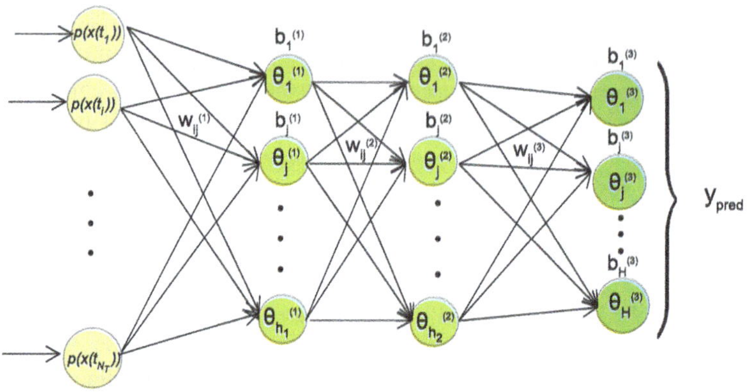

Fig. 7.1 A two-hidden-layer feedforward neural network structure with inputs $p(\bar{x})$ being a nonlinear function of state measurements within the detection window N_T, and output being the probability of each class label that indicates the status and/or type of cyber-attack

variations in real-time process state measurements and mitigate the impact of cyber-attacks before triggering the safety systems. Without explicit knowledge on the process model, a data-based detection approach that uses machine learning algorithms for solving classification problems provides a promising path for the detection of unknown intelligent cyber-attacks. The integration of real-time machine-learning-based detection algorithms and existing advanced process control schemes (e.g., MPC) adds another protective safeguard to the multi-layer cyber-defense strategy that is standard to next-generation smart manufacturing. Cyber-attack detection carried out using machine learning methods have been studied in many literature [1, 81, 145]. Using data-based methods to train a detection algorithm for cyber-attacks separates the detector from the physical process model, and therefore makes the detector resilient to both process changes and intelligent stealthy attacks designed based on process behavior. Among advanced machine learning methods, neural networks (NNs) have been successfully implemented in a wide range of applications for both unsupervised and supervised classification problems [72]. In a supervised classification problem, new data is classified by the neural network into classes that share similar characteristics based on the training dataset with labeled data corresponding to each target class. For example, the neural network can distinguish between two (i.e., "no attack" or "attack") or multiple classes, where each class represents a type of cyber-attack depending on the training data.

A feedforward artificial neural network is used to solve the supervised classification problem in this study. Each layer in the neural network consists of a series of nonlinear functions of the weighted sum of inputs or neurons (i.e., activation functions), yielding values for the neurons in the subsequent layer from the previous layer. The structure of a two-hidden-layer neural network model used in this study is shown in Fig. 7.1, with each input unit representing a nonlinear function $p(\cdot)$ of the full state measurements at each sampling time, and an output vector representing the probability of each class label. The two-hidden-layer feedforward neural network is mathematically formulated as follows:

$$\theta_j^{(1)} = g_1\left(\sum_{i=1}^{N_T} w_{ij}^{(1)} p(\bar{x}(t_i)) + b_j^{(1)}\right) \tag{7.12a}$$

$$\theta_j^{(2)} = g_2\left(\sum_{i=1}^{h_1} w_{ij}^{(2)} \theta_i^{(1)} + b_j^{(2)}\right) \tag{7.12b}$$

$$\theta_j^{(3)} = g_3\left(\sum_{i=1}^{h_2} w_{ij}^{(3)} \theta_i^{(2)} + b_j^{(3)}\right), \quad y_{pred} = [\theta_1^{(3)}, \theta_2^{(3)}, \ldots, \theta_H^{(3)}]^T \tag{7.12c}$$

where $\theta_j^{(l)}$, $j = 1, \ldots, h_l, l = 1, 2$ are the neurons in the first ($l = 1$) and second ($l = 2$) hidden layers, respectively. The output node is represented by $\theta_j^{(3)}$, $j = 1, \ldots, H$, where H is the number of class labels. In this study, two hidden layers were used in the neural network for the cyber-attack detector design as it achieves the best closed-loop performance and computational efficiency; however, in general, there is no restriction on the number of layers, and a multiple-hidden-layer neural network can be developed using similar formulations for a more complex problem. The input node $p(\bar{x}(t_i))$ receives the state measurement at time t_i, where $i = 1, \ldots, N_T$ is the length of the time-varying trajectory. $w_{ij}^{(l)}$ and $b_j^{(l)}$ represent the weights connecting neurons i and j in consecutive layers (from $l - 1$ to l), and the bias term on the jth neuron in the lth layer, respectively. Based on the information received from the previous layer as well as the optimized biases, weights, and the nonlinear activation function g_l, each layer calculates an output and sends it to the next layer. Examples of the activation functions include softmax function $g(z_j) = \frac{e^{z_j}}{\sum_{i=1}^{H} e^{z_i}}$, hyperbolic tangent sigmoid transfer function $g(z) = \frac{2}{1+e^{-2z}} - 1$, and some other common functions such as sigmoid, radial basis functions, and ReLu. The interested readers may refer to [179] for the analysis of their performances. The output node y_{pred} computes the probabilities of each class label, from which the class with the highest probability will indicate the status (i.e., no attack or under attack), or the type of the cyber-attack, depending on the requirement for the machine learning detector.

To optimize the weights and biases in Eq. 7.12, we use the Levenberg–Marquardt algorithm [107, 120] to minimize a Bayesian regularized mean squared error cost function of the following form:

$$S(w) = \mu \sum_{k=1}^{N_s} (y_{\text{pred},k} - y_{\text{true},k})^2 + \zeta \sum_{p=1}^{N_w} w_p^2 \tag{7.13}$$

where $p = 1, \ldots, N_w$ and $k = 1, \ldots, N_s$ represent the number of biases and weights in the neural network, and the number of samples in the training dataset, respectively. y_{pred} and y_{true} are the vector of the predicted probabilities associated with each class label, and the vector of target class labels of each sample. ζ and μ are the two regularization hyper-parameters. The Hessian matrix and the gradient of $S(w)$ are calculated using the backpropagation method in Levenberg–Marquardt algorithm. We assume that the data and the weights follow Gaussian prior probability distributions. Then, the regularization hyper-parameters, μ and ζ, are updated by maximizing their posterior probability distribution provided the data, which is equivalent to maximizing the likelihood of evidence by Bayes' Theorem. Within each training epoch, the cost function $S(w)$ is first minimized with respect to w, and then, the likelihood of evidence is maximized with respect to μ and ζ. Readers may refer to [41] for detailed formulation of this procedure. Lastly, we calculate the testing accuracy, which is the ratios between the number of correctly classified samples and total number of samples in the testing sets.

To develop an NN detector, a dataset needs to be first developed with state measurement data collected during the operation under feedback controllers, i.e., the LMPC of Eq. 7.4 or the LEMPC of Eq. 7.5. To build a high-quality dataset, extensive simulations for a broad range of initial conditions within the stability region Ω_ρ are carried out to generate a large number of state evolutions within the stability region. We record the full state measurements $\bar{x}(t)$ along the time-varying trajectory for $t \in [t_0, t_{N_T}]$, and feed it to the NN input layer. It is noted that we use a nonlinear function $p(\bar{x})$ to provide an effective one-dimensional input feature that captures the dynamic behavior of all states for the NN detector. The selection of this input variable, $p(\bar{x})$, will be discussed in Sect. 7.3.1.

After adequate training using collected data, the NN detector is implemented online with the process controlled by MPCs with cyber-attack resilient control strategies that will be discussed in Sect. 7.4. The feedforward NN model is a static model receiving inputs of fixed dimension, N_T, which is the length of the time-varying trajectory. Therefore, the detection window of the NN detector matches the trajectory length of the training data, N_T. For example, the detector can be activated every time full state measurements become available, and uses a moving horizon detection window, receiving latest sequences of $x(t_k)$ of fixed length N_T. Alternatively, the NN detector can be activated at the end of each material constraint period (under EMPC), where $N_T = N_p$. In this case, the detector will receive the entire sequence of full state measurements $\bar{x}(t_k)$ over the latest material constraint period with a fixed length N_T. Each data sample consists of a two-dimensional matrix $n \times N_T$, where n and N_T are the full state dimension, and the length of each state trajectory within the detection window, respectively. Each training sample is obtained from the closed-loop system simulation under a different set of initial conditions, and equal number of samples are collected within each class labels to ensure training accuracy.

7.3.1 Choice of Detection Input Variable

We first consider the case of LMPC. Since the control objective of the LMPC of Eq. 7.4 is to stabilize the system at the origin, for any initial condition in the operating region Ω_ρ, the closed-loop state profiles ultimately converge to their steady-state values if no attacks occur. Therefore, the closed-loop state profiles provide a good measure of system dynamic operations under LMPC, and thus, can be directly used as the NN input. However, unlike the case of operation under tracking MPC where the Lyapunov function decreases as the process states approach the origin, the time-varying operation of LEMPC results in a state trajectory that remains on the boundary of the operating region Ω_ρ where $V(\bar{x}) = \rho$ to maximize process economic benefits. Therefore, the exact state profile of each state variable does not follow a general expected trend under the nominal operation, which means the detection methods based on the assessment of state trajectory might not be effective for detecting cyber-attacks in EMPC systems. Moreover, if a cyber-attack is designed to destabilize the closed-loop system within the shortest amount of time, the current state measurement will be set to the minimum/maximum allowable attack value (i.e., the boundary of the operating region Ω_ρ) while not triggering the alarm system. In this case, the falsified sensor measurements will also yield a Lyapunov function that is equal to ρ, which is similar to the dynamic behavior under nominal operation of EMPC. For these reasons, although the Lyapunov function value of the full-state measurements $V(\bar{x})$ is a good candidate for the input variable of the NN-based detector used in LMPC, it might not be a good measure of input for the NN-based detector developed for LEMPC.

Given that the economic benefit is optimized under EMPC via its cost function, the progression of economic benefit is a measure that can effectively reflect the time-varying operation under LEMPC; thus, information derived from the cost function provides a good comparison for not-attacked and attacked scenarios. In this study, the evolution of economic benefits will be continuously monitored during closed-loop operation. Specifically, it is noticed that as operation time progresses, the cumulative economic benefit increases monotonically. If we calculate the first derivative of cumulative economic benefit, we will be able to obtain the incremental economic benefit, which is analogous to the reaction rate at each time instance. From extensive closed-loop simulations, the first derivative of cumulative economic benefit shows varying patterns depending on the material consumption constraint and the initial conditions. The second derivative of the cumulative economic benefit represents the rate of change in the incremental economic benefit (i.e., the rate of change in the optimized cost function l_e in Eq. 7.5a). We will use this rate of change as the input $p(\bar{x})$ for the NN-based detector in this study.

7.3.2 *Sliding Detection Window*

As the NN detector may not have perfect classification accuracy, a one-time detection may lead to false alarms when large oscillatory data within normal ranges is misclassified as a cyber-attack. To reduce false alarm rates, we develop a sliding alarm verification window, within which we count the number of positive attack detections to determine whether the system is under attack or not. Specifically, a detection indicator D_i generated by a sliding detection window with length N_s and

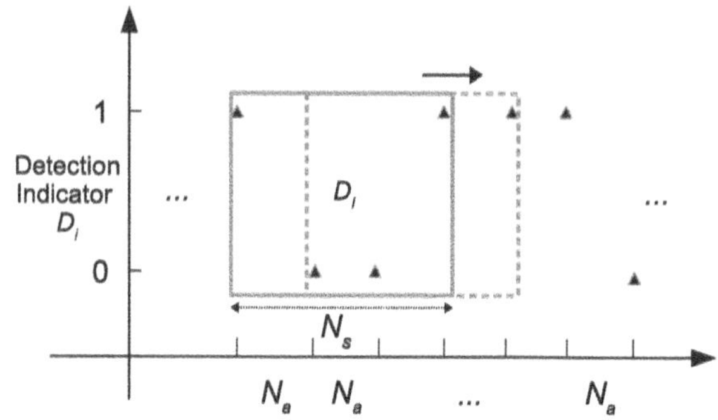

Fig. 7.2 The sliding detection window with a length of N_s, where D_i is the indicator for the detection triggered every N_a sampling steps

each sub-model M_i and are developed as follows:

$$D_i = \begin{cases} 1, & \text{if attack is detected by } M_i \\ 0, & \text{if no attack is detected by } M_i \end{cases} . \tag{7.14}$$

Based on the detection indicator D_i at every N_a sampling steps, the weighted sum of detection indicators within the sliding detection window D_I shown in Fig. 7.2 at $t = t_k = k\Delta$ is calculated as follows:

$$D_I = \sum_{j=\lceil (k-N_s+1)/N_a \rceil}^{\lfloor k/N_a \rfloor} \lambda^{\lfloor \frac{k}{N_a} \rfloor - j} D_j \tag{7.15}$$

where λ is a detection factor that gives more weight to more recent detections within the sliding window as the classification is more accurate with more data being collected. If $D_I \geq D_{TH}$, where D_{TH} is a threshold for the sliding alarm verification window, then the cyber-attack is confirmed by the NN-based detector; otherwise, the detection system remains silent and the sliding window will move one sampling time forward. To balance missed detections and false alarm rates, we determine the threshold D_{TH} via extensive closed-loop simulations under cyber-attacks to derive a desired detection rate while not triggering the alarm systems too frequently.

Additionally, since recursive feasibility is not guaranteed for the optimization problem of LMPC/LEMPC if the state of the system of Eq. 7.1 leaves the stability region Ω_ρ, it is also necessary to check whether the state is inside Ω_ρ, especially when cyber-attacks occur but have not been detected yet. Therefore, to ensure closed-loop stability and recursive feasibility, we can use a redundant, secure sensor (if there is any) to check whether the closed-loop state is still bounded in Ω_ρ at the time when $D_i = 1$. If the state x is shown to be outside of Ω_ρ, safety systems (e.g., alarm systems, emergency shutdown systems, and relief systems) need to be activated to ensure operational safety as closed-loop stability is no longer guaranteed $\forall x \notin \Omega_\rho$ [237].

Remark 7.2 Given that the classification accuracy may not be perfect, we confirm the occurrence of a cyber-attack based on multiple detections instead of a one-time detection. To achieve a desired performance in terms of high detection efficiency and low false alarm rates, the sliding window length N_s and the threshold will be carefully chosen through extensive closed-loop simulations. Specifically, a higher detection threshold D_{TH} ($D_I \geq D_{TH}$ represents the presence of cyber-attacks) and a larger N_s would lead to a lower false alarm rate but a longer detection time, while a lower D_{TH} and a smaller N_s would have the opposite effect.

Remark 7.3 The NN-based detector in this section is developed based on the supervised learning algorithm with a large amount of labeled data to distinguish the abnormal operation under cyber-attacks from the nominal operation. However, supervised learning algorithms may not work for unknown cyber-attacks since no labeled data can be collected for training off-line. Therefore, to detect unknown cyber-attacks, we can utilize unsupervised learning-based detection method by clustering unknown cyber-attack data into a new class. However, if the unknown cyber-attack shares similar properties (e.g., similar attack mechanism) with a trained (known) cyber-attack, it may still be detected as one of the known cyber-attacks by the NN detector. This broadens the applications of the NN-based detector to practical systems, where many cyber-attacks are not exactly the same as those reported in the literature, but share very similar properties in their designs.

7.4 Cyber-Attack Resilient Control Systems

In this section, we focus on the development of cyber-attack resilient control systems that can mitigate the impact of cyber-attacks upon detection. Several resilient control strategies are discussed for the closed-loop system of Eq. 7.1 under LMPC and LEMPC.

7.4.1 *Redundant Sensors*

When the cyber-attack is detected by $D_i = 1$ but not confirmed by $D_I \geq D_{TH}$ yet, the LMPC (LEMPC) optimization problem can use the state measurement from redundant, secure sensors instead of the original sensors as the initial condition $x(t_k)$ for the optimization problem of Eq. 7.4 (Eq. 7.5) until the next instance of detection. However, if the cyber-attack is finally confirmed by $D_I \geq D_{TH}$, the misbehaving sensor will be isolated, and the LMPC (LEMPC) optimization problem starts to use the state measurement from secure sensors instead of the compromised state measurement as the initial condition $x(t_k)$ for the optimization problem of Eq. 7.4 (Eq. 7.5) for the remaining time of process operation. The structure of the integrated cyber-attack-detection-control system for LMPC is shown in Fig. 7.3.

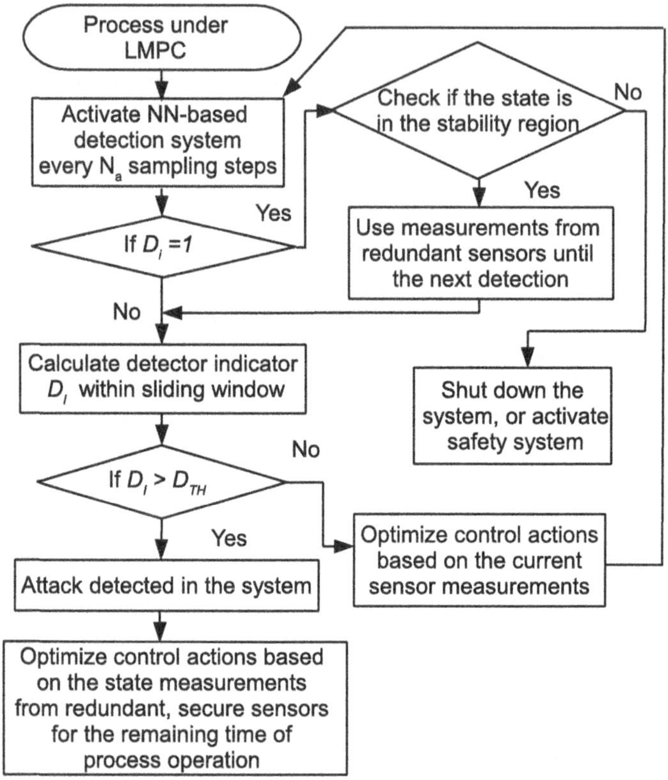

Fig. 7.3 Basic structure of the proposed integrated NN-based detection and LMPC control method

If the cyber-attack is detected and confirmed before the closed-loop state is driven out of the stability region, it follows that the closed-loop state is always bounded in the stability region Ω_ρ thereafter and ultimately converges to a small neighborhood $\Omega_{\rho_{\min}}$ around the origin for any $x_0 \in \Omega_\rho$ under the LMPC of Eq. 7.4. An example trajectory is shown in Fig. 7.4, where it is demonstrated that starting from an initial condition in Ω_ρ, the trajectory first moves away from the origin due to cyber-attacks and finally re-converges to a small neighborhood $\Omega_{\rho_{\min}}$ around the origin under LMPC once the cyber-attack is detected by the proposed NN-based detection scheme.

7.4.2 *Attack-Resilient Operation Combining Open-Loop and Closed-Loop Control*

Upon the successful detection of cyber-attacks in sensors, one strategy that we have shown in Sect. 7.4.1 is to utilize the response plan that involves physical replacements

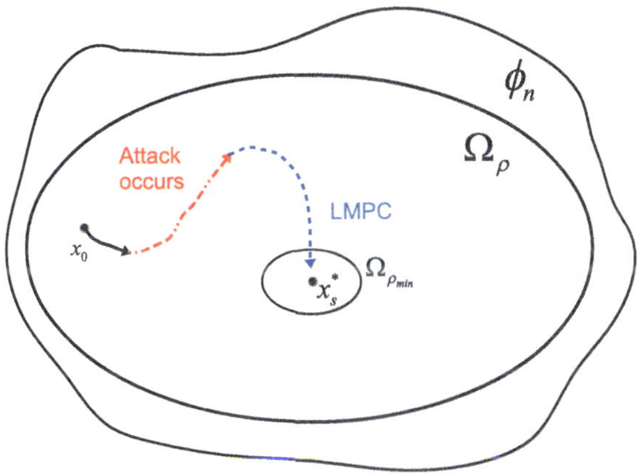

Fig. 7.4 A schematic showing an example state trajectory under the integrated cyber-attack detection and control scheme

of problematic sensors with their redundant back-up sensors. However, sensor device replacement may not be an effective measure for all circumstances. For example, if redundant sensors cannot be immediately deployed upon detection of cyber-attacks, instead of using falsified state measurements, it would be better to operate the process in open loop without reliable feedback measurements. Specifically, in this section, we consider the case of LEMPC, under which the state of the nominal system of Eq. 7.1 (i.e., no disturbances or cyber-attack) is bounded in Ω_ρ for all times. Additionally, since the LEMPC operates the system at the boundary of the operating region for the majority of operating time, we define a smaller level set $\Omega_{\rho_{\text{secure}}} := \{x \in \Omega_\rho \mid V(x) \leq \rho_{\text{secure}}\}$ inside the the stability region Ω_ρ as the new operating region such that the state may leave $\Omega_{\rho_{\text{secure}}}$ due to cyber-attacks but still remains in Ω_ρ before detection. Specifically, as EMPC optimizes the economic benefit of the process, it is likely that the optimized states will reach, and evolve along the boundary of the secure region $\Omega_{\rho_{\text{secure}}}$ during the operating period. Assuming that the attacker has knowledge on the secure region and the stability region for LEMPC, the tampered state measurements will be set to a value near or on the boundary of the secure region $\Omega_{\rho_{\text{secure}}}$ to induce maximum destructive impact on the system (e.g., surge or min-max cyber-attack) without triggering any alarms. Therefore, regardless of the presence of a cyber-attack, it is very likely that the measured process states will reach the boundary of $\Omega_{\rho_{\text{secure}}}$ (i.e., $V(\bar{x}) = \rho_{\text{secure}}$) during the operation of one material constraint period. In other words, there could be two reasons when measured process states yield $V(\bar{x}) = \rho_{\text{secure}}$: 1) under the normal operation with no cyber-attacks, the measured process state is driven to the boundary of the bounded secure region $\Omega_{\rho_{\text{secure}}}$ at time t_k under the optimized control actions $u^*(t_k)$ computed by EMPC, or 2) the measured states are compromised by a cyber-attack (e.g., surge or min-max attack) at time t_k. Therefore, when the measured states $\bar{x}(t_k)$ shows $V(\bar{x}(t_k)) = \rho_{\text{secure}}$, we can no longer trust this measurement due to the ambiguous cause of this observation.

When the measured states reach the boundary of $\Omega_{\rho_{\text{secure}}}$, the control system will switch to open-loop control mode to combat the ambiguity of state measurements.

Assuming that the measured states are secure and correct at the beginning of each material constraint period, $t = t_{N_0}$, (the LEMPC can operate the system in multiple periods where the material constraint of Eq. 7.6 is satisfied in each operating period), we solve the following nonlinear optimization problem to obtain the open-loop control actions at the beginning of the material constraint period:

$$\max_{u' \in S(\Delta)} \int_{t_{N_0}}^{t_{N_0+N_p}} l_e(\tilde{x}(t), u'(t))\mathrm{d}t \tag{7.16a}$$

$$\text{s.t.} \quad \dot{\tilde{x}}(t) = f(\tilde{x}(t), u'(t)) \tag{7.16b}$$

$$u'(t) \in U, \ \forall\, t \in [t_{N_0}, t_{N_0+N_p}) \tag{7.16c}$$

$$\tilde{x}(t_{N_0}) = \bar{x}(t_{N_0}) \tag{7.16d}$$

$$V(\tilde{x}(t)) \leq \rho_{\text{secure}}, \ \forall\, t \in [t_{N_0}, t_{N_0+N_p}), \ \ \text{if } \bar{x}(t_{N_0}) \in \Omega_{\rho_{\text{secure}}} \tag{7.16e}$$

$$\dot{V}(\bar{x}(t_{N_0}), u') \leq \dot{V}(\bar{x}(t_{N_0}), \Phi(\bar{x}(t_{N_0}))), \ \ \text{if } \bar{x}(t_{N_0}) \in \Omega_{\rho} \backslash \Omega_{\rho_{\text{secure}}} \tag{7.16f}$$

where N_p is the prediction horizon for open-loop control, which is also the number of sampling periods in one material constraint period. At the start of a new material constraint period (i.e., time t_k), the EMPC in open-loop control mode computes the optimal trajectory of N_p control actions based on the state measurement received at $t = t_k$. The open-loop control actions are implemented in a sample-and-hold fashion and will be applied until the end of this material constraint period. In the case that there are no process disturbances or cyber-attacks, this optimal control action profile would yield maximum economic benefits while meeting all state and input constraints.

When the feedback measurement becomes unreliable for the controller to calculate control actions, we will use the open-loop control actions that are calculated at the start of the material constraint period for the remaining time of the material constraint period as a substitute. At the end of the material constraint period, we activate the cyber-attack detector to re-assess reliability of the control system by examining the past state measurements and determining whether the system is under attack or not. The security status of the state measurements over the last material constraint period will be provided by the detector. Once the security of the control system is confirmed, or the impact of cyber-attacks are mitigated, we will reactivate closed-loop control to optimize control actions in real time with with secure feedback measurement.

Although a minor performance degradation may be observed due to the modeling error and process disturbances when we switch to open-loop control mode, this attack-resilient operation strategy guarantees that the impact of a surge or min-max cyber-attack is fully eliminated. Figure 7.5 outlines the implementation strategy using a logic flow diagram, and and the specific steps are states as follows:

1. At the beginning of a material constraint period ($t = t_{N_0}$), we compute open-loop control actions within one material constraint period using Eq. 7.16. Meanwhile, the optimal control action for the next sampling is calculated by the LEMPC of Eq. 7.5 with $\Omega_{\rho_{\text{secure}}}$ replacing Ω_{ρ_e}.

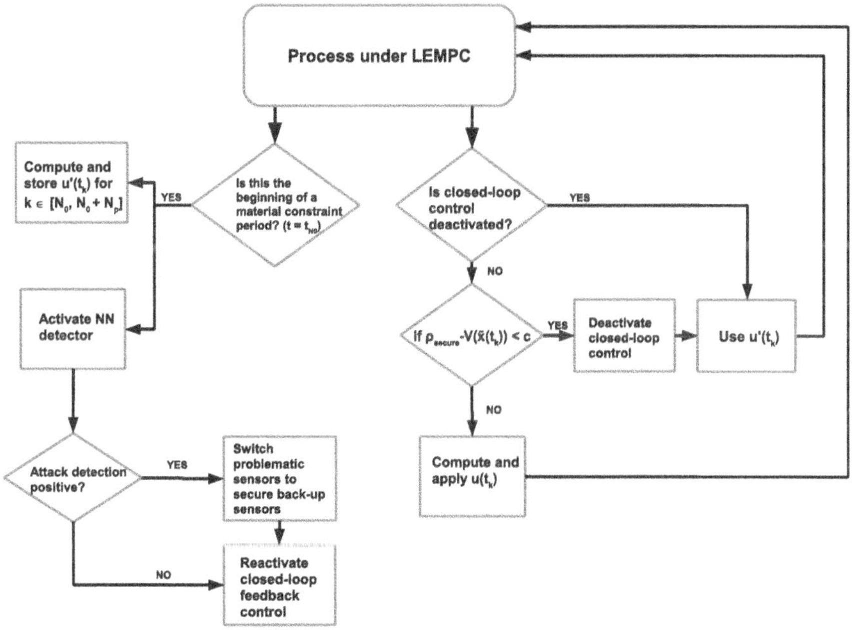

Fig. 7.5 Logic flowchart showing the implementation steps of the attack-resilient operation of LEMPC that combines open-loop and closed-loop control actions together for the system operated in a secure region $\Omega_{\rho_{\text{secure}}}$

2. If $\rho_{\text{secure}} - V(\bar{x}(t_k)) \leq c$, (where c is a positive real number that quantifies the distance from the boundary of secure region to categorize a state measurement as being untrustworthy), then we deactivate the LEMPC of Eq. 7.5, and implement the open-loop control action $u'(t_k)$ calculated by the LEMPC of Eq. 7.16 as an substitute.
3. The control system will implement open-loop control actions $u'(t_k)$ until $t_{N_0+N_p}$.
4. At $t_{N_0+N_p}$, we activate the cyber-attack detector using the past full-state measurements $\bar{x}(t_k)$ for $k \in [N_0, N_0 + N_p]$. If a cyber-attack is detected, the tampered sensors will be disconnected and, these measurement signals will be rerouted to a set of secure back-up sensors. Then, go to Step 5. If no attack is detected, go to Step 5.
5. At $t_{N_0+N_p}$, the closed-loop control is reactivated with a renewed material constraint. Repeat Steps 1–4.

Remark 7.4 Note that the closed-loop system may never reach the boundary of $\Omega_{\rho_{\text{secure}}}$ in some cases, depending on the length of the material constraint period, the size of $\Omega_{\rho_{\text{secure}}}$, and the initial condition. If this is the case, and the measured states are set to be on the boundary of $\Omega_{\rho_{\text{secure}}}$ by cyber-attacks, then we will follow the implementation strategy of Step 2 to deactivate closed-loop control and use open-loop control instead.

7.4.3 *Post Cyber-Attack State Reconstruction*

Measurement reconstruction has been of interest for many decades in fault detection field, e.g., [9, 74, 97, 162, 198, 206]. In addition to redundant sensors and integrated closed-loop and open-loop control, in this section, we present a state reconstruction method to estimate true state values based on the compromised sensor measurements and continue closed-loop control following the successful detection of cyber-attacks. As it is important to develop accurate detectors to promptly report the intrusion of a cyber-attack as well as building robust frameworks to mitigate the impact of cyber-attacks before the detector is activated, it is equally important to have recuperating measures in place to maintain controllability of the system in the absence of reliable sensors. A state reconstructor can be developed to estimate the true state values based on the control actions u and past state measurements $\bar{x}$ [215]. In this section, recurrent neural network is used to develop the state reconstructor using extensive open-loop simulation data of the nonlinear system of Eq. 7.1 under cyber-attacks. Subsequently, the RNN-based state reconstructor is implemented in closed-loop simulation to estimate the true state values that will be used in MPC.

7.4.3.1 Recurrent Neural Network

Recurrent neural network (RNN) has been one of the most popular machine learning methods in developing nonlinear dynamic functions using time-series data. Figure 7.6 shows the RNN structure and its mathematical formulation can be found in Eq. 6.4. Due to the existence of a feedback loop in the hidden-layer neurons, RNN models exhibit temporal behavior and can be used to capture the dynamic behavior of nonlinear systems. In Sect. 6, RNN models have been developed for nonlinear systems and incorporated in MPC to provide prediction of future states. However, since state measurements are now under attack, to estimate true state values in real time, in this chapter, we develop an RNN-based state reconstructor based on control actions u and faulty measurements $\bar{x}$. Specifically, the RNN inputs are $u(t)$ and $\bar{x}(t)$, $\forall t \in [t_k, t_{k+r})$,

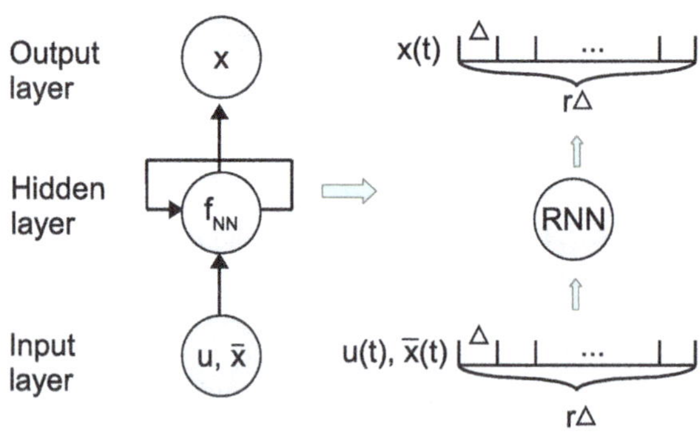

Fig. 7.6 Structures of recurrent neural network (left) and of a restruction window (right), where the input vectors are $\bar{x}$, u, the output vector is x, and f_{NN} represents the hidden neurons that incorporate nonlinear activation functions

and the RNN output is the estimate of the true state x over $t \in [t_k, t_{k+r})$, where r represents the length of the reconstruction window (i.e., the number of sampling periods within the window).

To generate datasets for the training of RNN-based state reconstructors for different types of cyber-attacks (i.e., surge, geometric, and min-max attacks) on sensor measurements that have been introduced in the previous section, we first carry out extensive open-loop simulations for the nonlinear system of Eq. 7.1 under various combinations of $u \in U$ and $x \in \Omega_\rho$ and under each of the different cyber-attacks, respectively. Specifically, a set of open-loop input sequences are applied to the nonlinear system of Eq. 7.1 that initiated inside the stability region Ω_ρ. Then, the aforementioned cyber-attacks are imposed on sensor measurements from the second sampling time. Since then, the true state trajectory will deviate from the measured state trajectory, and we will record both of them for each simulation run. Finally, we split the entire dataset into training, validation and testing datasets, and train the RNN models following the standard procedure as introduced in Sect. 6.2.2 (i.e., minimize the error between the actual true state and the predicted trajectories). To achieve a reliable and accurate state estimation, a constraint on the error between the actual states x and the estimated states $\hat{x}$ is imposed as follows: $|x - \hat{x}| \leq \gamma$, where $\gamma > 0$ is a sufficiently small bound. As will be discussed below (Fig. 7.11b), the RNN models are demonstrated to well capture the attacking behavior (e.g., the zigzag pattern for state measurements under a min-max cyber-attack), and provide a desired estimate of true state trajectory.

Remark 7.5 The RNN-based state reconstruction method can be applied to the nonlinear system of Eq. 7.1 under cyber-attacks provided that cyber-attacks on sensor measurements are sparse attacks (i.e., only a part of process state measurements are under attack, while others remain secure). Under the worst-case scenario that the attack targets all the state measurements, for example, the attack sets all the state measurement to constant values for all times, it is barely possible for RNN-based state reconstructor to estimate the true states without any reliable information of secure sensors. In this case, the attack-resilient operation using open-loop control could be an alternative approach to mitigating the impact of cyber-attacks to the greatest extent.

Remark 7.6 Note that the RNN-based state reconstructor developed in this section is not restricted to the LEMPC of Eq. 7.5 or the LMPC of Eq. 7.4 as the dataset is developed from open-loop simulations that do not depend on any control laws. For example, the RNN-based state reconstructor can be directly applied to the closed-loop system of Eq. 7.1 under a proportional–integral–derivative (PID) controller to estimate true state values using the (falsified) state measurement at each sampling step. Therefore, the machine learning-based state reconstruction provides a general approach to handling sparse sensor attacks in the closed-loop control of the system of Eq. 7.1.

7.4.3.2 Online Reconstruction

Upon the detection of cyber-attacks by the NN-based detectors developed in Sect. 7.3, we will implement online state reconstruction within MPC from the last secure checkpoint. Specifically, the following steps are carried out to implement the RNN-based state reconstruction. (1) To ensure that the initial state measurement for the RNN reconstructor is not attacked, we set the secure checkpoint to be the sampling step before the first detection as the NN-based detectors are implemented with a moving detection window that improves detection accuracy based on multiple detections. (2) Then, the state reconstructor predicts the state evolution from the last secure checkpoint to the current time step $t = t_k$ based on the control actions and (falsified) sensor measurements during this period. Since a reconstruction window of length $r\Delta$ is used in the development of the RNN model, we will use the estimated state in the second sampling period in the window as the initial condition for the next reconstruction as the window rolls one sampling time forward. (3) Then, we will send the estimated state $x(t_k)$ at the current time step to the controller (e.g., the LEMPC of Eq. 7.5 or the LMPC of Eq. 7.4) to compute the control action $u(t_k)$ for the next sampling period $t \in [t_k, t_{k+1})$. 4) After we apply the control action $u(t_k)$ to the nonlinear system and receive new state measurements $\bar{x}(t)$, $t \in [t_k, t_{k+1})$, the state reconstruction window will move one sampling time forward, and use new control actions and state measurements to estimate the true state value at $t = t_{k+1}$.

Remark 7.7 The RNN-based state reconstruction method is a data-driven approach that does not depend on any knowledge of cyber-attacks or of the process model. As a result, it is not restricted to the cyber-attacks discussed in this chapter, and can be applied to many other deception attacks such as scheduled attacks, randomly injected attacks , and optimization-based deception attacks on sensor measurements. However, there is one restriction for the implementation of the RNN-based state reconstruction, that is the communication networks between controllers and sensors are not blocked by cyber-attacks, such that (falsified) state measurements can be continuously received to make estimation. Therefore, the data-driven estimation approach may not be able to handle attacks such as denial-of-service (DOS) attack that temporarily disrupts network services to make sensor measurement unavailable to its intended users.

7.4.3.3 Closed-Loop Control with Reconstructed States

If the feed measurements are no longer reliable, the LEMPC of Eq. 7.5 and the LMPC of Eq. 7.4 will switch to the estimated state $\hat{x}(t_k)$ provided by RNN-based reconstructor. However, considering that a non-zero state estimation error may exist, in this section, we will show that to ensure closed-loop stability under LMPC/LEMPC that uses state estimation from RNN-based reconstructor, the RNN models need to be well trained with a desired estimation accuracy. To proceed, we first develop the following proposition to demonstrate that under the same control actions, the error between the estimated state trajectories $\hat{x}$ and the true state trajectories x of the nonlinear system of Eq. 7.1 is bounded for finite time.

Proposition 7.1 *Consider the solution $\hat{x}(t)$ of the nonlinear system $\dot{\hat{x}} = f(\hat{x}, u, 0)$ based on the estimated state $\hat{x}$ and the solution $x(t)$ of the nominal system $\dot{x} = f(x, u, 0)$ of Eq. 7.1 based on the actual state x with the initial condition $|x_0 - \hat{x}_0| \leq \gamma$, where $\gamma > 0$. If $x(t), \hat{x}(t) \in \Omega_\rho$ holds for all $t \geq 0$, then there exists a positive real number $\kappa > 0$ such that the following inequalities hold $\forall x, \hat{x} \in \Omega_\rho$:*

$$|x(t) - \hat{x}(t)| \leq \gamma e^{L_x t} \tag{7.17a}$$

$$V(x) \leq V(\hat{x}) + \alpha_4(\alpha_1^{-1}(\rho))|x - \hat{x}| + \kappa|x - \hat{x}|^2. \tag{7.17b}$$

Proof Let $e(t) = x(t) - \hat{x}(t)$ represent the state error vector. Using using Eq. 7.3c , the time-derivative of $e(t)$, $\forall x, \hat{x} \in \Omega_\rho$ and $u \in U$ is derived as follows:

$$|\dot{e}| = |f(x, u, 0) - f(\hat{x}, u, 0)| \leq L_x|e(t)|. \tag{7.18}$$

Since the error between x_0 and $\hat{x}_0$ is bounded (i.e., $|x_0 - \hat{x}_0| \leq \gamma$), the upper bound for $|e(t)|$ is derived for all $x(t), \hat{x}(t) \in \Omega_\rho$ as follows:

$$|e(t)| = |x(t) - \hat{x}(t)| \leq \gamma e^{L_x t}. \tag{7.19}$$

Additionally, using Eqs. 7.2a and 7.2c, and applying the Taylor series expansion of $V(x)$ around $\hat{x}$ for all $x, \hat{x} \in \Omega_\rho$, the upper bound for $V(x)$ is derived as follows:

$$\begin{aligned} V(x) &\leq V(\hat{x}) + \frac{\partial V(\hat{x})}{\partial x}|x - \hat{x}| + \kappa|x - \hat{x}|^2 \\ &\leq V(\hat{x}) + \alpha_4(\alpha_1^{-1}(\rho))|x - \hat{x}| + \kappa|x - \hat{x}|^2 \end{aligned} \tag{7.20}$$

where κ is a positive real number.

Then, the following proposition is developed to demonstrate that the sample-and-hold implementation of the stabilizing controller $u = \Phi(\hat{x}) \in U$ based on estimated states $\hat{x}$ can drive the state x of the nonlinear system of Eq. 7.1 towards the origin by rendering the time-derivative of $V(x)$ (i.e., $\dot{V}(x)$ based on the state of the nonlinear system of Eq. 7.1) negative for all times.

Proposition 7.2 *Consider the nominal system of Eq. 7.1 with $w(t) \equiv 0$ under the controller $u = \Phi(\hat{x}) \in U$ (in sample-and-hold fashion) based on the estimated state $\hat{x}$ that satisfies $|\hat{x} - x| \leq \gamma$. Let $\varepsilon_s > 0$, $\Delta > 0$ and $\rho > \rho_s > 0$ satisfy*

$$-\alpha_3(\alpha_2^{-1}(\rho_s)) + L_x^{'}(\gamma + M\Delta) \leq -\varepsilon_s. \tag{7.21}$$

Then, $\dot{V}(x) \leq -\varepsilon_s$ holds for any $x(t_k) \in \Omega_\rho \backslash \Omega_{\rho_s}$.

Proof The time-derivative of $V(x(t_k))$ is obtained as follows:

$$\begin{aligned}\dot{V}(x(t_k)) =& \frac{\partial V(x(t_k))}{\partial x} f(x(t_k), \Phi(\hat{x}(t_k)), 0)\\ =& \frac{\partial V(\hat{x}(t_k))}{\partial x} f(\hat{x}(t_k), \Phi(\hat{x}(t_k)), 0) + \frac{\partial V(x(t_k))}{\partial x} f(x(t_k), \Phi(\hat{x}(t_k)), 0)\\ & - \frac{\partial V(\hat{x}(t_k))}{\partial x} f(\hat{x}(t_k), \Phi(\hat{x}(t_k)), 0).\end{aligned} \tag{7.22}$$

The following inequalities are further derived using Eqs. 7.2a and 7.2b and the Lip:cyberschitz condition of Eq. 7.3:

$$\begin{aligned}\dot{V}(x(t_k)) \leq& -\alpha_3(\alpha_2^{-1}(\rho_s)) + L'_x|x(t_k) - \hat{x}(t_k)|\\ \leq& -\alpha_3(\alpha_2^{-1}(\rho_s)) + L'_x\gamma.\end{aligned} \tag{7.23}$$

Therefore, we can prove that $\dot{V}(x) \leq -\varepsilon_s$ holds provided that Eq. 7.21 is satisfied as follows:

$$\begin{aligned}\dot{V}(x(t)) =& \frac{\partial V(x(t))}{\partial x} f(x(t), \Phi(\hat{x}(t_k)), 0) - \frac{\partial V(x(t_k))}{\partial x} f(x(t_k), \Phi(\hat{x}(t_k)), 0)\\ & + \frac{\partial V(x(t_k))}{\partial x} f(x(t_k), \Phi(\hat{x}(t_k)), 0)\\ \leq& L'_x|x(t) - x(t_k)| + \dot{V}(x(t_k))\\ \leq& L'_x M\Delta - \alpha_3(\alpha_2^{-1}(\rho_s)) + L'_x\gamma\\ \leq& -\varepsilon_s.\end{aligned} \tag{7.24}$$

Based on the above proposition showing that $\dot{V}$ can be rendered negative within each sampling period, closed-loop stability for the nonlinear system of Eq. 7.1 under the LMPC of Eq. 7.4 can be readily proved, and therefore, is omitted here. The interested reader is referred to the similar proof for LMPC with secure state measurement in Sect. 2.3.3.

The next proposition demonstrates that to ensure the invariance of the stability region Ω_ρ, we need to account for the estimation error when characterizing the set Ω_{ρ_e}.

Proposition 7.3 *Consider the nominal system of Eq. 7.1 with $w(t) \equiv 0$ under the sample-and-hold implementation of the LEMPC of Eq. 7.5. Let $\Delta > 0$ and $\rho > \rho_e > \rho_s > 0$ satisfy the following inequality:*

$$\rho_e \leq \rho - \alpha_4(\alpha_1^{-1}(\rho))\gamma e^{L_x\Delta} - \kappa(\gamma e^{L_x\Delta})^2. \tag{7.25}$$

If the state estimation error is bounded by a sufficiently small number γ for all times, i.e., $|\hat{x}(t) - x(t)| \leq \gamma$, $\forall t \geq 0$, then, it is guaranteed that for any $x_0 \in \Omega_\rho$, the true state of the nonlinear system of Eq. 7.1 remains inside the stability region Ω_ρ, $\forall t \geq 0$.

Proof Following the results of Proposition 7.1, we determine the value of ρ_e to make Ω_ρ an invariant set when a non-zero estimation error exists between the estimated state trajectories $\hat{x}$ and the true state trajectories x of the nonlinear system of Eq. 7.1 under the same control actions implemented in a sample-and-hold fashion. The proof follows closely to that for LEMPC with secure state measurement in Sect. 2.3.4 (see, also, the proof for Theorem 2 in [76]), and is omitted here.

Therefore, given that a sufficiently small estimation error, i.e., $|x - \hat{x}| \leq \gamma$, is achieved by the RNN model, the resilient LEMPC and LMPC using estimated state $\hat{x}$ guarantees closed-loop stability for the nonlinear system of Eq. 7.1 upon detection of cyber-attacks.

Remark 7.8 Note that the training dataset for the RNN-based state reconstructor developed in this section is generated from extensive open-loop simulations, where no measurement noise is introduced into the data. However, if there exists noise (e.g., white Gaussian noise) on sensor measurement in practical systems, the proposed state reconstruction method may still be used as long as the measurement noise with the same distribution is accounted for in the generation of the dataset. In this case, if a sufficiently small modeling error is achieved by the RNN reconstructor, we can still ensure closed-loop stability for system under the MPC using the data-driven state reconstructor.

7.5 Application to a Chemical Process Example

The training and online detection of NN cyber-attack detectors as well as the applications of the resilient control strategy presented in Sect. 7.4.2, and of the LEMPC of Eq. 7.5 are illustrated using the chemical reactor example that has been discussed in Chap. 6. Specifically, we consider an irreversible second-order reaction, $A \rightarrow B$, that transforms reactant A to product B at a reaction rate $r_B = k_0 e^{-E/RT} C_A^2$ in a well-mixed, non-isothermal continuous stirred tank reactor (CSTR). A heating jacket is used in the CSTR to remove or supply heat from or to the CSTR at a rate Q. Based on the material and energy balance equations, the dynamic model of this CSTR process is presented as follows:

$$\frac{\mathrm{d}C_A}{\mathrm{d}t} = \frac{F}{V}(C_{A0} - C_A) - k_0 e^{\frac{-E}{RT}} C_A^2 \tag{7.26a}$$

$$\frac{\mathrm{d}T}{\mathrm{d}t} = \frac{F}{V}(T_0 - T) + \frac{-\Delta H}{\rho_L C_p} k_0 e^{\frac{-E}{RT}} C_A^2 + \frac{Q}{\rho_L C_p V} \tag{7.26b}$$

where the description of process variables can be found in Sect. 6.3.6, and a complete list of the process parameter values is given in Table 6.1. The CSTR is initially operated at the unstable steady-state $[C_{As}, \ T_s] = [1.95\ \mathrm{kmol/m^3}, \ 402\ \mathrm{K}]$, and $[C_{A0_s} \ Q_s] = [4\ \mathrm{kmol/m^3}, \ 0\ \mathrm{kJ/h}]$. The manipulated inputs are the inlet concentration of reactant A and the heat input rate represented by the deviation variables, i.e., $\Delta C_{A0} = C_{A0} - C_{A0_s}$ and $\Delta Q = Q - Q_s$, respectively. Additionally, considering the physical limitations, the manipulated inputs are bounded as follows: $|\Delta C_{A0}| \leq 3.5$ kmol/m^3 and $|\Delta Q| \leq 5 \times 10^5$kJ/h. Both the state and the inputs of the closed-loop CSTR system are represented in deviation variable forms, i.e., $x^T = [C_A - C_{As}\; T - T_s]$ and $u^T = [\Delta C_{A0}\; \Delta Q]$, respectively. Therefore, the origin of the state-space is the equilibrium point of the system (i.e., $x_s^T = [0, 0]$, $u_s^T = [0, 0]$). At time $t = t_0$, the system is assumed to be at the equilibrium point (i.e., the initial condition is at the origin, $x_0 = [0, 0]^T$).

The control objective of LEMPC is to maximize the economic profit of the CSTR process of Eq. 7.26 and ensure closed-loop stability by maintaining the closed-loop state in the stability region Ω_ρ for all times. The heat input rate ΔQ and the inlet concentration ΔC_{A0} are the two manipulated inputs. The objective function of the LEMPC is developed using the following equation to optimize the production rate of B:

$$l_e(\tilde{x}, u) = r_B(C_A, T) = k_0 e^{-E/RT} C_A^2. \tag{7.27}$$

The explicit Euler method is used to numerically simulate the dynamic model of Eq. 7.26 with a sufficiently small integration time step of $h_c = 2.5 \times 10^{-5}$ h. The MATLAB OPTI Toolbox is used to solve the nonlinear optimization problem of the LEMPC of Eq. 7.5 with the sampling period $\Delta = 2.5 \times 10^{-3}$ h.

The following material constraint is utilized in the LEMPC of Eq. 7.5 to make the averaged reactant material available within one operating period $t_{N_p} = 0.06$ h to be at its steady-state value, C_{A0s} (i.e., the averaged reactant material in deviation form, u_1, is equal to 0).

$$\frac{1}{t_{N_p}} \int_0^{t_{N_p}} u_1(\tau) d\tau = 0 \text{ kmol/m}^3. \tag{7.28}$$

We design the control Lyapunov function in the form of $V(x) = x^T P x$ with the positive definite P matrix given as follows:

$$P = \begin{bmatrix} 1060 & 22 \\ 22 & 0.52 \end{bmatrix}. \tag{7.29}$$

Then, we characterize the closed-loop stability region Ω_ρ as a level set of Lyapunov function with $\rho = 320$ for the CSTR. The secure operating region $\Omega_{\rho_{\text{secure}}}$ with $\rho_{\text{secure}} = 90$ for the LEMPC in Eq. 7.5 is determined accounting for the impact of cyber-attacks. Specifically, the size of the secure operating region $\Omega_{\rho_{\text{secure}}}$ can be adjusted depending on desired threshold for economic benefits and system dynamics. If the process dynamics is very fast, then more room needs to be vacated between Ω_ρ and $\Omega_{\rho_{\text{secure}}}$ to accommodate for the fast changes in process states when under cyber-attacks. However, since the economic gain depends heavily on the size of operating region, a conservative secure operating region $\Omega_{\rho_{\text{secure}}}$ may reduce the economic

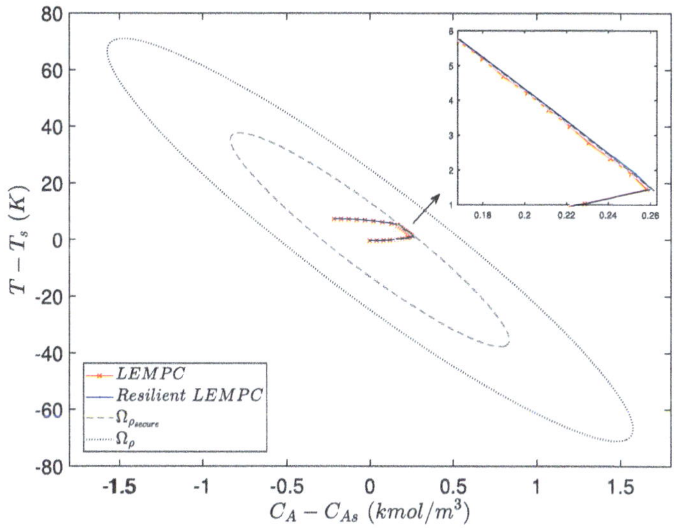

Fig. 7.7 Evolution of measured process states within one material constraint period under resilient LEMPC (blue trajectory) and under LEMPC (red trajectory)

benefits significantly. Therefore, the size of $\Omega_{\rho_{\text{secure}}}$ should be determined to balance economic performance and operational stability.

Resilient Operation of LEMPC

We carry out the closed-loop simulations for the CSTR of Eq. 7.26 under the resilient control of LEMPC using combined closed-loop and open-loop control actions (see Sect. 7.4.2), and under the LEMPC of Eq. 7.5 with the same initial conditions $x_0 = [0, 0]^T$, for one material constraint period t_{N_p}. Figure 7.7 shows the trajectories of the measured process states using the resilient LEMPC control strategy and the LEMPC of Eq. 7.5 when the process is under no attack. The control system switches to using open-loop control actions at $t_s = 0.0175$ h. For $t_0 \leq t_k < t_s$, the CSTR is operated under the LEMPC of Eq. 7.5 with real-time state measurements (i.e., closed-loop control), and it can be seen that the measured process states are well within the secure operating region $\Omega_{\rho_{\text{secure}}}$. Then, the measured process states first reach the boundary of the secure operating region (i.e., when $\rho_{\text{secure}} - V(\bar{x}(t_k)) \leq c$, where $c = 0.5$ for this case study) at $t_s = 0.0175$ h; since then, the feedback measurements are no longer trustworthy as this may also occur under a cyber-attack. We deactivate the LEMPC of Eq. 7.5 at $t_s = 0.0175$ h, and implement the control actions $u'(t_k)$ from the open-loop optimization of Eq. 7.16 that are solved based at the beginning of simulation for the remaining time of this operating period, i.e., $\forall t_s \leq t \leq t_{N_p}$.

Even in the case that no cyber-attack, no model mismatch, and no process disturbance is present, the resulting state trajectories under the resilient LEMPC (closed loop followed by open-loop control actions after the switching time t_s), and under LEMPC (closed-loop only) are slightly different. This is because the open-loop control actions are computed using an LEMPC with a prediction horizon of $N_p = 24$ that covers the entire material constraint period, while the prediction horizon used in the LEMPC with feedback measurements is much shorter, i.e., $N = 8$, and is rolled forward as new measurements are received. Therefore, the control actions $u(t_k)$ computed from online optimization will be slightly different from the open-loop control actions $u'(t_k)$, which lead to slightly different state trajectories over time.

Despite the subtle differences in the state trajectory, the process states are maintained within the secure operating region (hence the stability region) at all times under the closed-loop control of LEMPC followed by the open-loop control actions. It is important to note that the total economic benefits (i.e., $\int_{t_0}^{t_{N_p}} l_e(\bar{x}(t))dt$) for the steady-state operation (i.e., the process is operated at steady-state for all times) is 0.6397 kmol/m^3, which is much less than that under time-varying EMPC operation. The total economic benefits from t_0 to t_{N_p} using closed-loop-only control actions from the LEMPC of Eq. 7.5 is 0.7936 kmol/m^3, and using the resilient control strategy outlined in Sect. 7.4.2 is similarly 0.7947 kmol/m^3. Therefore, the effectiveness of the resilient control strategy is demonstrated for the system that is under no attack as closed-loop stability is still ensured and economic performance is close to that under the closed-loop control of LEMPC. Additionally, the similarity in the two state-space trajectories also suggests that in the absence of cyber-attacks, the evolution of true process states after implementing the resilient control strategy will highly resemble that under closed-loop control after a cyber-attack occurs.

Cyber-attack Resiliency Assessment

In Sect. 7.4.2, we developed the resilient control strategy to prevent true process states from exiting the stability region Ω_ρ in the presence of sensor cyber-attacks. Figure 7.8 shows the trajectories of attacked state measurements and true process states under LEMPC and under resilient LEMPC when the temperature sensor measurement is compromised by min-max, geometric, replay, and surge attacks, respectively. All the state trajectories are from the same initial conditions $x_0 = [0, 0]^T$ and are simulated for one material constraint period. In all cases, the cyber-attacks will be present in the system until it has been successfully detected. After the sensor is tampered by a cyber-attack, the resulting falsified state measurements will not exit the secure operating region $\Omega_{\rho_{secure}}$ so as to remain inconspicuous to the control engineer. The results of the NN-based detector and of the CSTR simulation after the detection will be shown in Sect. 7.5.

Min-max and surge cyber-attacks are added at $t = t_s = 0.0175$ h such that there will be no suspicious deviation in the Lyapunov function of the system. At $t = 0.0175$ h, both the attacked state measurement and the true process state reach the boundary of the secure operating region, i.e., $V(x(t_s)) = V(\bar{x}(t_s)) = \rho_{\text{secure}}$. As shown in Fig. 7.8a, d, the true process states exit $\Omega_{\rho_{secure}}$ and eventually Ω_ρ if we continue to use closed-loop control actions from the LEMPC of Eq. 7.5 after the temperature sensor is attacked by min-max and surge attacks, respectively. To mitigate the impact of cyber-attacks, we deactivate the closed-loop control and implement the resilient LEMPC control strategy at $t = 0.0175$ h such that the control system will not be affected by falsified feedback measurements. Specifically, since $t = 0.0175$ h,

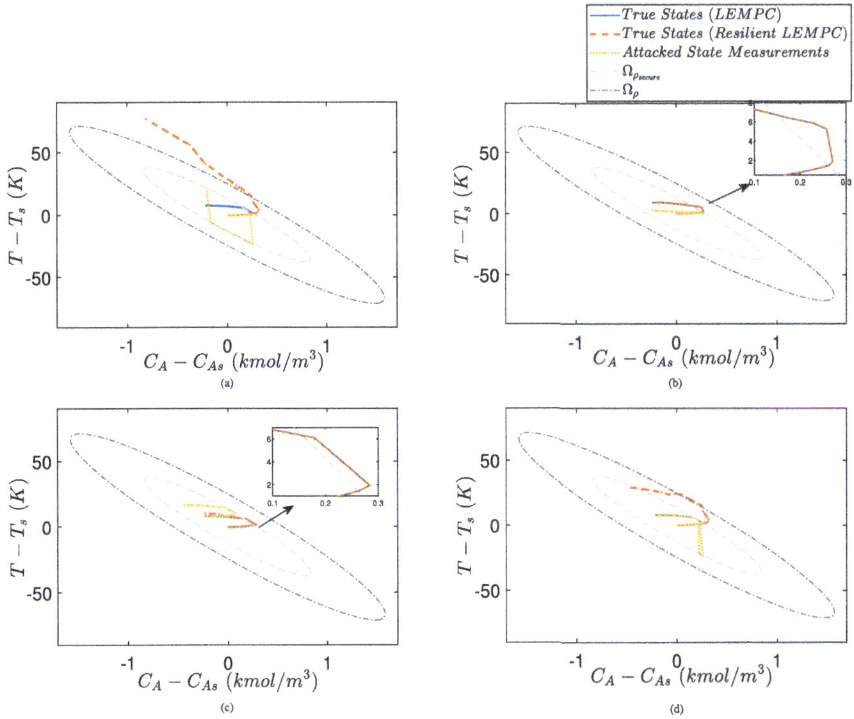

Fig. 7.8 Evolution of attacked state measurements (yellow trajectories) and true process states over one material constraint period under resilient LEMPC (red trajectories) and under LEMPC (blue trajectories) when **a** min-max, **b** geometric, **c** replay, and **d** surge attacks are targeting the temperature sensor, where the dashed ellipse is $\Omega_{\rho_{\text{secure}}}$ and the dash-dotted ellipse is the stability region Ω_{ρ}

we implement the open-loop control actions that are calculated based on a correctly measured set of initial conditions to the CSTR until the end of the material constraint period, i.e., $t = t_{N_p} = 0.06$ h. As a result, we can see that the true process states remain inside $\Omega_{\rho_{\text{secure}}}$ at all times, and the evolution of the true process states is almost identical to that under secure closed-loop control. Therefore, it is concluded that the system stays resilient to surge and min-max attacks, with guaranteed stability and desired control performance.

However, when other types of cyber-attacks occur, under which, the falsified state measurement does not approach the boundary of $\Omega_{\rho_{\text{secure}}}$, the resilient control strategy may not be as effective as it was in handling min-max and surge attacks. To illustrate this, we carry out the simulation study under a geometric attack of Eq. 7.8 with $\beta = x(t) * (1.001)$ and $\alpha = 0.1$ on the temperature measurements starting at $t = 0.01$ h, and show the simulation results in Fig. 7.8b. As cyber-attacks could happen at any time instant during operation, geometric attacks are designed and inserted as such to demonstrate the incapability of the resilient control strategy in handling geometric attacks or attacks alike. Specifically, as can been seen in Fig. 7.8b, the state measurements never reach the boundary of $\Omega_{\rho_{secure}}$ for the entire duration of cyber-attack, and therefore, the condition for deactivating closed-loop control is never met. Despite having the open-loop control actions computed at $t = 0\ hr$ using the correctly measured initial conditions, these control actions were never used in the presence of cyber-attacks. As a result, the control system implements the closed-loop control actions based on false measurements all the time, and ultimately drives the true state outside of $\Omega_{\rho_{secure}}$ during operation. In this case, the resilient control strategy fails to ensure closed-loop stability as the geometric cyber-attack influences the system in a different way than min-max or surge attack.

Moreover, even when the open-loop control actions are used once the measured states reach the boundary of $\Omega_{\rho_{secure}}$, there may be situations where the true process states still exit $\Omega_{\rho_{secure}}$ because the open-loop control actions are calculated based on false sensor measurements. To illustrate this scenario, we carry out the simulation study for the CSTR under a replay attack starting at $t_0 = 0$ h, and show the simulation results in Fig. 7.8c. The replayed signals in this example span the duration of one material constraint period, and are from real closed-loop state measurements that start from a different initial condition, $\bar{x}_0 = [-0.2107\ \text{kmol/m}^3; 7.8047\ \text{K}]$. Therefore, it is straightforward to show that the open-loop control actions computed based on $\bar{x}_0$ are also not correct for the prediction horizon of N_p because the initial condition $\bar{x}_0$ is incorrect. As a result, although we deactivate closed-loop control when the falsified state measurement reaches the boundary of $\Omega_{\rho_{secure}}$ at $t = 0.0175$ h, the true process states still leave the secure operating region due to incorrect open-loop control actions.

While the true process states did not leave the stability region Ω_ρ under replay and geometric attacks in this example, it should be pointed out that this may not be the case for a different replay attack that used more aggressive open-loop control actions, a different geometric attack with larger α (geometric factor), or for a faster process. In other words, using the resilient control strategy only may not always guarantee system stability, and thus, an effective cyber-attack detection mechanism should be included.

Detectors Training and Testing

To train NN-based detectors, we collect training data from closed-loop operation with the secure LEMPC outlined in Eq. 7.5. Simulation period is one material constraint period $t_{N_p} = 0.06$ h with $N_p = 24$. Cyber-attacks are added at random times and last until the end of the simulation period. The MATLAB Machine Learning and Deep Learning Toolboxes are used to construct and train the neural network models.

Based on the full-state measurement $\bar{x}(t)$, we can calculate the reaction rate $r_B(\bar{x})$ of product B at each time instant t_k from $k = 0$ to $k = N_p$ following Eq. 7.27, where $C_A = \bar{x}_1 + C_{A_s}$ and $T = \bar{x}_2 + T_s$. The neural network inputs, which are denoted as $p(\bar{x})$, are the time-varying trajectory of the rate of change in $r(\bar{x})$ over one material constraint period $N_p = 24$:

$$p(\bar{x}(t)) = \frac{\mathrm{d}r(\bar{x})}{\mathrm{d}t}. \tag{7.30}$$

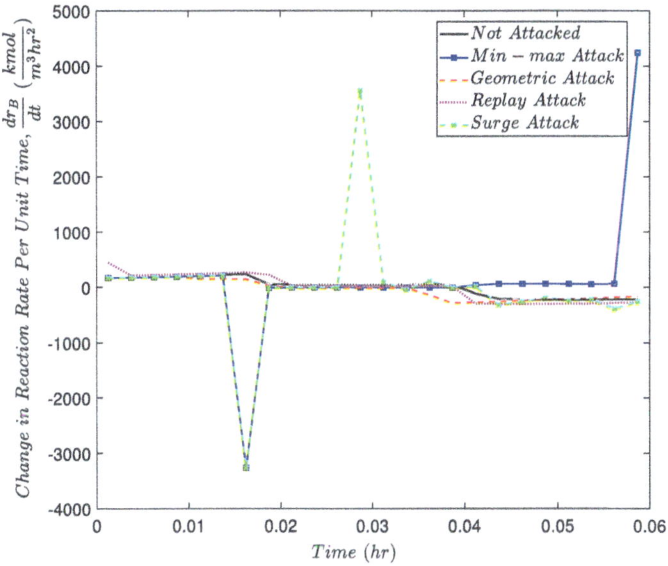

Fig. 7.9 Time-derivative of the reaction rate r_B of Eq. 7.27 based on measured process states over one material constraint period, when the temperature sensor is under no attack, and under min-max, geometric, replay, and surge attacks, respectively

Figure 7.9 shows the evolution of $p(\bar{x})$ when the temperature sensor is under min-max, geometric, replay, and surge attacks, and under no attack. Each sample starts from a different initial condition within Ω_ρ and consists of a 1×24 array of $p(\bar{x})$. Equal number of samples are collected for each output label from extensive closed-loop simulations, from which 30% are used for testing and 70% are used for training.

We first train a NN-based detector for detecting min-max attacks. We develop a feedforward NN model that has two hidden layers with 12 and 10 neurons in each layer, respectively. The activation function *tansig* in the form of $g_{1,2}(z) = \frac{2}{1+e^{-2z}} - 1$ are utilized in both hidden layers. The activation function *softmax* in the form of $g_3(z_j) = \frac{e^{z_j}}{\sum_{i=1}^{H} e^{z_i}}$, where H denotes the number of class labels, is used in the output layer to provide a predicted probability of the class labels. The training process optimizes the weights and biases to minimize the Bayesian regularized mean squared error cost function $S(w)$. The Levenberg–Marquardt algorithm is used for optimization, in which the gradient and the Hessian matrix of $S(w)$ are calculated through backpropagation method. We collect a total of 750 data samples for each class. The training accuracy for a 2-class detector is up to 98.9% after 70 training epochs, and the computation time is around 2.05 seconds. The testing accuracy of this detector against different types of cyber-attacks types is shown in Table 7.1. Due to the vast difference in the trends of $p(\bar{x})$ under min-max attacks, and under geometric attack (Fig. 7.9), it is noticed that geometric attacks are not identified as being attacked in this case.

Table 7.1 Detection accuracies of NN detectors in response to different types of cyber-attacks

	Detector 1 (Attacked versus Not attacked)	Detector 2 (Min-max versus Geometric versus Not attacked)
Min-max	98.3%	89.7%
Geometric	2.4% (Attacked)	71.1%
Surge	87.0% (Attacked)	71.0% (Min-max); 10.0% (Geometric)
Not Attacked	98.4%	95.6%

Then, we develop a second detector that can classify between three classes: attacked by geometric cyber-attacks, attacked by min-max cyber-attacks, and not attacked. Therefore, in addition to indicating the presence of cyber-attacks, the detector is capable of differentiating the types of cyber-attacks. We include the geometric attacks in the training as it exhibits very different behavior than min-max attacks, and thus, cannot be efficiently detected by the 2-class detector. We develop 3-class detector using a feedforward NN that has two hidden layers with 15 and 12 neurons each. The cost function and the activation functions are the same as those in the 2-class detector. The overall training accuracy for the 3-class detector is up to 91.8% after 300 training epochs, and the computation time is around 39.48 seconds. Table 7.1 shows the testing accuracies in response to min-max, geometric, and surge attacks, respectively. The detector identifies geometric and min-max attacks as their respective labels accurately, and it classifies 10.0% as geometric, 71.0% of surge attacks as min-max, and the remaining 19% are wrongly classified as "not attacked".

Remark 7.9 Since replay attacks use past signals of one entire material constraint period to mimic the secure operation starting at a different initial condition, they are essentially a different sample that belongs to the class of "not attacked", and will be rightfully classified as being "not attacked". After one material constraint period, it will be seen that the falsified signals follow exactly the state trajectory of previous secure measurements, and thus, replay attacks remain undetectable by the NN detectors.

Online Detection

Detector 1 is developed to detect surge and min-max attacks, whereas detector 2 is developed to detect geometric attacks. At the end of each material constraint period, we activate the corresponding detector to examine the state measurements collected over the last material constraint period. As replay attacks cannot be detected (see Remark 7.9), the online detection results are also not shown. The profiles of the measured and the true process states for the system under the resilient LEMPC and different types of cyber-attacks (i.e., min-max, geometric, and surge cyber-attacks) are shown in Fig. 7.10. Two material constraint periods are simulated, in which NN-based detector is triggered at the end of the first period and of the second period. It

Fig. 7.10 Evolution of attacked state measurements (red trajectories) and true process states (blue trajectories) over two material constraint periods under the resilient LEMPC when **a** min-max, **b** geometric, and **c** surge attacks, targeting the temperature sensor are successfully detected by a NN detector at the end of the first material constraint period, $t = 0.06$ h, where the dashed ellipse is $\Omega_{\rho_{\text{secure}}}$ and the dash-dotted ellipse is the stability region Ω_ρ

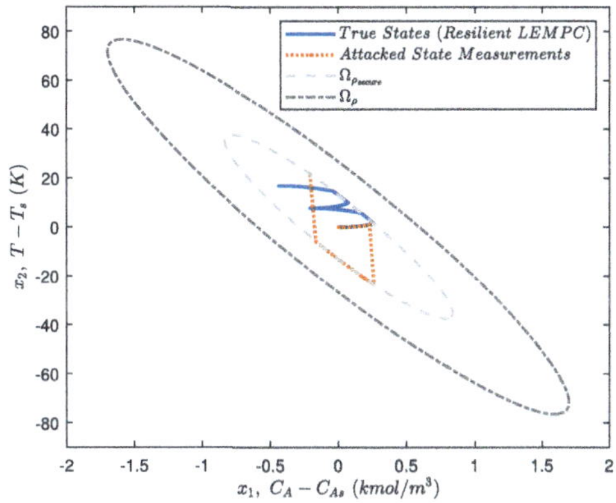

(a)

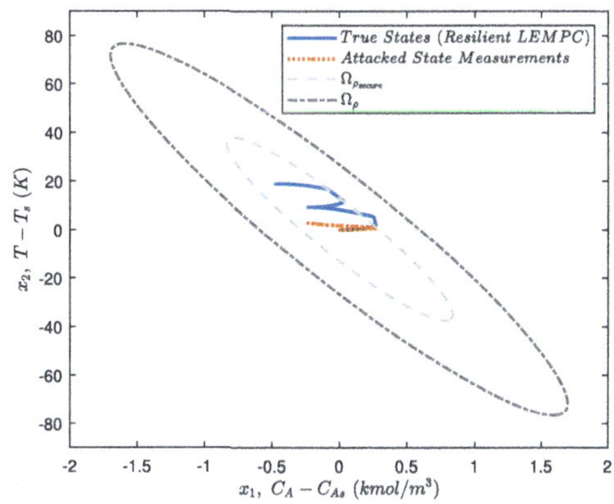

(b)

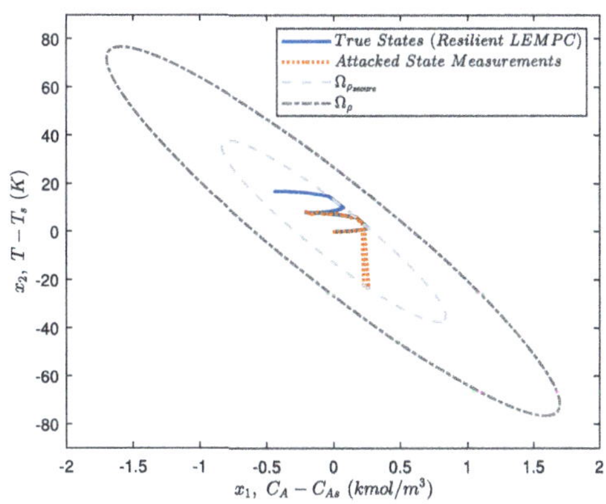

(c)

is demonstrated that detector 1 correctly detects surge and min-max attacks at the end of the first constraint period $t = 0.06$ h based on the measured state trajectory of $p(\bar{x}(t))$ from $t = 0$ h to $t = 0.06$ h. After the detection of min-max attack, the old set of sensors are disconnected from the control system, and a secure set of redundant sensors that are not tampered by cyber-attacks will be used for the second period. The operation continues with the secure sensor measurements, and at the end of the second material constraint period $t = 0.12$ h, detector 1 is activated again to correctly classify the secure measurements as "not attacked".

Moreover, the type of cyber-attack can be identified by the detector if the neural network is designed to train a particular cyber-attack type as a separate class (i.e., "geometric") from other attack types (i.e., "min-max"). In Fig. 7.10b, it is shown that although the closed-loop control based on false feedback signals was not deactivated, and the true process states exited $\Omega_{\rho_{\text{secure}}}$ during the first material constraint period, detector 2 still correctly identifies the falsified state measurements as geometric attacks at the end of the first material constraint period. After the attack is detected and the sensor devices are switched to the respective secure back-up sensors, the $p(\bar{x}(t))$ profile over the second material constraint period (i.e., from $t = 0.06$ h to $t = 0.12$ h) is correctly identified by detector 2 as "not attacked". This implies that the cyber-attacks can still be efficiently detected at the end of each material constraint period, even if the attacked measurement deliberately avoids approaching the boundary of $\Omega_{\rho_{\text{secure}}}$, and closed-loop stability may not be ensured under the resilient control strategy. Following the successful detection, mitigation measures can be taken to reduce the impact of the cyber-attacks. This also suggests that a conservative secure region and a shorter material constraint period could be used in safety-critical systems to activate cyber-attack detection more frequently.

Real-time State Reconstruction

In addition to the integrated closed-loop and open-loop control, we also carry out the closed-loop simulations for the CSTR system of Eq. 7.26 under LEMPC with state reconstruction that was discussed in Sect. 7.4.3. In this case, we assume that under no attacks, the CSTR system of Eq. 7.26 is operated in the region Ω_{ρ_e} with $\rho_e = 280$, and the cyber-attack detection is implemented in real time, i.e., at each sampling period, instead of after each material constraint period. When cyber-attacks are present in the closed-loop system, the true state may exit Ω_{ρ_e} under LEMPC. Thus, to maintain the state within the stability region Ω_ρ before the detection of cyber-attacks, the size of Ω_{ρ_e} needs to be carefully chosen.

We develop multiple RNN models for the state reconstructors under min-max, surge, and geometric cyber-attacks, respectively. All of them have two hidden layers with 60 neurons in each layer. The datasets are generated from open-loop simulations and contain around 150,000 data sequences. The state-of-the-art machine learning library, Keras, is used to train the RNN models. The averaged mean square errors are calculated for the three state reconstructors (i.e., for min-max, surge, and geometric cyber-attacks), and the results based on training and validation datasets are demonstrated to be below 10^{-5}. The averaged computation time for training one neural network is around 2.5 h. Note that the training is done completely off-line, while

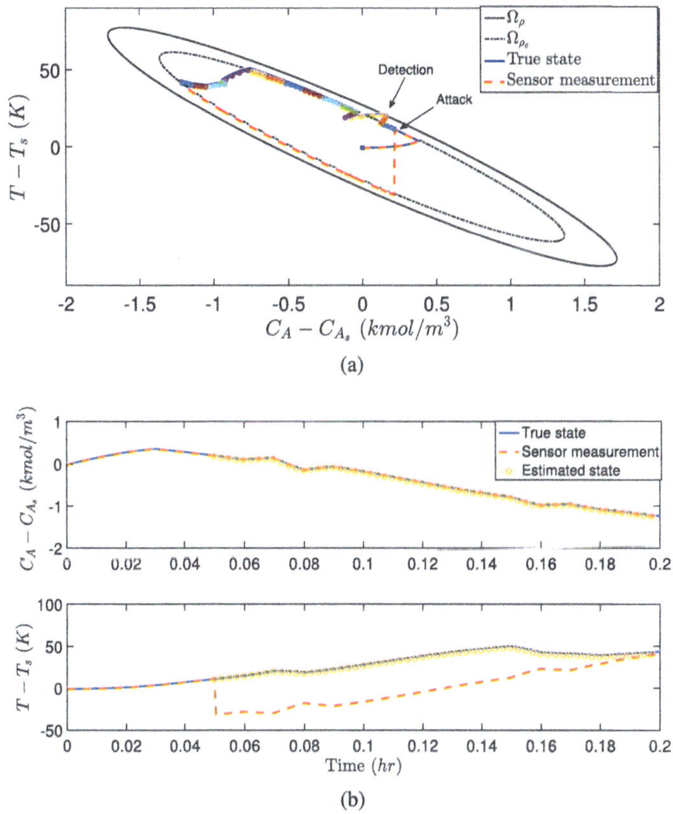

Fig. 7.11 **a** State-space trajectories, and **b** closed-loop profiles of reconstructed state (marked by colored circles), measured state (red), and true state (blue) for the CSTR system of Eq. 7.26 under LEMPC when the temperature sensor is attacked by a min-max cyber-attack at $t = 0.05$ h

the obtained RNN model is used within MPC on-line for state estimation. Since the neural network model is essentially a nonlinear function that predicts outputs (i.e., estimated state values) based on inputs (i.e., past state measurements), the computation of state estimation within MPC is done almost instantaneously, which is different from the training process that takes hours to finish. Therefore, the RNN-based state reconstructor can be implemented in practical systems as the computational time for RNN-based state estimation within controller is negligible compared to the sampling time.

Figures 7.11a, b, 7.12a, b, and 7.13a, b show the state-space trajectories and state profiles for the closed-loop system under min-max, surge, and geometric cyber-attacks, respectively. Specifically, the system of Eq. 7.26 is initially operated under no attacks from an initial condition $x_0 = (0,\ 0)$. As shown in Fig. 7.11a, at $t = 0.05$ h, the min-max cyber-attack is imposed on the temperature sensor to render the falsified state measurement (dashed red trajectory) on the lower boundary of Ω_{ρ_e}. After a

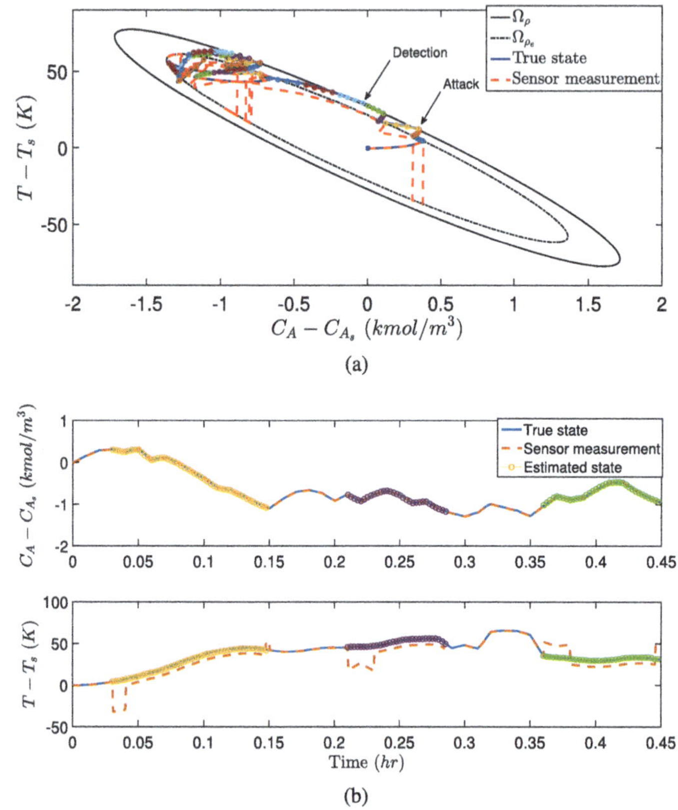

Fig. 7.12 **a** State-space trajectories, and **b** closed-loop profiles of reconstructed state (marked by colored circles), measured state (red), and true state (blue) for the CSTR system of Eq. 7.26 under LEMPC when the temperature sensor is attacked by surge cyber-attacks at $t = 0.03$ h, $t = 0.21$ h, and $t = 0.36$ h

few sampling steps, the true state trajectory (blue trajectory) exits the Ω_{ρ_e} from the upper boundary. Once the cyber-attack is detected at $t = 0.07$ h, the true states (colored dotted trajectories) are reconstructed based on past control actions and falsified sensor measurements. As we can see from Fig. 7.11a, the LEMPC of Eq. 7.5 using the estimated state re-stabilizes the CSTR system by maintaining the state in Ω_ρ. Additionally, it is demonstrated in Fig. 7.11b that the reconstructed temperature and concentration provide reliable state estimation for the feedback control with LEMPC as they both are very close to the true state values. Ideally, since detection is implemented in real time and reports the occurrence of a cyber-attack promptly, we will activate state reconstruction after the cyber-attack detector gives the first positive detection during online implementation to save computational power. In this study, to demonstrate the effectiveness of the proposed state reconstruction method using RNN models, the reconstructed states are plotted right after the occurrence of attacks.

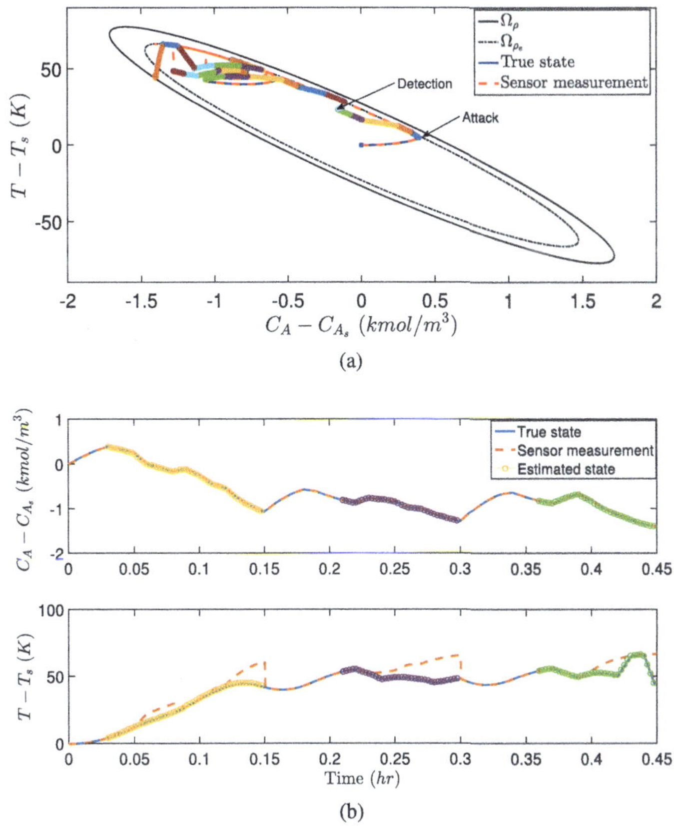

Fig. 7.13 **a** State-space trajectories, and **b** closed-loop profiles of reconstructed state (marked by colored circles), measured state (red), and true state (blue) for the CSTR system of Eq. 7.26 under LEMPC when the temperature sensor is attacked by geometric cyber-attacks at $t = 0.03$ h, $t = 0.21$ h, and $t = 0.36$ h

Furthermore, it is observed that in this example the RNN-based state reconstructor can successfully predict the true state values with a desired accuracy, even when the sensor measurements are not compromised. Therefore, we can activate state reconstruction at the start of the operation period, as long as the sensor measurements prior to which time remain secure.

Then we carry out the closed-loop simulation for system under surge cyber-attack over multiple EMPC operating periods, and show the simulation results in Fig. 7.12a, b. The surge cyber-attacks are added on temperature measurements in each material constraint period (i.e., from $t = 0$ h to $t = 0.15$ h, from $t = 0.15$ h to $t = 0.3$ h, and from $t = 0.3$ h to $t = 0.45$ h with $t_{N_p} = 0.15$ h). As shown in Fig. 7.12b, under a surge attack, the compromised sensor measurement initially reaches the upper bound (i.e., the maximum allowable value) and uses a small deviation from true states afterwards such that it remains undetectable by conventional detection schemes, e.g., CUSUM. Similarly, RNN-based state reconstructor accurately estimates the true state trajectory and the LEMPC using estimated state ensures the boundedness of the process state within the stability region. Additionally, Fig. 7.13a, b shows the closed-loop simulation results under geometric cyber-attack. The analysis is omitted here as it is very similar to the above.

7.6 Conclusions

In this chapter, the secure operation of nonlinear chemical processes under MPC and EMPC was presented via the design of a neural-network-based cyber-attack detector and of resilient control strategies. Considering a general class of nonlinear systems, the NN-based detection system was first developed with the sliding detection window to detect intelligent cyber-attacks. Subsequently, resilient control systems were developed with several control strategies including redundant sensors, combined closed-loop and open-loop control, and post cyber-attack state reconstruction. Through the simulation of a CSTR process, we demonstrated that the process stability was maintained against particular types of malicious cyber-attacks, namely min-max, geometric and surge attacks, under the proposed control strategy, and comparable economic performance was achieved compared to nominal operation without any attacks. Additionally, the RNN-based state reconstructor successfully estimated the true states in real-time implementation of LEMPC such that stability was guaranteed for the nonlinear system upon detection of cyber-attacks.

Chapter 8
A Two-Tier Control Architecture For Cybersecurity and Operational Safety

8.1 Introduction

While the resilient control systems in Chap. 7 were shown to successfully mitigate the impact of cyber-attacks and re-stabilize the system upon detection, the control systems themselves are not inherently cyber-secure, which means they have to rely on redundant sensors or accurate state estimation in the presence of cyber-attacks. Although many advances have been made in improving efficiency of data-based detectors and in the development of resilient control schemes in response to cyber-attacks, it is possible that the control system has to be shut down due to unavailability of state measurements from redundant sensors or of reliable state estimation.

This chapter presents a detector-integrated control system with a two-tier control architecture that can ensure closed-loop stability of nonlinear processes upon detection of cyber-attacks without having to switch to secure redundant sensors or state estimation. Traditionally, control systems are developed based on a small number of actuators and sensors with point-to-point wired communication. Hybrid communication networks that incorporate additional (wired or wireless) networked sensor and actuator devices into the existing point-to-point communication networks may benefit the operation of chemical processes. In the study of hybrid communication networks that use both point-to-point and networked sensors, cybersecurity is a key issue for secure and stable operation of chemical processes. Working with a general class of nonlinear systems, cyber-secure lower-tier controllers that stabilize a multivariable nonlinear process at the steady-state based on point-to-point dedicated sensor measurements are coupled with an upper-tier model predictive controller (MPC) that uses networked sensor measurements to improve closed-loop performance. The two-tier control system guarantees that the process stays immune to malicious cyber-attacks that target the networked sensor measurements to destabilize the system. Additionally, the safety systems discussed in the previous chapters are also integrated with cyber-secure control systems to ensure safe operation upon successful detection of

Z. Wu and P. D. Christofides, *Process Operational Safety and Cybersecurity*, Advances in Industrial Control, https://doi.org/10.1007/978-3-030-71183-2_8

cyber-attacks. In this chapter, we first develop machine-learning-based attack detectors that identify the occurrence of cyber-attacks based on real-time sensor measurements. Simulation results of the application to a multivariable nonlinear process example, e.g., [156, 185, 232], demonstrate that the detection algorithms can efficiently detect and distinguish between multiple types of intelligent cyber-attacks. A reactor-reactor-separator process will be used to illustrate the application of the machine-learning-based attack detectors and of the two-tier control architecture. It will be demonstrated in the simulation studies that the cyber-secure control architecture ensures closed-loop stability through reconfiguration of the control system once the cyber-attacks are successfully detected.

8.1.1 Class of Nonlinear Systems

The class of continuous-time nonlinear systems considered is described by the following state-space form:

$$\dot{x}(t) = f(x(t), u_c(t), u_a(t), u_s(t)) \tag{8.1a}$$

$$y_c(t) = h_c(x(t)), \;\; y_a(t) = h_a(x(t)) \tag{8.1b}$$

where $x \in D \subset \mathbf{R}^{n_x}$ is the state vector, $y_c(t) \in \mathbf{R}^{n_{y_c}}$ represents the vector of measured states that are continuously sampled (e.g., $y_c(t)$ is the reactor temperature in the benchmark chemical process example in this chapter), and $y_a(t) \in \mathbf{R}^{n_{y_a}}$ represents the vector of networked state measurements that are asynchronously sampled at $t = t_k$ (e.g., $y_a(t)$ is product concentration in the reactor example); u_c and u_a are the manipulated input vectors, which are constrained by $[u_c \in \mathbf{R}^{m_{u_c}}, u_a \in \mathbf{R}^{m_{u_a}}] \in U := \{u_i^{\min} \leq u_i \leq u_i^{\max}, i = 1, \ldots, m_{u_c} + m_{u_a}\}$. $u_s \in \mathbf{R}^{m_{u_s}}$ represents the vector of inputs to the process controlled by the safety system, with each component $u_{s,i}$, $i = 1, \ldots, m_{u_s}$, restricted to a set of discrete (on-off) values. For simplicity, we use $u_{s,i} = 0$ and $u_{s,i} = 1$ to represent deactivation and activation of the safety system, respectively, throughout this chapter. The safety system is then incorporated within the two-tier control system under many different operating conditions to allow the impacts of various safety system actions based on different states and threshold crossings to be evaluated so that: (1) Appropriate recommended procedures can be developed for operators given the alarms, (2) Appropriate automatic actions can be taken by the emergency shutdown system (ESS), and (3) Appropriate equipment choices can be made for the relief system (e.g., the safety relief valves that are designed to open/close to maintain chemical plant pressure within a desired range; see, also, Chap. 5 for the integration of safety systems with control systems).

Through y_c and y_a, we assume that the full-state measurements are available at t_k. Without loss of generality, the vector function $f(\cdot, \cdot, \cdot, 0)$, $h_c(\cdot)$ and $h_a(\cdot)$ are assumed to be sufficiently small with $h_a(0) = 0$, $h_c(0) = 0$, and $f(0, 0, 0, 0) = 0$

such that the origin is an equilibrium point of the system of Eq. 8.1 under $u_c(t) = 0$, $u_a(t) = 0$ and $u_s(t) = 0$. Additionally, the initial time t_0 is taken to be zero ($t_0 = 0$).

Remark 8.1 To simplify the notation, we assume that (asynchronous) full-state measurements are available for the controller $u_a(t)$. The results can be extended to the case where only partial state information is available.

8.2 Cyber-Secure Two-Tier Control Architecture

This section presents a cyber-secure control architecture that integrates a lower-tier controller that stabilizes the nonlinear system at the steady-state based on the dedicated sensor measurements, $y_c(t)$, with an upper-tier, advanced control system (e.g., model predictive control) that improves closed-loop performance significantly above what could be achieved with the lower-tier control system using both networked ($y_a(t)$) and dedicated ($y_c(t)$) sensor measurements. Due to the asynchronous nature of networked sensor measurements $y_a(t)$, the unknown time interval between two asynchronous measurements should be taken into account in the design of upper-tier controller. The main objective of two-tier control architecture is to enhance closed-loop performance with the additional state information from networked sensors while maintaining closed-loop stability properties achieved by lower-tier controller. The formulations of upper-tier and lower-tier control systems are presented in detail below.

8.2.1 *Lower-Tier Control System*

Throughout this chapter, it is assumed that an explicit feedback controller $u_c(t) = \Phi_c(y_c) \in U$ exists to stabilize the nominal system of Eq. 8.1 (i.e., in the absence of cyber-attacks) at the steady-state. This stabilizing controller will be used as in the lower-tier control system as it only uses the continuous measurements $y_c(t)$ to compute control actions. Additionally, we assume that the origin of the nominal system of Eq. 8.1 can be rendered asymptotically stable under $u_c(t) = \Phi_c(y_c) \in U$ and $u_a(t) = 0$ (i.e., u_a is set to its steady-state value at all times). This stabilizability assumption implies that there exist a $\mathscr{C}^1$ control Lyapunov function $V(x) : D \to \mathbf{R}_+$ and class $\mathscr{K}$ functions $\alpha_i(\cdot)$, $i = 1, 2, 3, 4$, that satisfy the following conditions:

$$\alpha_1(|x|) \leq V(x) \leq \alpha_2(|x|), \tag{8.2a}$$

$$\frac{\partial V(x)}{\partial x} f(x, \Phi_c(y_c), 0, 0) \leq -\alpha_3(|x|), \tag{8.2b}$$

$$\left| \frac{\partial V(x)}{\partial x} \right| \leq \alpha_4(|x|) \tag{8.2c}$$

for all $x \in D \subset \mathbf{R}^{n_x}$, where D is an open neighborhood around the origin. Then, using the stabilizing controller $u_c = \Phi_c(y_c) \in U$, we characterize the closed-loop stability region Ω_ρ as a level set of $V(x)$ within the set D, i.e., $\Omega_\rho := \{x \in D \mid V(x) \leq \rho, \rho > 0\}$, from which the state trajectory of the closed-loop system remains within Ω_ρ and asymptotically converges to the origin under $u = \Phi_c(y_c) \in U$ for any initial condition in Ω_ρ. Therefore, for any initial condition inside Ω_ρ, closed-loop stability is guaranteed for the process using the lower-tier controller only, provided that secure and reliable sensor measurements are available to the lower-tier controller.

Remark 8.2 Static lower-tier controllers are considered in this section to simplify the discussion of stabilization of the nonlinear system of Eq. 8.1 at the steady-state. However, the formulation of lower-tier controllers can be extended to dynamic control schemes. For example, in Sect. 8.4, we use proportional–integral (PI) controllers as the lower-tier controllers to stabilize a nonlinear chemical process at the operating steady-state.

8.2.2 Upper-Tier Model Predictive Control System

While the continuous dedicated sensor measurements $y_c(t)$ are used to ensure closed-loop stability, the (potentially asynchronous) networked state measurements $y_a(t)$ can be used in the optimization of the control actions $u_a(t)$ in the upper-tier controller to improve the closed-loop performance than using the lower-tier controller only. In this study, to take advantage of the process model in optimizing process performance, and to control the process when the feedback is unavailable between two consecutive (asynchronous) measurements, we use model predictive control scheme in upper-tier control system. Specifically, the Lyapunov-based MPC (LMPC) with the contractive constraint designed based on the stability region characterized by the lower-tier controller is used as the upper-tier controller, such that the calculation of $u_a(t)$ will not affect the asymptotic stability of the closed-loop system. The LMPC is represented as the following optimization problem:

$$\min_{u_a \in S(\Delta)} \int_{t_k}^{t_{k+N}} l_t(\tilde{x}(t), \tilde{u}_c(t), u_a(t))dt \tag{8.3a}$$

$$\text{s.t.} \quad \dot{\tilde{x}}(t) = f(\tilde{x}(t), \Phi_c(h_c(\tilde{x}(t)), u_a(t), 0) \tag{8.3b}$$

$$\dot{\hat{x}}(t) = f(\hat{x}(t), \Phi_c(h_c(\hat{x}(t)), 0, 0) \tag{8.3c}$$

$$\tilde{x}(t_k) = \hat{x}(t_k) = x(t_k) \tag{8.3d}$$

$$[u_c(t), u_a(t)] \in U, \ \forall\, t \in [t_k, t_{k+N}) \tag{8.3e}$$

$$V(\tilde{x}(t_k)) \leq V(\hat{x}(t_k)), \quad \text{if } V(\tilde{x}(t_k)) > \rho_{\min} \tag{8.3f}$$

$$V(\tilde{x}(t)) \leq \rho_{\min}, \ \forall\, t \in [t_k, t_{k+N}), \quad \text{if } V(\tilde{x}(t_k)) \leq \rho_{\min} \tag{8.3g}$$

where Δ, N, and $S(\Delta)$ are the sampling period, the number of sampling periods in the prediction horizon, and a family of piecewise constant functions with the time interval of Δ, respectively. The optimal control actions $u_a^*(t)$ over the prediction horizon $t \in [t_k, t_{k+N})$ are calculated at time t_k by the LMPC optimization problem of Eq. 8.3 based on a full-state measurement, from which the first control action, i.e., $u_a(t) = u_a^*(t_k|t_k)$, is applied in open loop to the nonlinear system until the next full-state measurement x of y_c and y_a in Eq. 8.1b is available to the optimization problem of LMPC. Then, the LMPC optimization problem will be solved with the new state measurements, and the above process is repeated until the end of the operating period. Note that if the time between two consecutive asynchronous measurements is longer than the prediction horizon $N \cdot \Delta$, then we set u_a to its steady-state value (i.e., $u_a = 0$) for the remaining time in the asynchronous sampling interval past the prediction horizon, such that it will not affect the closed-loop stability achieved by the lower-tier controller. Meanwhile, the lower-tier control actions $u_c = \Phi_c(y_c)$ are still continuously calculated based on continuous measurement feedback y_c to drive the process state towards the steady-state. $\tilde{x}(t)$ and $\hat{x}(t)$ in Eqs. 8.3b and 8.3c are the predicted states of the nominal system under the two-tier control system (i.e., $u_c = \Phi_c(y_c)$ where $y_c = h_c(\tilde{x})$ and u_a is optimized by LMPC), and under the lower-tier controller only (i.e., $u_c = \Phi_c(y_c)$, and u_a is set to 0), respectively. Equation 8.3d defines the initial condition for the optimization problem of LMPC, which is the full-state measurement received at the time t_k. Equation 8.3e defines the constraints on control actions for both lower-tier and upper-tier controllers.

Since the origin of the nonlinear system of Eq. 8.1 is rendered asymptotically stable under the lower-tier controller that satisfies the conditions in Eq. 8.2, the constraint of Eq. 8.3f ensures that the Lyapunov function value of the closed-loop system under two-tier control, $V(\tilde{x}(t_k))$, is not greater than that under lower-tier control alone $V(\hat{x}(t_k))$. Hence, the controller forces the closed-loop state to move towards the origin under the contractive constraint of Eq. 8.3f, and thus, is also bounded in the stability region Ω_ρ for all times under two-tier control. When the state approaches the steady-state and enters a small region around the steady-state, i.e., $x(t_k) \in \Omega_{\rho_{\min}}$, where $\Omega_{\rho_{\min}}$, $0 < \rho_{\min} < \rho$ is a level set of Lyapunov function, the constraint of Eq. 8.3g requires that the future states remain inside $\Omega_{\rho_{\min}}$ for the entire prediction horizon. Since the closed-loop state is ultimately bounded in $\Omega_{\rho_{\min}}$ that is very close to the origin under the LMPC of Eq. 8.3, the system is considered practically stable. Additionally, the two-tier control system improves the overall closed-loop performance through the optimization problem of LMPC, while maintaining the system stability by using the stabilizing constraint based on lower-tier control actions. It should be pointed out that the upper-tier MPC is executed only when a full-state measurement is available from both the asynchronous and continuous sensor measurements. Specifically, the continuous measurements are readily used by LMPC as they are continuously measured in a point-to-point sensor network and sent to the lower-tier controller to compute the control actions u_c (i.e., the stabilizing controller). Therefore, the execution of the LMPC optimization problem basically depends on the availability of (asynchronous) networked sensor measurements. Figure 8.1 shows

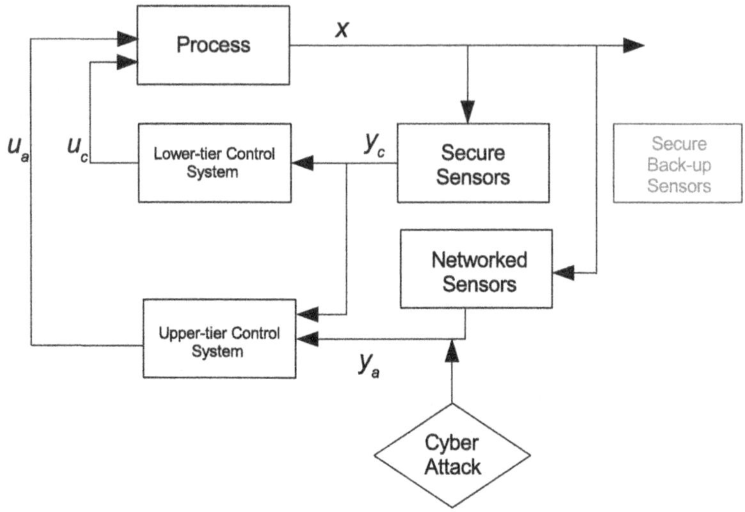

Fig. 8.1 Two-tier control-detector architecture with the upper-tier controller (i.e., MPC) using both networked and continuous (secure) sensor measurements, and the lower-tier controllers using only continuous (secure) sensor measurements, where the networked sensors are vulnerable to cyber-attacks

a structure of the two-tier control system, in which the networked sensors, $y_a(t)$ that will be used in LMPC are vulnerable to cyber-attack.

Remark 8.3 The lower-tier controller views the input $u_a(t)$ as a disturbance to the process if the upper-tier controller that manipulates $u_a(t)$ is designed improperly. Therefore, to improve closed-loop performance while maintaining system stability, the upper-tier controller should be designed accounting for the decisions that are made by the lower-tier controller. Specifically, in the formulation of upper-tier MPC system, the upper-tier controller is switched off when the system starts operating in open loop (i.e., all the control actions optimized over the prediction horizon have been implemented before the next asynchronous measurements are available). In this case, the last received optimal control actions from upper-tier controllers are no longer useful for the lower-tier controller to improve the closed-loop performance, and may even act as a disturbance to the process. As the two-tier control architecture is inherently stable (due to the stability properties of lower-tier controllers), the main challenge for the upper-tier controller is to improve closed-loop performance using non-reliable communications in a way such that closed-loop stability is not compromised. Therefore, when implementing upper-tier MPC, we will set the control action of the upper-tier controller to zero after a certain time to maintain the stability properties.

Remark 8.4 Note that the control systems using dedicated, local control networks have already been implemented in many chemical plants for years, and these controllers will not be replaced by networked control systems. Instead, to improve closed-loop performance and maintain system stability, we develop the networked control systems by augmenting the pre-existing control systems with networked sensor measurements. This supports the assumption we made at the beginning of this chapter that a stabilizing lower-tier controller exists for the nonlinear system based on the continuous sensor measurements.

Remark 8.5 The two-tier control system with stabilizing lower-tier controllers and upper-tier model predictive controllers can guarantee closed-loop stability in the sense that the closed-loop state remains within the stability region for all times, and is ultimately bounded in a small neighborhood around the origin for any initial condition in the stability region under nominal operating conditions (i.e., in the absence of process disturbances and cyber-attacks). The interested readers may refer to [111] for the stability analysis of the two-tier control architecture. Additionally, it should be noted that although the closed-loop performance is improved in general using two-tier control architecture as the cost function accounts for process performance index, we may not be able to derive quantitative results for guaranteed improvement of closed-loop performance using two-tier control architecture over other controllers, unless an infinite horizon is utilized.

8.3 Cyber-Attack Design and Detection

In this section, we consider intelligent cyber-attacks that are adaptive to the process and control system behavior. The intelligent cyber-attacks are assumed to be process-aware in the sense that they have access to process information such as the control command signals (actuator attack), and the measurement feedback signals (sensor attack), or auxiliary information such as the bias and threshold parameters in conventional detection methods, e.g., cumulative sum (CUSUM) [43, 132] (see, also, Eq. 7.10 for the formulation of CUSUM). Particularly, in this study, the attacks are designed with the information on the existing alarms that indicate normal operating conditions for the output and input variables, as well as the stability region characterized for the closed-loop system under two-tier control. Additionally, in this study, we only consider attacks on sensor measurements. Under nominal operation (i.e., under no attack), the closed-loop system is operated normally under the two-tier control system as the sensor feedback measurements remain secure and reflect the true process state accurately. However, in the presence of cyber-attacks on sensor measurements, closed-loop stability is no longer guaranteed as the process state may be driven away from the equilibrium point and eventually outside of the stability region Ω_ρ under falsified states measurements. Additionally, the falsified state measurement under intelligent cyber-attacks will be set to a value inside the closed-loop stability region Ω_ρ such that feasible control actions still exist, but will have large enough magnitude of variations to disrupt the control objective. Specifically, the four most important types of cyber-attacks, i.e., min-max, geometric, replay, and surge attacks that have been discussed in Sect. 7.2.1 are considered in this chapter.

8.3.1 Attack Scenarios

Since part of the measurement feedback (i.e., networked sensor information) is asynchronous in the upper-tier control system, the networked sensor measurements are vulnerable to cyber-attacks. Due to the sparse and irregular measurements as well as the possibility that multiple states are attacked by intelligent cyber-attacks, the resulting deviations on the process may be undetectable by conventional fault-detection schemes or by control engineers. While the asynchronous networked measurements are vulnerable to cyber-attacks, we assume that the dedicated sensor measurements received by both lower-tier and upper-tier controllers remain secure based on the following reasons. Firstly, the two-tier control system is developed based on the assumption that the system is stabilizable under lower-tier controllers. In fact, the closed-loop stability is guaranteed under two-tier control using the constraints based on lower-tier control actions, for which, secure and reliable continuous measurements are required. Secondly, when a cyber-attack targeting networked measurements is successfully identified by detector, the control structure will be reconfigured and the lower-tier controller (i.e., secure stabilizing controller) can quickly mitigate the impact of cyber-attacks. Since the closed-loop system can be stabilized at the operating steady-state using the lower-tier controller only, we will shut off the upper-tier controller and stop using the corrupted networked measurements after the detection of cyber-attack is confirmed. However, in the worst-case scenario that the continuous measurements are also attacked, cybersecurity is no longer guaranteed for the two-tier control system because the lower-tier controllers are also unable to stabilize the system at the steady-state. Due to the above considerations, it is instrumental to have secure continuous sensor measurements to ensure cybersecurity for the nonlinear system under a two-tier control system.

To distinguish between normal device fluctuations and cyber-attacks, and to capture realistic sensor variance, we also consider bounded sensor noise in this study. Therefore, we consider the following two scenarios in this study:

1. Nominal model refers to the nonlinear system of Eq. 8.1 where no sensor noise is added on sensor measurements.
2. Noise model adopts the same nonlinear model of Eq. 8.1a, where sensor measurements are corrupted by Gaussian noise $w(t) \in W$ that is bounded by the set $W = \{w \in \mathbf{R}^{n_{yc}+n_{ya}} \mid |w| \leq w_{\max}\}$. We adjust the noise distribution (i.e., standard deviation) for different sensors based on the range of the measured process variables. Therefore, we modify Eq. 8.1b to the following form to account for the sensor noise :

$$y_c(t) = h_c(x(t)) + w(t), \ \ y_a(t) = h_a(x(t)) + w(t), \ \ w(t) \in W. \tag{8.4}$$

The training dataset is generated from extensive closed-loop simulations with attacks being introduced at random times i_0 with varying durations L_a during the simulation period. The reader is referred to Sect. 7.2.1.5 for a detailed simulation design guide. In both the cases of the nominal model and of the noise model, we classify the signals without attack as "no attack". Additionally, we consider both the attacks that target single and multiple sensors, where the training process will utilize the data from single-sensor attack, and the multiple-sensor attacks are used for online testing to demonstrate the effectiveness of detection methods. Additionally, the machine-learning-based detector trained for a single-sensor attack can also be used for sensor isolation once the cyber-attack is detected. For clarity, in this study, we only consider the scenario that one type of cyber-attack occurs at a time, i.e., during each attack duration, the system is not attacked by a hybrid of multiple types of cyber-attacks.

Remark 8.6 Note that upper-tier controller and lower-tier controllers share the same continuous state measurements despite the asynchronous execution frequency of the upper-tier controller. This implies that the continuous state measurements sent to the upper-tier control system remain intact during the entire operating period. It also implies that even when a multiple-sensor attack occurs, it will only target the sensors that send sampled asynchronous state measurements to upper-tier controller, and not attack the continuous sensor measurements. Additionally, it is not meaningful for the intelligent cyber-attack to attack the two separate communication channels in the two tiers of controllers, under which the continuous measurements for the upper-tier controller are compromised while those sent to lower-tier controllers remain unchanged. The reason is that we can always develop a simple tracker that identifies the presence of this abnormality by examining the error between the same measurements sent to the lower and upper-tier controllers. Therefore, the assumption that both controllers receive secure continuous state measurements is valid in the two-tier control system.

8.3.2 Mitigation Measures via Reconfiguration of Control System

A feedforward artificial neural network is developed following the construction method in Sect. 7.3 to detect cyber-attacks by solving supervised classification problems. Compared to conventional detection schemes, data-based approaches have many advantages in the develop to development of the cyber-attack detector [1, 81, 145]. Firstly, physical-model-based detection methods that identify cyber-attacks based on false alarm thresholds may become ineffective for the intelligent cyber-attacks that have access to process information (e.g., variable operating window and stability region) [43]. Secondly, in a chemical plant, plant model parameters and structure may be modified sometimes to adapt to the variation of operating environment. In this case, the data-based detection method that does not depend on physical models is resilient to intelligently designed attacks and process changes. In this study, neural networks are developed to distinguish between two classes, i.e., "no attack" and "attack", or to distinguish between multiple classes, i.e., types of attacks and "no attack", depending on the training data collected from simulations. Additionally, a neural-network-based isolator can also be developed using the data collected from individual sensors to locate the corrupted sensor upon the detection of cyber-attacks.

In this case, this neural network model is developed with multiple labeled classes, where every class represents one problematic sensor.

Once an attack on the sensors is detected based on the (asynchronous) networked state measurements provided to two-tier control system, control system reconfiguration is executed using the following steps. First, the upper-tier controller should be deactivated completely, and the stabilizing controller (i.e., lower-tier controller) with secure, dedicated sensor measurements will be used to operate the system. Since the continuous measurements remain secure all the time, and system stability is ensured using the lower-tier controllers, the impact of the cyber-attacks can be fully eliminated after the reconfiguring the control system. Second, once we confirm sensor attack in the system and use sensor isolation detector (if there is any) to locate the compromised sensor(s), the upper-tier controller needs to abandon the corrupted sensors and use the secure, redundant back-up sensors. In this case, the upper-tier controller remains functional and an improved closed-loop performance can be achieved under the two-tier controller.

In the worst-case scenario that both asynchronous and continuous sensor measurements are attacked, we will shut off the upper-tier controller and continue to use the lower-tier controllers with secure back-up sensor measurements replacing the compromised continuous measurements. A reactor-reactor-separator process will be utilized in Sect. 8.4 to demonstrate the robustness of the two-tier control architecture against different types of attacks.

Remark 8.7 By reconfiguring the control system, we have shown that closed-loop stability is maintained for the system subject to cyber-attacks. This implies that the two-tier control architecture is inherently cyber-secure due to stabilizing lower-tier controller, and does not rely on redundant senors or accurate state estimation to re-stabilize the system at the steady-state. However, as upper-tier controller (i.e., model predictive controller) is deactivated after detection of cyber-attacks, closed-loop performance degradation may be observed afterwards by using lower-tier controller only. Therefore, to improve closed-loop performance by continuing using the upper-tier controller, the redundant back-up sensors and state reconstruction methods that have been discussed in Chap. 7 can be utilized within upper-tier controller. Specifically, if redundant back-up sensors are available, a straightforward approach to continuing closed-loop control is to replace the problematic sensors with the redundant sensors. However, as redundant sensors may not be immediately deployed upon detection of cyber-attacks, sensor device replacement is not an effective measure for all circumstances. Therefore, instead of using falsified state measurements, state reconstruction method provides an alternative way to optimize closed-loop performance via MPC with reconstructed process states (i.e., estimated true state values).

8.3.3 *Integration of Safety Systems with Two-Tier Control Systems*

As cyber-attacks on safety-critical systems have the potential to cause real harm in the physical world, the scope of cybersecurity goes beyond the use of cybersecurity software. In addition to the cyber-secure control systems introduced in the previous chapter and this chapter, process safety systems such as alarms systems, emergency shutdown systems and safety relief devices that have been discussed in the first half of the book can provide the last line of defense in the event of an abnormal situation due to cyber-attacks. It is noted that although the two-tier control system is able to maintain stable and safe operation using the lower-tier controller only with dedicated, secure sensor measurements y_c, the remaining process states (i.e., measured by networked sensors y_a) may leave their safety limits prior to the successful detection of cyber-attacks. To reduce physical risks of cyber-attacks ranging from simple unplanned downtime in operations to a plant explosion or release of hazardous materials, we integrate safety system with two-tier control system to maintain all process states within their safety operating limits once the cyber-attack is detected by the neural-network-based detector. Specifically, Safeness Index functions $S(x)$, a function of the (closed loop) process states that characterizes the "safeness" of a process operation, is initially adopted as a safety metric for activation/deactivation of safety systems (see Chap. 3 for the definition and application of Safeness Index in chemical process control problems). Safe and unsafe operations can then be evaluated by comparing the value of $S(x)$ with the threshold value S_{TH} that is pre-determined using process first-principles knowledge or past plant data (i.e., $S(x) < S_{TH}$ and $S(x) > S_{TH}$ represent safe and unsafe operations, respectively). Additionally, because $S(x)$ can provide information on both measured and estimated states, its use in the alarm system can help manage the trade-off between measuring fewer states (which may lead to missed alarms) and more states (which leads to instrumentation expenses and possibly more occurrences of alarm overloading).

In the traditional process safety paradigm, process variables are stabilized at their set-points by basic process control systems (BPCS) under normal operation; when the control system (BPCS) fails to operate the process in a safe operating region in the presence of disturbances or cyber-attacks, the safety systems (e.g., alarm systems, emergency shutdown systems (ESS), and safety relief devices) are activated to prevent further unsafe operation. However, since the process dynamics is changed after the activation of safety systems (e.g., opening of a pressure relief valve to prevent high pressure in a chemical reactor), the actions taken by the safety systems should be taken into account in the reconfiguration of control systems. In this section, we develop an integrated safety and cyber-secure control system that takes appropriate actions based on measured variables and $S(x)$ thresholds being crossed is developed based on the following requirements: (1) Secure, redundant sensors or reliable state estimation are available to the control, alarm, emergency shutdown, and relief systems with standard industrial practice; (2) To simplify the discussion, the actions initiated as a result of the alarm, ESS, and relief systems are on-off type actions (i.e., an operator or the relief system can fully open or fully close a valve; the ESS can turn a pump on or off); (3) The safety-based (lower-tier) control system continues to regulate the process state even when the safety system is triggered. Specifically, since networked sensor measurements y_a are untrustworthy under cyber-attacks, we measure/estimate the true state x value through redundant, secure sensors or state estimator based on secure and dedicated sensor measurements (see, also, Sect. 7.4 for a detailed description of the two methods). Additionally, it should be noted that the system dynamics is changed after the activation of the safety system, i.e., $u_s = 1$ in Eq. 8.1. Therefore, the two-tier control system needs to account for the change in system dynamics by updating the prediction model of Eqs. 8.3b–8.3c in the upper-tier MPC. After process states move into the safe operating region, the safety system is taken off-line and the two-tier control system switches to the initial process model where $u_s = 0$.

8.4 Application to a Chemical Process Example

We consider a chemical process network consisting of two CSTRs followed by a flash tank separator, in which the species concentration and the temperature in the three vessels are regulated by manipulating multiple inputs in both the lower-tier and upper-tier controllers [235]. A schematic diagram of the reactor-reactor-separator process is given in Fig. 8.2. Two reactions (A $\rightarrow$ B $\rightarrow$ C) take place in series in both reactors, and the overhead vapor from the flash tank is recycled to the first CSTR.

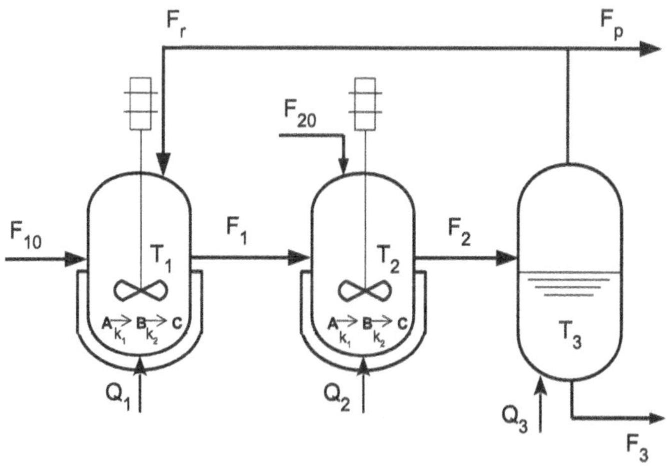

Fig. 8.2 Schematic of the reactor-reactor-separator process with two CSTRs and a flash drum separator

It is assumed that all three vessels have constant holdup. Based on mass and energy balances, the following nine nonlinear ordinary differential equations are developed to describe the process dynamics:

$$\frac{dx_{A1}}{dt} = \frac{F_{10}}{V_1}(x_{A10} - x_{A1}) + \frac{F_r}{V_1}(x_{Ar} - x_{A1}) - k_1 e^{\frac{-E_1}{RT_1}} x_{A1} \tag{8.5a}$$

$$\frac{dx_{B1}}{dt} = \frac{F_{10}}{V_1}(x_{B10} - x_{B1}) + \frac{F_r}{V_1}(x_{Br} - x_{B1}) + k_1 e^{\frac{-E_1}{RT_1}} x_{A1} - k_2 e^{\frac{-E_2}{RT_1}} x_{B1} \tag{8.5b}$$

$$\begin{aligned}\frac{dT_1}{dt} =& \frac{F_{10}}{V_1}(T_{10} - T_1) + \frac{(-\Delta H_1)}{\rho C_p} C_M k_1 e^{\frac{-E_1}{RT_1}} x_{A1} + \frac{(-\Delta H_2)}{\rho C_p} C_M k_2 e^{\frac{-E_2}{RT_1}} x_{B1} \\ &+ \frac{Q_1}{\rho C_p V_1} + \frac{F_r}{V_1}(T_3 - T_1)\end{aligned} \tag{8.5c}$$

$$\frac{dx_{A2}}{dt} = \frac{F_1}{V_2}(x_{A1} - x_{A2}) + \frac{F_{20}}{V_2}(x_{A20} - x_{A2}) - k_1 e^{\frac{-E_1}{RT_2}} x_{A2} \tag{8.5d}$$

$$\frac{dx_{B2}}{dt} = \frac{F_1}{V_2}(x_{B1} - x_{B2}) + \frac{F_{20}}{V_2}(x_{B20} - x_{B2}) + k_1 e^{\frac{-E_1}{RT_2}} x_{A2} - k_2 e^{\frac{-E_2}{RT_2}} x_{B2} \tag{8.5e}$$

$$\begin{aligned}\frac{dT_2}{dt} =& \frac{F_{20}}{V_2}(T_{20} - T_2) + \frac{(-\Delta H_1)}{\rho C_p} C_M k_1 e^{\frac{-E_1}{RT_2}} x_{A2} + \frac{(-\Delta H_2)}{\rho C_p} C_M k_2 e^{\frac{-E_2}{RT_2}} x_{B2} \\ &+ \frac{Q_2}{\rho C_p V_2} + \frac{F_1}{V_2}(T_1 - T_2)\end{aligned} \tag{8.5f}$$

$$\frac{dx_{A3}}{dt} = \frac{F_2}{V_3}(x_{A2} - x_{A3}) - \frac{F_r + F_p}{V_3}(x_{Ar} - x_{A3}) \tag{8.5g}$$

$$\frac{dx_{B3}}{dt} = \frac{F_2}{V_3}(x_{B2} - x_{B3}) - \frac{F_r + F_p}{V_3}(x_{Br} - x_{B3}) \tag{8.5h}$$

$$\begin{aligned}\frac{dT_3}{dt} =& \frac{F_2}{V_3}(T_2 - T_3) + \frac{(F_r + F_p)C_M}{\rho C_p V_3}(x_{Ar}\Delta H_{vapA} + x_{Br}\Delta H_{vapB} + x_{Cr}\Delta H_{vapC}) \\ &+ \frac{Q_3}{\rho C_p V_3}\end{aligned} \tag{8.5i}$$

where the state variables include the mass fractions of species A and B, i.e., x_{A1}, x_{A2}, x_{A3} and x_{B1}, x_{B2}, x_{B3}, as well as the temperatures of the three vessels, i.e., T_1, T_2, T_3. Specifically, the temperatures are measured continuously and securely, while the species mass fractions are measured asynchronously. The upper-tier control system receives the asynchronous networked measurements through a digital network that is vulnerable to cyber-attacks. The LMPC scheme is used as the upper-tier controller to optimize closed-loop performance based on both continuous and asynchronous state measurements. It needs to be mentioned that the LMPC optimization problem is executed only when the upper-tier controller receives a full-state measurement. In this example, each of the three vessels has an external heat input. To control the temperatures (in three vessels) at their desired set-points, we use three PI controllers to manipulate the heat inputs (i.e., Q_1, Q_2, and Q_3) to the three vessels. To speed up the closed-loop response, the manipulated input in LMPC is chosen to be the flow rate of the feed stream to second CSTR, F_{20}. Assuming that the relative volatility of each species remains constant within the operating temperature range and the reaction in the separator tank is negligible, the composition of the recycle stream is as follows:

$$x_{Ar} = \frac{\alpha_A x_{A3}}{\alpha_A x_{A3} + \alpha_B x_{B3} + \alpha_C x_{C3}} \tag{8.6a}$$

$$x_{Br} = \frac{\alpha_B x_{B3}}{\alpha_A x_{A3} + \alpha_B x_{B3} + \alpha_C x_{C3}} \tag{8.6b}$$

$$x_{Ar} = \frac{\alpha_C x_{C3}}{\alpha_A x_{A3} + \alpha_B x_{B3} + \alpha_C x_{C3}} \tag{8.6c}$$

where the relative volatility of species i is represented by α_i, $i = A, B, C$. As discussed in Sect. 8.3, the sensor measurements of six mass fraction (i.e., networked measurements) are vulnerable to the cyber-attacks that can tamper the sensor readings based on the current true states values. The control objective of the two-tier cyber-secure control system is to stabilize all 9 states at an unstable steady-state while staying immune to potential cyber-attacks. Table 8.1 lists the values of process parameter and of the steady-state used in this example.

We use deviation variables to present the input vector and the state vector as the deviation from their steady-state values, such that the origin of state-space is the equilibrium point of the process. The following operating constraints are imposed on the manipulated inputs (in deviation variable form): $-4.04\ \text{m}^3/\text{h} \leq \Delta F_{20} \leq 3.96\ \text{m}^3/\text{h}$, $|\Delta Q_1| \leq 5 \times 10^7\ \text{kJ/h}$, $|\Delta Q_2| \leq 5 \times 10^7\ \text{kJ/h}$, $|\Delta Q_3| \leq 5 \times 10^7\ \text{kJ/h}$.

As proportional–integral (PI) controllers are easy to implement and can be solved instantaneously, they are used to regulate the vessel temperatures in lower-tier controller. The formulation of PI controller is given below

$$u_{c_i}(t) = K_{c_i}(e_{c_i}(t) + \frac{1}{\tau_i}\int_0^t e_{c_i}(\tau)\mathrm{d}\tau), \quad e_{c_i}(t) = y_{c_i}^{REF}(t) - y_{c_i}(t) \tag{8.7a}$$

where τ_i and K_{c_i}, $i = 1, 2, 3$, are the integral time constant and proportional gain of each PI controller, respectively. $e_{c_i}(t)$ is the error between the set-points $y_{c_i}^{REF}$ (i.e., the operating steady-state in this example), and the measured output values y_{c_i}. In order to guarantee stability for the closed-loop system under PI control, we linearize the nonlinear model of Eq. 8.1 around the operating steady-state we considered, and evaluate the eigenvalues of this linearized model $\dot{x} = Ax + Bu_c$ to determine the value of K_{c_i} and τ_i. The values of K_{c_i} and τ_i used in this example are reported below

$$[K_{c_1}\ K_{c_2}\ K_{c_3}]^T = [-8 \times 10^5,\ -8 \times 10^5,\ -8 \times 10^5]^T \tag{8.8a}$$

$$[\tau_1\ \tau_2\ \tau_3]^T = [5000,\ 5000,\ 5000]^T. \tag{8.8b}$$

Table 8.1 Descriptions and values of process parameters.

Parameter/Value	Description
$F_{10} = 5.04\ m^3/h$	Feed flow rate of CSTR 1
$F_r = 50.4\ m^3/h$	Flow rate of recycle stream
$F_p = 5.04\ m^3/h$	Flow rate of purge stream
$T_{10} = 300$ K, $T_{20} = 300$ K	Feed temperatures of CSTR 1 & 2
$V_1 = 1.0\ m^3$, $V_2 = 0.5\ m^3$, $V_3 = 1.0\ m^3$	Volume of 3 vessels
$\Delta H_{vapA} = -3.53 \times 10^4$ kJ/kmol,	Heat of vaporization for A, B, C
$\Delta H_{vapB} = -1.57 \times 10^4$ kJ/kmol,	
$\Delta H_{vapC} = -4.068 \times 10^4$ kJ/kmol	
$\Delta H_1 = -1.2 \times 10^5$ kJ/kmol,	Heat of reaction for reactions 1 & 2
$\Delta H_2 = -1.4 \times 10^5$ kJ/kmol	
$E_1 = 5.0 \times 10^4$ kJ/kmol,	Activation energy of reactions 1 & 2
$E_2 = 6.0 \times 10^4$ kJ/kmol	
$k_1 = 9.972 \times 10^6\ h^{-1}$,	Pre-exponential constants of reactions 1 & 2
$k_2 = 9.36 \times 10^6\ h^{-1}$	
$Q_{1s} = 2.9 \times 10^9$ kJ/h, $Q_{2s} = 1.9 \times 10^9$ kJ/h,	Input steady-state values
$Q_{3s} = 2.9 \times 10^9$ kJ/h, $F_{20s} = 5.04\ m^3/h$	
$x_{A1s} = 0.1762, x_{A2s} = 0.1965, x_{A3s} = 0.0651,$	Process state steady-state values
$x_{B1s} = 0.6731, x_{B2s} = 0.6536, x_{B3s} = 0.6703,$	
$T_{1s} = 480.32$ K, $T_{2s} = 472.79$ K, $T_{3s} = 474.89$ K	
$C_M = 2\ kmol/m^3$	Total molar concentration
$\alpha_A = 3.5, \alpha_B = 1.0, \alpha_C = 0.5$	Relative volatility of A, B, C
$\rho = 1000\ kg/m^3$	Liquid solution density
$R = 8.314$ kJ/(kmol K)	Gas constant
$C_p = 4.2$ kJ/(kg K)	Heat capacity

The Cohen-Coon tuning method is initially utilized to choose the PI controller parameters. Then, the optimal PI parameters that lead to smooth and reasonable control actions are determined via closed-loop simulations. It is shown that P-only control guarantees closed-loop stability using the above parameters since the eigenvalues of model $\dot{x} = Ax + Bu_c$ all have negative real parts, which is shown below

$$\begin{aligned}\Lambda = [&-2.599,\ -56.97,\ -758.8,\ -257.8 - 26.93i,\ -257.8 + 26.93i,\\ &-27.93 + 149.2i,\ -27.93 - 149.2i,\ -99.98 + 26.28i,\ -99.98 - 26.28i].\end{aligned} \tag{8.9}$$

Additionally, an integral term is added to eliminate the offset. To prevent integral windup, we implement an anti-windup strategy within PI controllers to temporarily eliminate the integral term after the system input hits the lower/upper bound. In this simulation, the LMPC of Eq. 8.3 is used for the upper-tier control system with the objective function in the form of $l_t(x, u_a) = x^T Q_c x + u_a^T R_c u_a$, where $Q_c = diag([5000,\ 10,\ 0.001,\ 5000,\ 10,\ 0.001,\ 5000,\ 10,\ 0.001])$ and $R_c = 1.0$ are weighting matrices to penalize x and u_a, respectively. The control Lyapunov function is developed in a quadratic form, i.e., $V(x) = x^T Px$, where P is a positive definite matrix:

$$P = diag([3228.31,\ 220.79,\ 4.334 \times 10^{-4},\ 2576.72,\ 233.80,\ 4.474 \times 10^{-4},\ 23675.92,\ 222.77,\ 4.434 \times 10^{-4}]). \tag{8.10}$$

The LMPC is simulated with the following settings: the prediction horizon is $N = 10$ and the sampling period is $\Delta = 0.02\ hr$. The OPTI-Toolbox in MATLAB is used to solve the optimization problem of LMPC. The explicit Euler method is used to numerically simulate the system of Eq. 8.5 with a sufficiently small integration step of $h_c = 10^{-4}$ h. A lower-bounded random Poisson process is used to choose the time sequence at which the upper-tier controller receives the asynchronous measurements. The following sequence of asynchronous intervals is used for the execution of the LMPC calculations in the simulation for every 1.5 h:

$$\Delta_a = [0.04,\ 0.08,\ 0.1,\ 0.06,\ 0.12,\ 0.08,\ 0.02].$$

Note that the unequal interval between two consecutive asynchronous measurements should satisfy $\Delta_{a_k} \geq \Delta$ for all $k \in [1, N_T]$. After we design the Lyapunov function to be in the form of $V(x) = x^T Px$, the stability region Ω_ρ, and the small neighborhood $\Omega_{\rho_{\min}}$ that the state will be ultimately bounded in are characterized as the level sets of V with $\rho = 120$ and $\rho_{\min} = 0.1$, respectively. The safe operating region for all the states in this example is designed as follows:

$$\begin{aligned} x_u &= [0.7237,\ 0.2269,\ 50,\ 0.7035,\ 0.2464,\ 50,\ 0.8349,\ 0.2297,\ 50]^T \\ x_l &= [-0.1763,\ -0.6731,\ -50,\ -0.1965, \\ &\quad -0.6536,\ -50,\ -0.0651,\ -0.6703,\ -50]^T \end{aligned} \tag{8.11}$$

where x_u and x_l denote the upper and lower bounds for the process states in deviation variable form, respectively. The operating envelope and the stability region are the two key parameters in designing intelligent cyber-attacks.

To generate the training dataset, we run extensive closed-loop simulations for 3 h, within which LMPC is executed 42 times and PI controllers are executed 150 times. As a result, each state trajectory contains $N_T = 43$ state measurements, accounting for the initial condition and the state measurements received by LMPC that was executed 42 times. The following initial condition is used in the closed-loop simulations under PI-only and under two-tier control for comparison of the closed-loop performances:

$$x_0 = [0.0176,\ 0.067299,\ 48.032,\ 0.0197,\ 0.0654,\ 47.279, 0.006499,\ 0.067,\ 47.489]^T. \tag{8.12}$$

We compare the closed-loop performances under the two-tier PI/LMPC control scheme and under the lower-tier PI controllers only by evaluating their normalized cumulative mean squared errors (MSEs) along the state trajectories and settling times. The simulation results show that it takes 0.6 h for two-tier PI/LMPC, and 2.46 h for lower-tier PI controllers, to settle to the set-point. The normalized cumulative MSEs are calculated to be 0.8014 and 4.1203 for two-tier PI/LMPC and lower-tier PI, respectively. Therefore, it is demonstrated that the two-tier control system outperforms the lower-tier PI control system as it eliminates process overshoots and offset more effectively, and stabilizes the system within shorter time.

8.4.1 Cyber-Attacks and Detector Training

We train the neural-network-based detector for detecting min-max cyber-attack with and without sensor noise. When only one type of attack is considered, the neural network (NN) detector is developed to solve a binary classification problem, where the resulting output has 2 classes—no attack and under attack. Additionally, as replay attacks are challenging to detect (see discussion in Remark 7.9), we also incorporate replay attack in the neural network detector such that it is capable of identifying different types of attack (i.e., the neural network output consists of three labeled classes: attacked by replay attacks, by min-max attacks, and not attacked). Specifically, state feedback shows extreme oscillations with larger magnitudes in the first five sampling steps. Therefore, we design replay attacks by collecting the aggressively oscillatory measurements with attack duration $L_a = 5$. Considering that cyber-attacks can occurr at various times and with various durations during operation, other attacks can be randomly introduced at the sampling time between $i_0 \in [6, 42]$ with varying attacking lengths. We collect equal number of samples for each output class from extensive closed-loop simulations. In the case of attack identification, each sample consists of a 1×43 array of $V(x)$ values, while in the case of sensor isolation, each sample will consist of x values with the dimension of 9×43 along the dynamic trajectory. To generate the input to the NN-based detector, we then collapse the 9×43 matrix into a 1×387 array for each data sample. We develop the following NN-based detectors to detect cyber-attacks: (1) to detect a min-max cyber-attack in a nominal system, a 2-class NN model is developed with nominal operation, where 12000 samples are collected for each class label, and the training time is 24.05 seconds, (2) to detect a min-max cyber-attack in the presence of sensor noise, a 2-class NN model is developed with noisy sensor measurements, where 1044 samples are collected for each class label. The training time is 4.332 seconds, (3) to detect min-max and replay attacks with noisy sensors, a 3-class NN model is developed, where 1044 samples are collected for each class label, and the training time is 5.265, and (4) To train the compromised sensor isolation detector, a 6-class NN model is developed under min-max attack with noisy sensors, where 2800 samples are collected for each class label. The training time is 6211.52 seconds. To develop the NN-based detector that can distinguish between sensor noise and cyber-attacks, we add bounded white noise of

Gaussian distribution on sensors to simulate the sensor noise. Additionally, the sensor noises are bounded as follows: $|w_1| \leq 7.5 \times 10^{-5}$, $|w_2| \leq 5.5 \times 10^{-5}$, $|w_3| \leq 0.032$ K, $|w_4| \leq 7.5 \times 10^{-5}$, $|w_5| \leq 5.5 \times 10^{-5}$, $|w_6| \leq 0.032$ K, $|w_7| \leq 3.5 \times 10^{-5}$, $|w_8| \leq 5.5 \times 10^{-5}$, $|w_9| \leq 0.032$ K. These white Gaussian noises have standard deviations $\sigma_1 = \sigma_4 = 0.0002$, $\sigma_2 = \sigma_5 = \sigma_8 = 0.001$, $\sigma_3 = \sigma_6 = \sigma_9 = 0.1\,K$, and $\sigma_7 = 0.0001$, and a mean of $\mu = 0$. The MATLAB Machine Learning and Deep Learning Toolboxes are used to develop the two-layer feedforward neural networks (FNNs) with 12 and 10 neurons, respectively, in each hidden layer. The activation function $tansig$, i.e., $g_{1,2}(z) = \frac{2}{1+e^{-2z}} - 1$ is used in both hidden layers. The activation function $softmax$, i.e., $g_3(z_j) = \frac{e^{z_j}}{\sum_{i=1}^{H} e^{z_i}}$, where H represents the number of class labels, is used in the output layer to provide a predicted probability of the class labels.

The training and testing accuracy are 99.6% and 92.2%, respectively, for the NN detector trained with nominal conditions, and those of the NN detector trained with sensor noise are 99.9% and 100%, respectively. The reason for the NN algorithm trained with noisy sensors achieving a higher accuracy than the NN model trained using the nominal model is that the NN detector is rendered more robust by introducing more variance (i.e., the contributions of sensor noise) into the training dataset. Moreover, the NN detector trained with two different types of cyber-attacks and accounting for sensor noise has accuracy of 98.2% and 91.4% for training and testing datasets, respectively, and the NN-based sensor isolation detector has an accuracy of 99.6% and 99.0% on training and testing datasets, respectively.

8.4.2 Cyber-Attack Detection Results

We carry out the closed-loop simulation for the system under an detector-integrated two-tier control system with initial conditions $x_0 = [0,\ 0,\ 0,\ 0,\ 0,\ 0,\ 0,\ 0,\ 0]^T$. We introduce the cyber-attacks after the control system stabilizes the process at the steady-state. Since the NN detectors in this simulation example are developed with a fixed input dimension of 43 (i.e., 42 sampling steps), we activate the detector at the $k = 42$th sampling time in the asynchronous time sequence to ensure that there is sufficient input data for executing the NN prediction. The NN-based detector computes an output showing the status of system cybersecurity (i.e., attacked or not attacked) based on the previous 42 state measurements. As the attack detector and the upper-tier LMPC are executed in real time, this fixed-length window of time-series data also rolls forward in time. In this example, we designed a three-sampling-period alarm verification window for the upper-tier LMPC. The presence of a cyber-attack is confirmed if two positive detections are observed within the alarm verification window (i.e., within three consecutive sampling periods). Once an attack is confirmed, it triggers the detection alarm and at the same time, deactivates the LMPC. Furthermore, to examine whether not-attacked signals will be misclassified as attack by detector, we introduce attacks a few sampling periods after the activation of detector at $t = 3.0$ h. In this way, the first few outputs are from nominal process for the detector. Specifically, in this example, we introduce cyber-attacks at time instant $i_0 = 45$ corresponding to the simulation time $t = 3.22$ h. The attack will last for $L_a = 40$ sampling periods. The sensor is under attack till the end of the simulation period (i.e., $t = 6$ h). The sensor measurements and the true state values of state 1 for the closed-loop system of Eq. 8.5 under surge, geometric, replay, and min-max cyber-attacks on sensor measurements of mass fraction x_{A1} with sensor noise are shown in Fig. 8.3, from which the pattern and impact of different types of cyber-attacks are illustrated. Specifically, it is shown that the true state settles at an offset under min-max attack; the true state shows aggressive oscillations under replay attack; the process state is driven away from the steady-state and settles at an offset when the geometric attack hits the boundary of Ω_ρ; the true state shows an initial jump similar to that under min-max attack, and is ultimately driven closer to the set-point with a smaller offset than min-max attack because the surge cyber-attack reduce attack severity to avoid being detected by conventional detection methods. Although we only show one of the states in Fig. 8.3, we observe that the deviating patterns of all 9 states are similar under the cyber-attacks. The simulation results of all 9 states are not shown here due to the space limitation. After the sensor measurements are tampered by cyber-attacks and received by the upper-tier controller, the LMPC is unable to compute correct control actions that drive the true state back to the steady-state. However, since three PI controllers (in lower-tier control system) that utilize secure sensor measurements play a dominant role in the stabilization of the system, it is shown in Fig. 8.3 that the true states do not diverge and can be bounded in Ω_ρ. Regardless, when no data-based detectors are utilized, the attack successfully disrupt the closed-loop performance by driving the state away from its steady-state.

Since we use an alarm verification window to reduce false alarm rates by requiring two or three positive detections in three detection instances, time delays are observed when we implement NN detectors online. The time delay in this simulation study is defined as the number of sampling periods between the time of the attack being added, and the time of the attack being confirmed. Although the first two detectors are trained for the system under min-max attack only with noise and with nominal operations, respectively, all four types of cyber-attacks (i.e., min-max, replay, geometric, and surge attacks) are identified by the neural-network-based detector trained with nominal model. Specifically, the min-max, surge, and replay attacks are detected by the NN detector with a time delay of 1 sampling period, at which time the control system receives the second consecutive positive detection to confirm the occurrence of cyber-attaks in the alarm verification window. The geometric attack is detected with a time delay of 2 sampling periods due to the small bias implemented at the early stage, which is challenging for the NN detector to classify it as an attack. As time progresses, the geometric cyber-attack exponentially increases towards an attacking value, at which the detector recognizes this abnormal behavior that is on par with the other three attacks. The potential time delay for the case of geometric attacks will vary according to different geometric parameters, i.e., α and β in Eq. 7.8, used by attackers. Then, we test the NN detector whose training process accounts for sensor noise. It is demonstrated that this NN detector can successfully detect

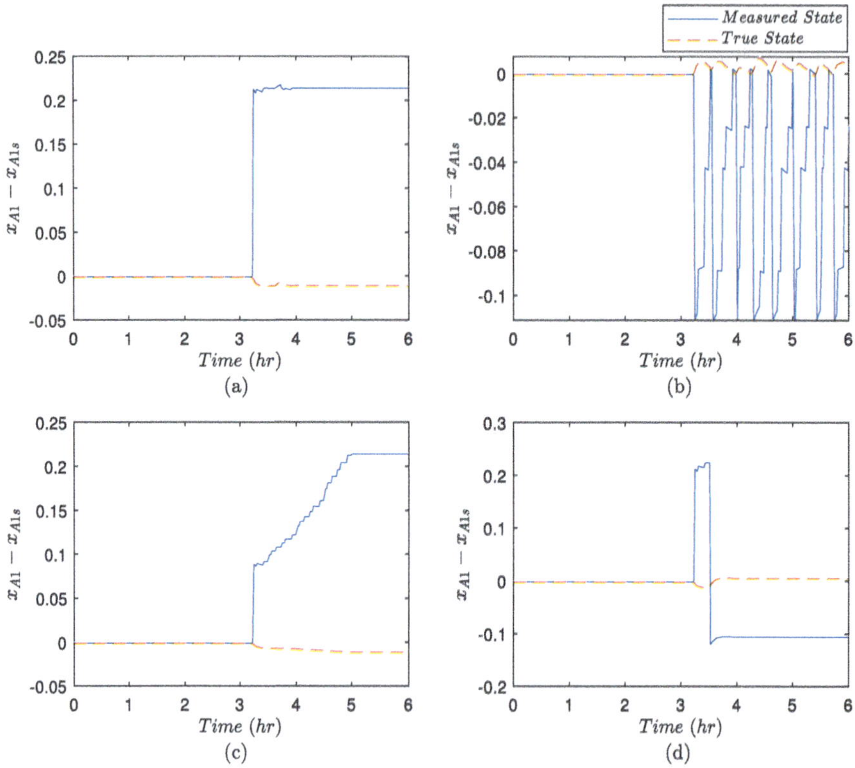

Fig. 8.3 Measured and true state values (in deviation variable form) of x_{A1} when **a** min-max, **b** replay, **c** geometric, and **d** surge cyber-attacks are added on the sensor measurement of concentration x_{A1} at 3.22 h, and no detection or mitigation mechanisms are used

surge, min-max, and geometric attacks, with a longer time delay for detecting the geometric attack (i.e., it takes 7 sampling periods). However, replay attacks cannot be detected by this NN detector because of the oscillatory pattern of replay signals. It is observed that replay attacks are oscillating over time, which is significantly different from the other 3 attacks that share similar attacking pattern (i.e., the attacked measurement remains at the attack target for at least 2 sampling periods). As replay attacks show oscillatory behavior that is similar to the system with sensor noise, the NN detector trained with sensor noise is unable to distinguish between sensor noise and replay attacks. Therefore, we train a third NN detector to account for both min-max and replay cyber-attacks. In this case, the min-max and replay attacks are correctly classified, and the detection is confirmed with a 1-sampling-period time delay.

Subsequently, we test the detection algorithms on cyber-attacks targeting multiple sensors at once. An extreme case where all 9 sensors are attacked by min-max cyber-attack and no online detectors are implemented is first simulated to demon-

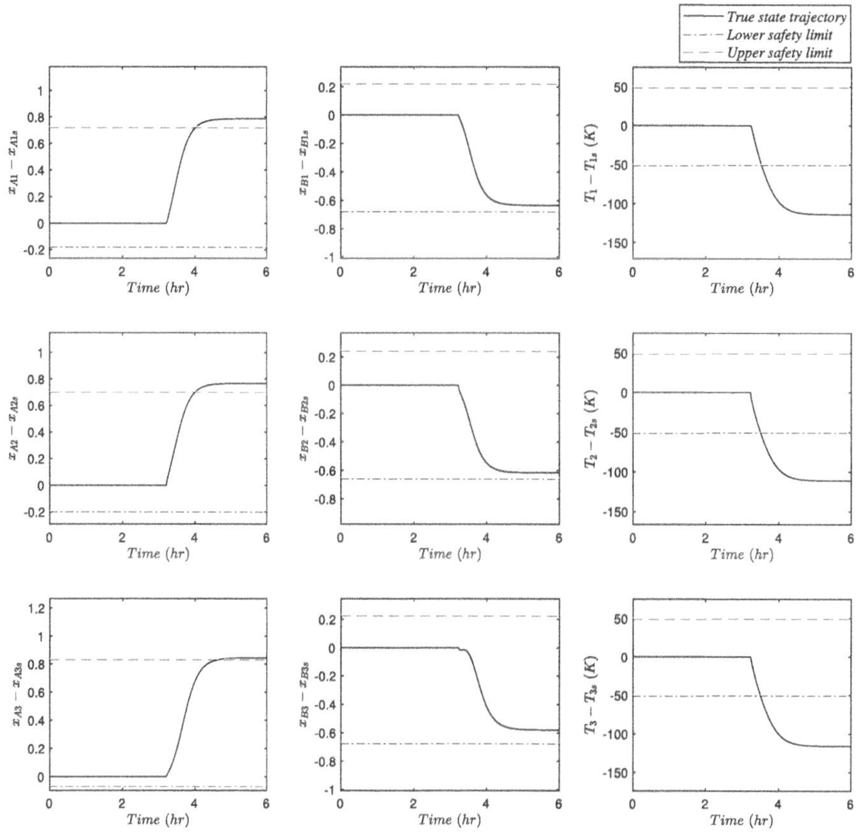

Fig. 8.4 Profiles of true process states when all 9 state measurement sensors are attacked at 3.22 h by min-max cyber-attacks, and no detection or reconfiguration of the two-tier control architecture are implemented

strate the effect of cyber-attacks that target all sensor measurements; this scenario helps demonstrate the motivation for the two-tier control architecture. Fig. 8.4 shows the true state trajectories, where a min-max attack is added at 3.22 h and lasts until the end of simulation. As the continuous measurements of temperature are attacked, system stability properties are no longer achieved by the lower-tier controllers. As a result, the true state trajectory leaves the stability region without the implementation of cyber-attack detectors. Additionally, the temperatures and the mass fractions of species A in CSTRs and separator violate the safety limits (in deviation variable form), and exceed their operating boundaries as well. Under this worst-case scenario that the attack also jeopardizes continuous temperature measurements, the only way to ensure cybersecurity is to abandon the corrupted sensors, and use measured temperature signals from a set of redundant sensors with secure readings (if there are any) in the lower-tier controllers. Moreover, safety systems such as safety relief valve

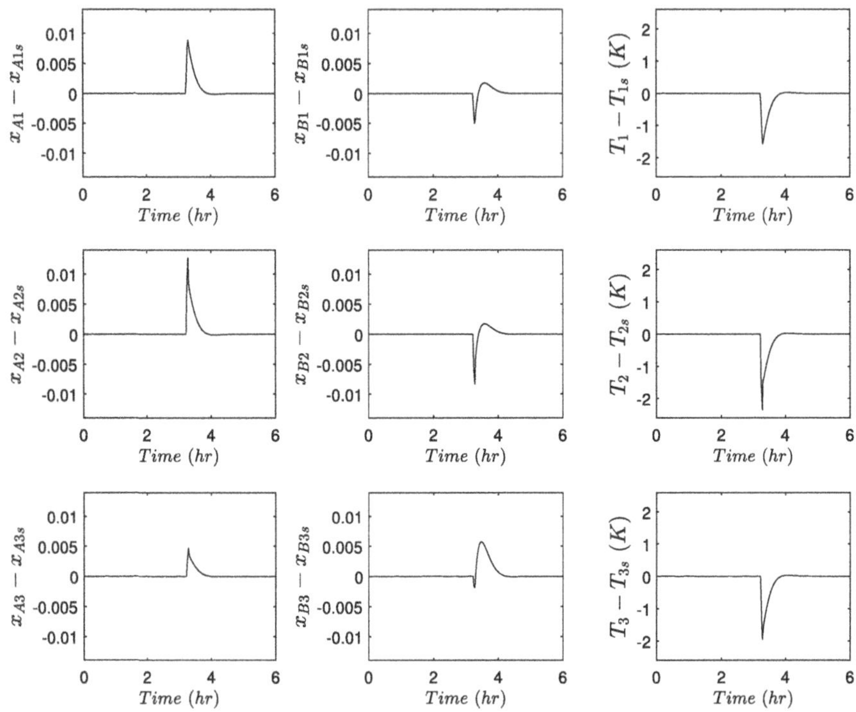

Fig. 8.5 Profiles of true process states when the six sensors of mass fraction are attacked at 3.22 h by min-max cyber-attacks; the attacks are detected at 3.28 h, and the process is re-stabilized at the steady-state by turning off upper-tier LMPC and using lower-tier PIs

and cold water injection can be adopted to discharge material from the reactor and cool down reaction mixture's temperature when the reactor temperature exceeds its safety limit (see, also, the case study of MIC reaction in a CSTR in Sect. 5.2.1.1 for the implementation of the aforementioned safety systems). Through this extreme scenario, we demonstrate the destabilizing impacts of cyber-attacks that target nine sensor measurements. It also implies that to maintain cybersecurity of the two-tier control system, reliable and secure feedback measurements should be available for lower-tier control system all the time.

To efficiently detect the above cyber-attack that targets all sensor measurements, we implement the NN detector trained with two cyber-attack types and noisy measurements. It is shown that the min-max attack added at 3.22 h can be detected within 0.06 h (i.e., at $t = 3.28$ h) with the use of an alarm verification window. After that, we turn off the upper-tier LMPC, replace the corrupted sensors with secure back-up sensors, and then control the system using the lower-tier PI controllers. In this way, stability properties hold for the closed-loop system under lower-tier control in the sense that the system is re-stabilized at the steady-state.

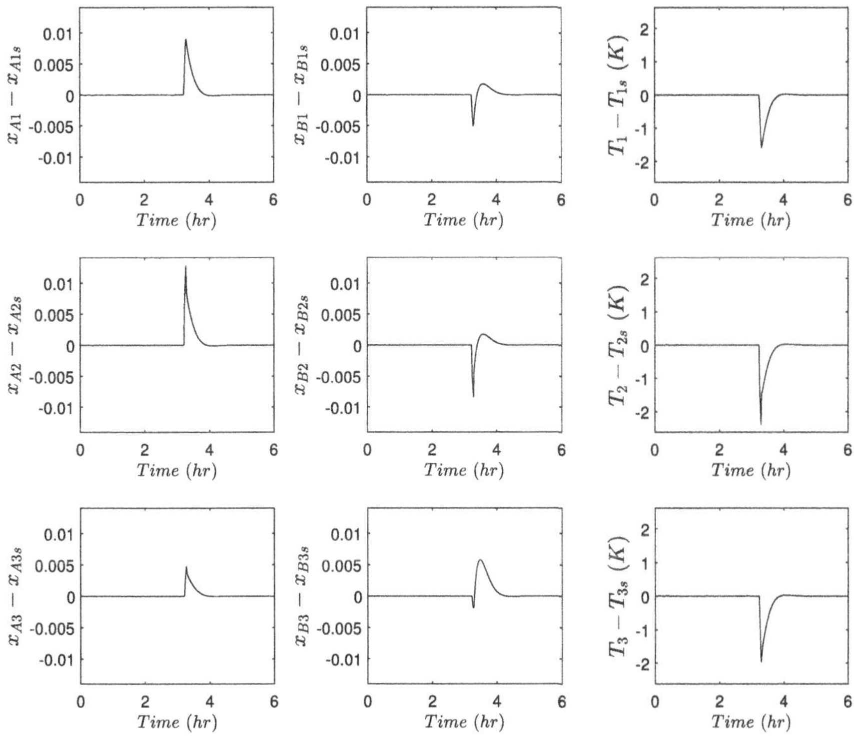

Fig. 8.6 Profiles of true process states when the six sensors of mass fraction are attacked at 3.22 h by replay cyber-attacks; the attacks are detected at 3.28 h, and the process is re-stabilized at the steady-state by turning off upper-tier LMPC and using lower-tier PIs

Then, we consider a more practical case where the measurements of temperature received by lower-tier PIs and upper-tier LMPC remain secure, and the cyber-attacks only target the networked mass fraction measurements received by the upper-tier LMPC. Therefore, once the detection of an attack is confirmed, to prevent the erroneous control actions by LMPC from acting as a unnecessary disturbance to the overall system, we will turn off the upper-tier LMPC (i.e., $u_a = 0$) for the remaining time of simulation. Similarly, we use the lower-tier PI controllers only with secure continuous temperature measurements to stabilize the system at the steady-state. The simulation results illustrate the effectiveness of the mitigation method, from which it is seen that the true process states ultimately converge to the steady-states under lower-tier PIs despite the gradual deviations or the sudden jumps as a result of cyber-attacks. Figures 8.5, 8.6, 8.7, and 8.8 show the closed-loop state trajectories

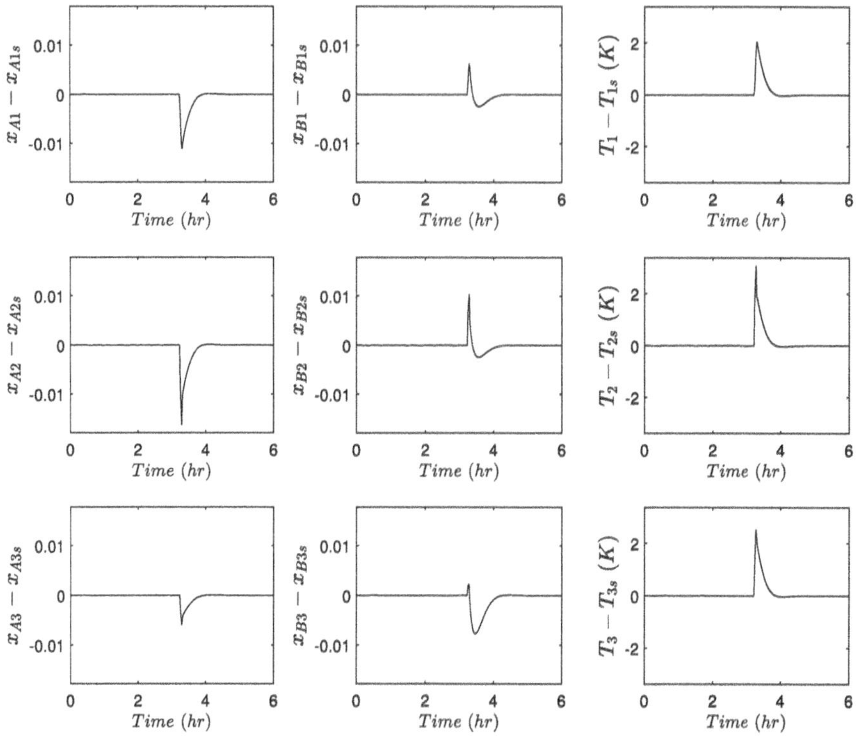

Fig. 8.7 Profiles of true process states when the six sensors of mass fraction are attacked at 3.22 h by geometric cyber-attacks; the attacks are detected at 3.28 h, and the process is re-stabilized at the steady-state by turning off upper-tier LMPC and using lower-tier PIs

under min-max, replay, geometric, and surge attacks, respectively. Specifically, for the attacks that occur at $t = 3.22$ h, the detector successfully identifies the occurrence of cyber-attacks at 3.28 h. Then, we turn off the LMPC and use the lower-tier PI controllers to re-stabilize the process state at the steady-state. Despite the minor degradation in closed-loop performance due to the use of lower-tier controllers only, closed-loop stability is successfully maintained by the reconfigured control system in the presence of intelligent cyber-attacks. Additionally, as cyber-attacks are detected in time, the closed-loop state does not leave the stability region, and thus, the safety system is not activated in this case.

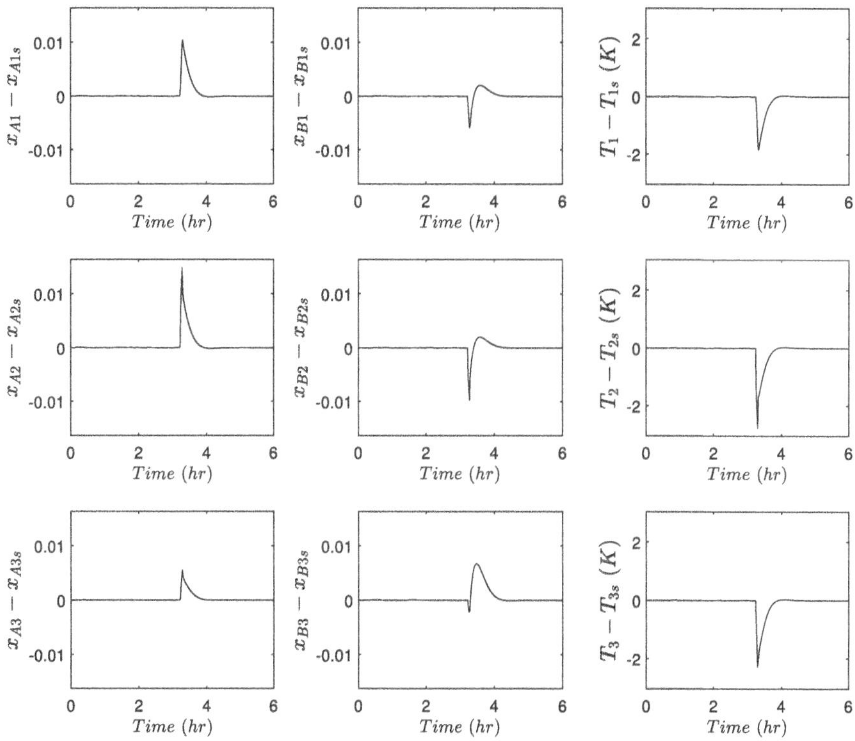

Fig. 8.8 Profiles of true process states when the six sensors of mass fraction are attacked at 3.22 h by surge cyber-attacks; the attacks are detected at 3.28 h, and the process is re-stabilized at the steady-state by turning off upper-tier LMPC and using lower-tier PIs

8.5 Conclusions

In this chapter, we presented a cyber-secure control architecture for nonlinear chemical processes that incorporates an upper-tier MPC and lower-tier explicit feedback controllers. The lower-tier controllers contributed to the stabilization of nonlinear processes, while the upper-tier LMPC used asynchronous networked sensor measurements that are susceptible to sensor cyber-attacks to improve closed-loop performance. An integrated framework for safety and cyber-secure control systems was also discussed to ensure safe operation upon successful detection of cyber-attacks. A neural-network-based detector was designed to efficiently identify the occurrence of cyber-attacks and integrated with the two-tier control architecture to reconfigure the control system to stabilize the system at the steady-state. Neural-network-based detection algorithms were developed and implemented online for detecting common types of cyber-attacks on single or multiple sensors. Specifically, we trained four FNN models and implemented them in the closed-loop systems under nominal operating conditions and operations with sensor noise. All of them were demonstrated to achieve a detection accuracy no less than 91%. Finally, we applied the proposed detector-integrated two-tier control system to a reactor-reactor-separator process, and demonstrated improved cybersecurity and robustness under the machine-learning-based detection methods and the two-tier control architecture.

References

1. Agrawal, S., Agrawal, J.: Survey on anomaly detection using data mining techniques. Procedia Comput. Sci. **60**, 708–713 (2015)
2. Ahlén, A., Akerberg, J., Eriksson, M., Isaksson, A.J., Iwaki, T., Johansson, K.H., Knorn, S., Lindh, T., Sandberg, H.: Toward wireless control in industrial process automation: a case study at a paper mill. IEEE Control Syst. Mag. **39**, 36–57 (2019)
3. AIChE: Dow's Chemical Exposure Index Guide, 1st edn. AIChE, New York (1994)
4. AIChE: Dow's Fire and Explosion Index Hazard Classification Guide, 7th edn. AIChE, New York (1994)
5. Al-Malah, K.I.: Aspen Plus: Chemical Engineering Applications. Wiley, New York (2016)
6. Albalawi, F., Alanqar, A., Durand, H., Christofides, P.D.: A feedback control framework for safe and economically-optimal operation of nonlinear processes. AIChE J. **62**, 2391–2409 (2016)
7. Albalawi, F., Durand, H., Christofides, P.D.: Distributed economic model predictive control for operational safety of nonlinear processes. AIChE J. **63**, 3404–3418 (2017)
8. Albalawi, F., Durand, H., Christofides, P.D.: Process operational safety using model predictive control based on a process safeness index. Comput. Chem. Eng. **104**, 76–88 (2017)
9. Alcala, C.F., Qin, S.J.: Reconstruction-based contribution for process monitoring with kernel principal component analysis. Ind. Eng. Chem. Res. **49**, 7849–7857 (2010)
10. Alhabdan, F., Elnashaie, S.: Simulation of an ammonia plant accident using rigorous heterogeneous models: effect of shift converter disturbances on the methanator. Math. Comput. Model. **21**, 85–106 (1995)
11. Ali, J.M., Hussain, M.A., Tade, M.O., Zhang, J.: Artificial intelligence techniques applied as estimator in chemical process systems-a literature survey. Exp. Syst. Appl. **42**, 5915–5931 (2015)
12. Allen, J.T., El-Farra, N.H.: A model-based framework for fault estimation and accommodation applied to distributed energy resources. Renew. Energy **100**, 35–43 (2017)
13. Almasi, G.S., Gottlieb, A.: Highly Parallel Computing (1988)
14. Alrowaie, F., Gopaluni, R.B., Kwok, K.E.: Alarm design for nonlinear stochastic systems. In: Proceeding of the 11th World Congress on Intelligent Control and Automation, pp. 473–479. Shenyang, China (2014)
15. Ames, A.D., Coogan, S., Egerstedt, M., Notomista, G., Sreenath, K., Tabuada, P.: Control barrier functions: theory and applications. In: Proceedings of the 18th European Control Conference, pp. 3420–3431. Saint Petersburg, Russia (2019)
16. Ames, A.D., Grizzle, J.W., Tabuada, P.: Control barrier function based quadratic programs with application to adaptive cruise control. In: Proceedings of the 53rd IEEE Conference on Decision and Control, pp. 6271–6278. Los Angeles, California (2014)

Z. Wu and P. D. Christofides, *Process Operational Safety and Cybersecurity*, Advances in Industrial Control, https://doi.org/10.1007/978-3-030-71183-2

17. Ames, A.D., Xu, X., Grizzle, J.W., Tabuada, P.: Control barrier function based quadratic programs for safety critical systems. IEEE Trans. Autom. Control **62**, 3861–3876 (2017)
18. Appl, M.: Ammonia. Ullmann's Encyclopedia of Industrial Chemistry, Wiley Online Library (2000)
19. Argyle, M., Bartholomew, C.: Heterogeneous catalyst deactivation and regeneration: a review. Catalysts **5**, 145–269 (2015)
20. Arifin, B.M.S., Choudhury, M.A.A.S.: An alternative approach of risk analysis for multivariable alarm system. J. Chem. Eng. IEB **26**, 75–79 (2011)
21. Artstein, Z.: Stabilization with relaxed controls. Nonlinear Anal.: Theory, Methods Appl. **7**, 1163–1173 (1983)
22. Aspen Technology Inc: Aspen Plus User Guide. Aspen Technology Inc, Cambridge, MA (2003)
23. Aspen Technology Inc: Aspen Plus Ammonia Model. Aspen Technology Inc, Bedford, MA (2017)
24. Atherton, J., Gil, F.: Incidents that Define Process Safety. Wiley Online Library, Hoboken, New Jersey (2008)
25. Ba, Y., Zhao, G., Kadambi, A.: Blending diverse physical priors with neural networks (2019). arXiv:1910.00201
26. Babiceanu, R.F., Seker, R.: Big data and virtualization for manufacturing cyber-physical systems: a survey of the current status and future outlook. Comput. Ind. **81**, 128–137 (2016)
27. Bakolas, E., Saleh, J.H.: Augmenting the traditional defense-in-depth strategy with the concept of a diagnosable safety architecture. In: Briš, R., Soares, C.G., Martorell, S. (eds.) Reliability, Risk and Safety: Theory and Applications, vol. 3, pp. 2113–2122. CRC Press/Balkema, Leiden, Netherlands (2010)
28. Baldi, M.: Cybersecurity defense for industrial process-control systems. Chem. Eng. **123**, 36 (2016)
29. Ball, R.: Oscillatory thermal instability and the Bhopal disaster. Process Saf. Environ. Protect. **89**, 317–322 (2011)
30. Beck, D.A., Carothers, J.M., Subramanian, V.R., Pfaendtner, J.: Data science: accelerating innovation and discovery in chemical engineering. AIChE J. **62**, 1402–1416 (2016)
31. Bertsekas, D.P.: Nonlinear Programming, 2nd edn. Athena Scientific, Belmont, MA (1999)
32. Biegler, L.T.: Nonlinear Programming: Concepts, Algorithms, and Applications to Chemical Processes. SIAM, Philadelphia, PA (2010)
33. Bishop, C.M.: Training with noise is equivalent to Tikhonov regularization. Neural Comput. **7**, 108–116 (1995)
34. Bishop, C.M.: Pattern Recognition and Machine Learning (Information Science and Statistics). Springer, New York, Inc (2006)
35. Bø, T.I., Johansen, T.A.: Dynamic safety constraints by scenario based economic model predictive control. In: Proceedings of the IFAC World Congress, pp. 9412–9418. Cape Town, South Africa (2014)
36. Boyd, S., Vandenberghe, L.: Convex Optimization. Cambridge University Press, Cambridge, UK (2004)
37. Braun, P., Kellett, C.M.: On (the existence of) Control Lyapunov Barrier Functions (2019). https://eref.uni-bayreuth.de/40899
38. Brooks, R., Thorpe, R., Wilson, J.: A new method for defining and managing process alarms and for correcting process operation when an alarm occurs. J. Hazardous Mater. **115**, 169–174 (2004)
39. Brown, N.: Alarm management/The EEMUA guidelines in practice. Meas. Control **36**, 114–119 (2003)
40. Buczak, A.L., Guven, E.: A survey of data mining and machine learning methods for cyber security intrusion detection. IEEE Commun. Surv. Tutorials **18**, 1153–1176 (2015)
41. Burden, F., Winkler, D.: Bayesian regularization of neural networks. In: Artificial Neural Networks, pp. 23–42. Springer, New York, NY (2008)
42. Camacho, E.F., Alba, C.B.: Model Predictive Control, 2nd edn. Springer, Berlin (2013)

43. Cárdenas, A.A., Amin, S., Lin, Z.S., Huang, Y.L., Huang, C.Y., Sastry, S.: Attacks against process control systems: risk assessment, detection, and response. In: Proceedings of the 6th ACM Symposium on Information, Computer and Communications Security, pp. 355–366. ACM, Hong Kong, China (2011)
44. Chang, Y., Khan, F., Ahmed, S.: A risk-based approach to design warning system for processing facilities. Process Saf. Environ. Protect. **89**, 310–316 (2011)
45. Chen, S., Wu, Z., Christofides, P.D.: Cyber-attack detection and resilient operation of nonlinear processes under economic model predictive control. Comput. Chem. Eng. **136**, 106806 (2020)
46. Chen, S., Wu, Z., Christofides, P.D.: A cyber-secure control-detector architecture for nonlinear processes. AIChE J. **66**, e16907 (2020)
47. Chen, Y., Peng, H., Grizzle, J.: Obstacle avoidance for low-speed autonomous vehicles with barrier function. IEEE Trans. Control Syst. Technol. **26**, 194–206 (2017)
48. Chollet, F., et al.: Keras (2015). https://www.keras.io
49. Christofides, P.D., Davis, J.F., El-Farra, N.H., Clark, D., Harris, K.R., Gipson, J.N.: Smart plant operations: vision, progress and challenges. AIChE J. **53**, 2734–2741 (2007)
50. Christofides, P.D., El-Farra, N.H.: Control of Nonlinear and Hybrid Process Systems: Designs for Uncertainty. Constraints and Time-Delays. Springer, Berlin, Germany (2005)
51. Christofides, P.D., Liu, J., Muñoz de la Peña, D.: Networked and Distributed Predictive Control: Methods and Nonlinear Process Network Applications. Advances in Industrial Control Series. Springer, London, England (2011)
52. Chylla Jr., R.W., Adomaitis, R.A., Cinar, A.: Stability of tubular and autothermal packed bed reactors using phase plane analysis. Ind. Eng. Chem. Res. **26**, 1356–1362 (1987)
53. Çinar, A., Palazoglu, A., Kayihan, F.: Chemical Process Performance Evaluation. CRC Press, Boca Raton, Florida (2007)
54. Cowlagi, R.V., Saleh, J.H.: Coordinability and consistency: application of systems theory to accident causation and prevention. J. Loss Prevent. Process Ind. **33**, 200–212 (2015)
55. Crowl, D.A., Louvar, J.F.: Chemical Process Safety: Fundamentals with Applications, 3rd edn. Pearson Education, Upper Saddle River, NJ (2011)
56. Crowl, D.A., Tipler, S.A.: Sizing pressure-relief devices. Chem. Eng. Progress 68–76 (2013)
57. Cruz, T., Rosa, L., Proença, J., Maglaras, L., Aubigny, M., Lev, L., Jiang, J., Simoes, P.: A cybersecurity detection framework for supervisory control and data acquisition systems. IEEE Trans. Ind. Inf. **12**, 2236–2246 (2016)
58. Dibaji, S., Pirani, M., Flamholz, D., Annaswamy, A., Johansson, K., Chakrabortty, A.: A systems and control perspective of CPS security. Ann. Rev. Control **47**, 394–411 (2019)
59. Ding, D., Han, Q.L., Xiang, Y., Ge, X., Zhang, X.M.: A survey on security control and attack detection for industrial cyber-physical systems. Neurocomputing **275**, 1674–1683 (2018)
60. Dubljević, S., Kazantzis, N.: A new Lyapunov design approach for nonlinear systems based on Zubov's method. Automatica **38**, 1999–2007 (2002)
61. EEMUA: EEMUA-191: Alarm Systems—A Guide to Design, Management and Procurement. Engineering Equipment and Materials Users Association, London, England (2013)
62. Ellis, M., Durand, H., Christofides, P.D.: A tutorial review of economic model predictive control methods. J. Process Control **24**, 1156–1178 (2014)
63. Ettouney, H.M., Shaban, H.I., Nayfeh, L.J.: Theoretical analysis of high and low temperature shift converters. Chem. Eng. Commun. **134**, 1–16 (1995)
64. Fawzi, H., Tabuada, P., Diggavi, S.: Secure estimation and control for cyber-physical systems under adversarial attacks. IEEE Trans. Autom. Control **59**, 1454–1467 (2014)
65. Gajjar, S., Palazoglu, A.: A data-driven multidimensional visualization technique for process fault detection and diagnosis. Chemometrics Intell. Lab. Syst. **154**, 122–136 (2016)
66. García, C.E., Prett, D.M., Morari, M.: Model predictive control: theory and practice-a survey. Automatica **25**, 335–348 (1989)
67. Genceli, H., Nikolaou, M.: Robust stability analysis of constrained L1-norm model predictive control. AIChE J. **39**, 1954–1965 (1993)
68. Gentile, M., Rogers, W.J., Mannan, M.S.: Development of an inherent safety index based on fuzzy logic. AIChE J. **49**, 959–968 (2003)

69. Goble, G., Stauffer, T.: Don't be alarmed: avoid unplanned downtime from alarm overload, use top techniques to improve alarm management. InTech Mag. **54**, 42–46 (2007)
70. Gong, J., You, F.: Optimal design and synthesis of algal biorefinery processes for biological carbon sequestration and utilization with zero direct greenhouse gas emissions: MINLP model and global optimization algorithm. Ind. Eng. Chem. Res. **53**, 1563–1579 (2014)
71. Gupta, J.P., Edwards, D.W.: Inherently safer design—present and future. Process Saf. Environ. Protect. **80**, 115–125 (2002)
72. Gurney, K.: An Introduction to Neural Networks. CRC Press (2014)
73. Hace, I.: The pressure relief system design for industrial reactors. J. Ind. Eng. **2013**, 1–14 (2013)
74. Harkat, M.F., Djelel, S., Doghmane, N., Benouaret, M.: Sensor fault detection, isolation and reconstruction using nonlinear principal component analysis. Int. J. Autom.Comput. **4**, 149–155 (2007)
75. He, C., You, F.: Shale gas processing integrated with ethylene production: novel process designs, exergy analysis, and techno-economic analysis. Ind. Eng. Chem. Res. **53**, 11442–11459 (2014)
76. Heidarinejad, M., Liu, J., Christofides, P.D.: Economic model predictive control of nonlinear process systems using Lyapunov techniques. AIChE J. **58**, 855–870 (2012)
77. Heikkilä, A.M., Hurme, M., Järveläinen, M.: Safety considerations in process synthesis. Comput. Chem. Eng. **20**, S115–S120 (1996)
78. Hink, R.C.B., Beaver, J.M., Buckner, M.A., Morris, T., Adhikari, U., Pan, S.: Machine learning for power system disturbance and cyber-attack discrimination. In: Proceedings of the 7th International Symposium on Resilient Control Systems, pp. 1–8. IEEE, Denver, CO, USA (2014)
79. Hoeting, J.A., Madigan, D., Raftery, A.E., Volinsky, C.T.: Bayesian model averaging: a tutorial. Stat. Sci. 382–401 (1999)
80. Hopfield, J.J.: Neural networks and physical systems with emergent collective computational abilities. Proc. Natl. Acad. Sci. **79**, 2554–2558 (1982)
81. Huang, L., Nguyen, X., Garofalakis, M.N., Hellerstein, J.M., Jordan, M.I., Joseph, A.D., Taft, N.: Communication-efficient online detection of network-wide anomalies. In: Proceedings of the 26th IEEE International Conference on Computer Communications—INFOCOM 2007, vol. 7, pp. 134–142. Anchorage, Alaska, USA (2007)
82. Huang, R., Biegler, L.T., Harinath, E.: Robust stability of economically oriented infinite horizon NMPC that include cyclic processes. J. Process Control **22**, 51–59 (2012)
83. Isidori, A.: Nonlinear Control Systems: An Introduction, 3rd edn. Springer, New York, NY (1995)
84. Jain, P., Pasman, H.J., Waldram, S., Pistikopoulos, E.N., Mannan, M.S.: Process resilience analysis framework (PRAF): a systems approach for improved risk and safety management. J. Loss Prevent. Process Ind. **53**, 61–73 (2018)
85. Jankovic, M.: Combining control Lyapunov and barrier functions for constrained stabilization of nonlinear systems. In: Proceedings of the American Control Conference, pp. 1916–1922. Seattle, Washington (2017)
86. Junejo, K.N., Goh, J.: Behaviour-based attack detection and classification in cyber physical systems using machine learning. In: Proceedings of the 2nd ACM International Workshop on Cyber-Physical System Security, pp. 34–43 (2016)
87. Karpatne, A., Watkins, W., Read, J., Kumar, V.: Physics-Guided Neural Networks (PGNN): An Application in Lake Temperature Modeling (2017). arXiv:1710.11431
88. Kellman, M., Bostan, E., Repina, N., Waller, L.: Physics-Based Learned Design: Optimized Coded-illumination for Quantitative Phase Imaging (2019). arXiv:1808.03571 (2019)
89. Kettunen, M., Zhang, P., Jämsä-Jounela, S.L.: An embedded fault detection, isolation and accommodation system in a model predictive controller for an industrial benchmark process. Comput. Chem. Eng. **32**, 2966–2985 (2008)
90. Khalil, H.K.: Nonlinear Syst., 3rd edn. Prentice Hall, Upper Saddle River, NJ (2002)

91. Khan, F.I., Abbasi, S.A.: Major accidents in process industries and an analysis of causes and consequences. J. Loss Prevent. Process Ind. **12**, 361–378 (1999)
92. Khan, F.I., Amyotte, P.R.: How to make inherent safety practice a reality. Canadian J. Chem. Eng. **81**, 2–16 (2003)
93. Kheradmandi, M., Mhaskar, P.: Data driven economic model predictive control. Mathematics **6**, 51 (2018)
94. Khorrami, F., Krishnamurthy, P., Karri, R.: Cybersecurity for control systems: a process-aware perspective. IEEE Design Test **33**, 75–83 (2016)
95. Khorsand, K., Marvast, M., Pooladian, N., Kakavand, M.: Modeling and simulation of methanation catalytic reactor in ammonia unit. Petroleum Coal **49**, 46–53 (2007)
96. Kidam, K., Hurme, M.: Analysis of equipment failures as contributors to chemical process accidents. Process Saf. Environ. Protecti. **91**, 61–78 (2013)
97. Kim, M., Liu, H., Kim, J.T., Yoo, C.: Sensor fault identification and reconstruction of indoor air quality (IAQ) data using a multivariate non-gaussian model in underground building space. Energy Build. **66**, 384–394 (2013)
98. Kletz, T.: What Went Wrong?—Case Histories of Process Plant Disasters and How They Could Have Been Avoided, 5th edn. Elsevier, Burlington, Massachusetts (2009)
99. Knowles, W., Prince, D., Hutchison, D., Disso, J., Jones, K.: A survey of cyber security management in industrial control systems. Int. J. Critical Infrastruct. Protect. **9**, 52–80 (2015)
100. Kohl, A.L., Nielsen, R.: Gas Purification. Gulf Publishing Co., Houston, USA (1997)
101. Kokotović, P., Arcak, M.: Constructive nonlinear control: a historical perspective. Automatica **37**, 637–662 (2001)
102. Kosmatopoulos, E.B., Polycarpou, M.M., Christodoulou, M.A., Ioannou, P.A.: High-order neural network structures for identification of dynamical systems. IEEE Trans. Neural Netw. **6**, 422–431 (1995)
103. Kravaris, C., Kantor, J.C.: Geometric methods for nonlinear process control. 1. Background. Ind. Eng. Chem. Res. **29**, 2295–2310 (1990)
104. Kravaris, C., Kantor, J.C.: Geometric methods for nonlinear process control. 2. controller synthesis. Ind. Eng. Chem. Res. **29**, 2310–2323 (1990)
105. Lao, L., Ellis, M., Christofides, P.D.: Proactive fault-tolerant model predictive control. AIChE J. **59**, 2810–2820 (2013)
106. Latham, D.A., McAuley, K.B., Peppley, B.A., Raybold, T.M.: Mathematical modeling of an industrial steam-methane reformer for on-line deployment. Fuel Proces. Technol. **92**, 1574–1586 (2011)
107. Levenberg, K.: A method for the solution of certain non-linear problems in least squares. Q. Appl. Math. **2**, 164–168 (1944)
108. Leveson, N.G.: Safeware: System Safety and Computers. Addison-Wesley Publishing Company, Reading, Massachusetts (1995)
109. Leveson, N.G., Stephanopoulos, G.: A system-theoretic, control-inspired view and approach to process safety. AIChE J. **60**, 2–14 (2014)
110. Lin, Y., Sontag, E.D.: A universal formula for stabilization with bounded controls. Syst. Control Lett. **16**, 393–397 (1991)
111. Liu, J., Muñoz de la Peña, D., Ohran, B.J., Christofides, P.D., Davis, J.F.: A two-tier control architecture for nonlinear process systems with continuous/asynchronous feedback. Int. J. Control **83**, 257–272 (2010)
112. Liu, P., Pistikopoulos, E.N., Li, Z.: A multi-objective optimization approach to polygeneration energy systems design. AIChE J. **56**, 1218–1234 (2010)
113. Long, Z., Lu, Y., Ma, X., Dong, B.: PDE-Net: Learning PDEs from Data (2017). arXiv:1710.09668
114. Lu, Y., Rajora, M., Zou, P., Liang, S.: Physics-embedded machine learning: case study with electrochemical micro-machining. Machines **5**, 4–15 (2017)
115. Luenberger, D.G.: Linear Nonlinear Program., 2nd edn. Kluwer Academic Publishers, Boston, MA (2003)

116. Mannan, M., Sachdeva, S., Chen, H., Reyes-Valdes, O., Liu, Y., Laboureur, D.: Trends and challenges in process safety. AIChE J. **61**, 3558 (2015)
117. Mannan, S.: Lees' Loss Prevention in the Process Industries—Hazard Identification, Assessment and Control, 4th edn. Elsevier, Waltham, Massachusetts (2012)
118. Marlin, T.: Process Control: Designing Process and Control Systems for Dynamic Performance. McGraw-Hill, New York (1995)
119. Marlin, T.: Operability in Process Design: Achieving Safe, Profitable, and Robust Process Operations. McMaster University, Ontario, Canada (2012)
120. Marquardt, D.W.: An algorithm for least-squares estimation of nonlinear parameters. J. Soc. Ind. Appl. Math. **11**, 431–441 (1963)
121. Marsh and McLennan Companies Inc.: The 100 largest losses 1974–2015: Large property damage losses in the hydrocarbon industry. Tech. rep., Marsh and McLennan Companies Inc. (2016)
122. Martínez, I., Armaroli, D., Gazzani, M., Romano, M.C.: Integration of the Ca-Cu process in ammonia production plants. Ind. Eng. Chem. Res. **56**, 2526–2539 (2017)
123. Massera, J.L.: Contributions to stability theory. Ann. Math. **64**, 182–206 (1956)
124. Mayne, D.Q., Rawlings, J.B., Rao, C.V., Scokaert, P.O.M.: Constrained model predictive control: Stability and optimality. Automatica **36**, 789–814 (2000)
125. Meel, A., Seider, W.D.: Plant-specific dynamic failure assessment using Bayesian theory. Chem. Eng. Sci. **61**, 7036–7056 (2006)
126. Mehra, A., Ma, W., Berg, F., Tabuada, P., Grizzle, J.W., Ames, A.D.: Adaptive cruise control: Experimental validation of advanced controllers on scale-model cars. In: Proceedings of the American Control Conference, pp. 1411–1418. Chicago, Illinois (2015)
127. Mendes-Moreira, J., Soares, C., Jorge, A.M., Sousa, J.F.D.: Ensemble approaches for regression: a survey. ACM Comput. Surv. **45**, 10 (2012)
128. Mendoza-Serrano, D.I., Chmielewski, D.J.: Smart grid coordination in building HVAC systems: EMPC and the impact of forecasting. J. Process Control **24**, 1301–1310 (2014)
129. Mhaskar, P., El-Farra, N.H., Christofides, P.D.: Predictive control of switched nonlinear systems with scheduled mode transitions. IEEE Trans. Autom. Control **50**, 1670–1680 (2005)
130. Mhaskar, P., El-Farra, N.H., Christofides, P.D.: Stabilization of nonlinear systems with state and control constraints using Lyapunov-based predictive control. Syst. Control Lett. **55**, 650–659 (2006)
131. Mhaskar, P., Liu, J., Christofides, P.D.: Fault-Tolerant Process Control: Methods and Applications. Springer, London, England (2013)
132. Mohanty, S.R., Pradhan, A.K., Routray, A.: A cumulative sum-based fault detector for power system relaying application. IEEE Trans. Power Deliv. **23**, 79–86 (2007)
133. Morari, M., Lee, J.H.: Model predictive control: past, present and future. Comput. Chem. Eng. **23**, 667–682 (1999)
134. Moskowitz, I.H., Seider, W.D., Arbogast, J.E., Oktem, U.G., Pariyani, A., Soroush, M.: Improved predictions of alarm and safety system performance through process and operator response-time modeling. AIChE J. **62**, 3461–3472 (2016)
135. Muñoz de la Peña, D., Christofides, P.D.: Lyapunov-based model predictive control of nonlinear systems subject to data losses. IEEE Trans. Autom. Control **53**, 2076–2089 (2008)
136. Murphey, Y.L., Masrur, M.A., Chen, Z.H., Zhang, B.: Model-based fault diagnosis in electric drives using machine learning. IEEE/ASME Trans. Mechatron. **11**, 290–303 (2006)
137. Naghoosi, E., Izadi, I., Chen, T.: Estimation of alarm chattering. J. Process Control **21**, 1243–1249 (2011)
138. Narendra, K.S., Annaswamy, A.M.: Stable Adaptive Systems. Courier Corporation (2012)
139. Niu, B., Zhao, J.: Barrier Lyapunov functions for the output tracking control of constrained nonlinear switched systems. Syst. Control Lett. **62**, 963–971 (2013)
140. Niziolek, A.M., Onel, O., Hasan, M.M.F., Floudas, C.A.: Municipal solid waste to liquid transportation fuels—Part II: process synthesis and global optimization strategies. Comput. Chem. Eng. **74**, 184–203 (2015)
141. Nocedal, J., Wright, S.: Numerical Optimization, 2nd edn. Springer, New York, NY (2006)

142. Noor, R.M., Ahmad, Z., Don, M.M., Uzir, M.H.: Modelling and control of different types of polymerization processes using neural networks technique: a review. Canadian J. Chem. Eng. **88**, 1065–1084 (2010)
143. Ojha, M., Dhiman, A.: Problem, failure and safety analysis of ammonia plant: a review. Int. Rev. Chem. Eng. **2**, 631–646 (2010)
144. de Oliveira, N.M., Biegler, L.T.: Constraint handing and stability properties of model predictive control. AIChE J. **40**, 1138–1155 (1994)
145. Omar, S., Ngadi, A., Jebur, H.H.: Machine learning techniques for anomaly detection: an overview. Int. J. Comput. Appl. **79**, 33–41 (2013)
146. Omell, B.P., Chmielewski, D.J.: IGCC power plant dispatch using infinite-horizon economic model predictive control. Ind. Eng. Chem. Res. **52**, 3151–3164 (2013)
147. Ozay, M., Esnaola, I., Vural, F.T.Y., Kulkarni, S.R., Poor, H.V.: Machine learning methods for attack detection in the smart grid. IEEE Trans. Neural Netw. Learn. Syst. **27**, 1773–1786 (2015)
148. Papachristodoulou, A., Prajna, S.: On the construction of Lyapunov functions using the sum of squares decomposition. In: Proceedings of the 41st IEEE Conference on Decision and Control, pp. 3482–3487. Las Vegas, NV (2002)
149. Pariyani, A., Seider, W.D., Oktem, U.G., Soroush, M.: Incidents investigation and dynamic analysis of large alarm databases in chemical plants: a fluidized-catalytic-cracking unit case study. Ind. Eng. Chem. Res. **49**, 8062–8079 (2010)
150. Pearlmutter, B.A.: Gradient calculations for dynamic recurrent neural networks: a survey. IEEE Trans. Neural Netw. **6**, 1212–1228 (1995)
151. de la Peña, D.M., Christofides, P.D.: Lyapunov-based model predictive control of nonlinear systems subject to data losses. IEEE Trans. Autom. Control **53**, 2076–2089 (2008)
152. Peng, C., Sun, H., Yang, M., Wang, Y.: A survey on security communication and control for smart grids under malicious cyber attacks. IEEE Trans. Syst. Man Cybern.: Syst. (2019)
153. Peng, P., Nguyen, H., Harold, M.P., Luss, D.: Spatio-temporal phenomena in monolithic reactors measured by combined spatially-resolved mass spectrometry and optical frequency domain reflectometry. In: Advances in Chemical Engineering, vol. 50, pp. 83–130. Elsevier (2017)
154. Polycarpou, M.M., Ioannou, P.A.: Identification and control of nonlinear systems using neural network models: Design and stability analysis. University of Southern California, Tech. rep. (1991)
155. Porfirio, C., Neto, E.A., Odloak, D.: Multi-model predictive control of an industrial C3/C4 splitter. Control Eng. Practice **11**, 765–779 (2003)
156. Pourkargar, D.B., Almansoori, A., Daoutidis, P.: Comprehensive study of decomposition effects on distributed output tracking of an integrated process over a wide operating range. Chem. Eng. Res. Design **134**, 553–563 (2018)
157. Prajna, S.: Barrier certificates for nonlinear model validation. Automatica **42**, 117–126 (2006)
158. Prajna, S., Jadbabaie, A.: Safety verification of hybrid systems using barrier certificates. In: Proceedings of the 7th International Workshop, HSCC, vol. 2993, pp. 477–492. Philadelphia, Pennsylvania (2004)
159. Qin, S.J.: Survey on data-driven industrial process monitoring and diagnosis. Ann. Rev. Control **36**, 220–234 (2012)
160. Qin, S.J., Badgwell, T.A.: A survey of industrial model predictive control technology. Control Eng.Practice **11**, 733–764 (2003)
161. Qin, S.J., Chiang, L.H.: Advances and opportunities in machine learning for process data analytics. Comput. Chem. Eng. **126**, 465–473 (2019)
162. Qin, S.J., Dunia, R.: Determining the number of principal components for best reconstruction. J. Process Control **10**, 245–250 (2000)
163. Rahimpour, M.R., Dehnavi, M.R., Allahgholipour, F., Iranshahi, D., Jokar, S.M.: Assessment and comparison of different catalytic coupling exothermic and endothermic reactions: a review. Appl. Energy **99**, 496–512 (2012)
164. Raiyn, J.: A survey of cyber attack detection strategies. Int. J. Secur. Appl. **8**, 247–256 (2014)

165. Rawlings, J.B.: Tutorial overview of model predictive control. IEEE Control Syst. Mag. **20**, 38–52 (2000)
166. Rawlings, J.B., Maravelias, C.T.: Bringing new technologies and approaches to the operation and control of chemical process systems. AIChE J. **65**, e16,615
167. Rawlings, J.B., Mayne, D.Q.: Model Predictive Control: Theory and Design. Nob Hill Publishing, Madison, WI (2009)
168. Reniers, G., Cozzani, V.: Domino Effects in the Process Industries: Modelling, Prevention and Managing. Newnes, Waltham, Massachusetts (2013)
169. Rodrigues, M., Theilliol, D., Adam-Medina, M., Sauter, D.: A fault detection and isolation scheme for industrial systems based on multiple operating models. Control Eng. Practice **16**, 225–239 (2008)
170. Romdlony, M.Z., Jayawardhana, B.: Stabilization with guaranteed safety using control Lyapunov-barrier function. Automatica **66**, 39–47 (2016)
171. Rönsch, S., Schneider, J., Matthischke, S., Schlüter, M., Götz, M., Lefebvre, J., Prabhakaran, P., Bajohr, S.: Review on methanation-from fundamentals to current projects. Fuel **166**, 276–296 (2016)
172. Rothenberg, D.H.: Alarm Management for Process Control: A Best-Practice Guide for Design, Implementation, and Use of Industrial Alarm Systems. Momentum Press, New York, New York (2009)
173. Samanta, B., Al-Balushi, K.R., Al-Araimi, S.A.: Artificial neural networks and support vector machines with genetic algorithm for bearing fault detection. Eng. Appl. Artif. Intell. **16**, 657–665 (2003)
174. Scokaert, P.O., Rawlings, J.B.: Feasibility issues in linear model predictive control. AIChE J. **45**, 1649–1659 (1999)
175. Seo, Y., Seo, D., Seo, Y., Yoon, W.: Investigation of the characteristics of a compact steam reformer integrated with a water-gas shift reactor. J. Power Sources **161**, 1208–1216 (2006)
176. Sepulchre, R., Janković, M., Kokotović, P.V.: Constructive Nonlinear Control. Communications and Control Engineering. Springer, London, England (1997)
177. Shen, Q., Jiang, B., Shi, P., Lim, C.C.: Novel neural networks-based fault tolerant control scheme with fault alarm. IEEE Trans. Cybern. **44**, 2190–2201 (2014)
178. Shi, G., Shi, X., O'Connell, M., Yu, R., Azizzadenesheli, K., Anandkumar, A., Yue, Y., Chung, S.: Neural lander: Stable drone landing control using learned dynamics. In: Proceedings of the International Conference on Robotics and Automation, pp. 9784–9790. Montreal, Canada (2019)
179. Sibi, P., Jones, S.A., Siddarth, P.: Analysis of different activation functions using back propagation neural networks. J. Theor. Appl. Inf. Technol. **47**, 1264–1268 (2013)
180. Singh, J., Nene, M.J.: A survey on machine learning techniques for intrusion detection systems. Int. J. Adv. Res. Comput. Commun. Eng. **2**, 4349–4355 (2013)
181. Smith, H., Howard, C., Foord, T.: Alarms management/Priority, floods, tears or gain? Introduction to the "problem". Meas. Control **36**, 109–113 (2003)
182. Sontag, E.D.: Neural nets as systems models and controllers. In: Proceedings of the Seventh Yale Workshop on Adaptive and Learning Systems, pp. 73–79. Yale University, 1992
183. Sontag, E.D.: A 'universal' construction of Artstein's theorem on nonlinear stabilization. Syst. Control Lett. **13**, 117–123 (1989)
184. Srinivasan, R., Liu, J., Lim, K.W., Tan, K.C., Ho, W.K.: Intelligent alarm management in a petroleum refinery. Hydrocarbon Proces. **83**, 47–53 (2004)
185. Stewart, B.T., Venkat, A.N., Rawlings, J.B., Wright, S.J., Pannocchia, G.: Cooperative distributed model predictive control. Syst. Control Lett. **59**, 460–469 (2010)
186. Takeda, K., Hamaguchi, T., Noda, M., Kimura, N., Itoh, T.: Use of two-layer cause-effect model to select source of signal in plant alarm system. In: Setchi, R., Jordanov, I., Howlett, R.J., Jain, L.C. (eds.) Knowledge-Based and Intelligent Information and Engineering Systems: 14th International Conference, KES 2010, Cardiff, UK, September 8–10, 2010, Proceedings, Part II, pp. 381–388. Springer, Berlin, Germany (2010)

187. Tatiya, R.R.: Elements of Industrial Hazards: Health, Safety. Environment and Loss Prevention. CRC Press/Balkema, Leiden, Netherlands (2011)
188. Tee, K.P., Ge, S.S., Tay, E.H.: Barrier Lyapunov functions for the control of output-constrained nonlinear systems. Automatica **45**, 918–927 (2009)
189. Tian, Y., Zhang, J., Morris, J.: Modeling and optimal control of a batch polymerization reactor using a hybrid stacked recurrent neural network model. Ind. Eng. Chem. Res. **40**, 4525–4535 (2001)
190. Toro, J.C.O., Dobrosz-Gómez, I., García, M.Á.G.: Dynamic modeling and bifurcation analysis for the methyl isocyanate hydrolysis reaction. J. Loss Prevent. Process Ind. **39**, 106–111 (2016)
191. Trischler, A.P., D'Eleuterio, G.M.: Synthesis of recurrent neural networks for dynamical system simulation. Neural Netw. **80**, 67–78 (2016)
192. Tsai, C.F., Hsu, Y.F., Lin, C.Y., Lin, W.Y.: Intrusion detection by machine learning: a review. Exp. Syst. Appl. **36**, 11994–12000 (2009)
193. Twigg, M.V.: Catalyst Handbook. Wolfe Publishing Ltd., London (1989)
194. Van Dijk, H., Boon, J., Nyqvist, R.N., Van Den Brink, R.W.: Development of a single stage heat integrated water-gas shift reactor for fuel processing. Chem. Eng. J. **159**, 182–189 (2010)
195. Venkatasubramanian, V.: Systemic failures: challenges and opportunities in risk management in complex systems. AIChE J. **57**, 2–9 (2011)
196. Venkatasubramanian, V.: The promise of artificial intelligence in chemical engineering: is it here, finally? AIChE J. **65**, 466–478 (2019)
197. Venkatasubramanian, V., Rengaswamy, R., Kavuri, S.N.: A review of process fault detection and diagnosis: Part II: qualitative models and search strategies. Comput. Chem. Eng. **27**, 313–326 (2003)
198. Venkatasubramanian, V., Rengaswamy, R., Kavuri, S.N., Yin, K.: A review of process fault detection and diagnosis: Part III: process history based methods. Comput. Chem. Eng. **27**, 327–346 (2003)
199. Venkatasubramanian, V., Zhao, J., Viswanathan, S.: Intelligent systems for HAZOP analysis of complex process plants. Comput. Chem. Eng. **24**, 2291–2302 (2000)
200. Vernières-Hassimi, L., Leveneur, S.: Alternative method to prevent thermal runaway in case of error on operating conditions continuous reactor. Process Saf. Environ. Protect. **98**, 365–373 (2015)
201. Wächter, A., Biegler, L.T.: On the implementation of an interior-point filter line-search algorithm for large-scale nonlinear programming. Math. Program. **106**, 25–57 (2006)
202. Walton, M., Southerton, T., Sharp, P.: Safety Improvements in a Methanation Reactor. Process Safety Progress, Wiley Online Library (2009)
203. Wang, J., Chen, T.: An online method for detection and reduction of chattering alarms due to oscillation. Comput. Chem. Eng. **54**, 140–150 (2013)
204. Wang, J., Yang, F., Chen, T., Shah, S.L.: An overview of industrial alarm systems: main causes for alarm overloading, research status, and open problems. IEEE Trans. Autom. Sci. Eng. **13**, 1045–1061 (2016)
205. Wang, L., Ames, A.D., Egerstedt, M.: Safety barrier certificates for collisions-free multirobot systems. IEEE Trans. Robot. **33**, 661–674 (2017)
206. Wang, S., Chen, Y.: Sensor validation and reconstruction for building central chilling systems based on principal component analysis. Energy Convers. Manag. **45**, 673–695 (2004)
207. Wang, Y.: A new concept using lstm neural networks for dynamic system identification. In: Proceedings of the American Control Conference, pp. 5324–5329. Seattle, Washington (2017)
208. West, S.R., Guo, Y., Wang, X.R., Wall, J.: Automated fault detection and diagnosis of HVAC subsystems using statistical machine learning. In: Proceedings of the 12th International Conference of the International Building Performance Simulation Association. Sydney, Australia (2011)
209. Widodo, A., Yang, B.: Support vector machine in machine condition monitoring and fault diagnosis. Mech. Syst. Signal Process. **21**, 2560–2574 (2007)
210. Wieland, P., Allgöwer, F.: Constructive safety using control barrier functions. IFAC Proc. Vol. **40**, 462–467 (2007)

211. Wilson, Z.T., Sahinidis, N.V.: The alamo approach to machine learning. Comput. Chem. Eng. **106**, 785–795 (2017)
212. Wu, Z., Albalawi, F., Zhang, J., Zhang, Z., Durand, H., Christofides, P.D.: Detecting and handling cyber-attacks in model predictive control of chemical processes. Mathematics **6**, 173 (2018)
213. Wu, Z., Albalawi, F., Zhang, Z., Zhang, J., Durand, H., Christofides, P.D.: Control Lyapunov-barrier function-based model predictive control of nonlinear systems. Automatica **109**, 108508 (2019)
214. Wu, Z., Albalawi, F., Zhang, Z., Zhang, J., Durand, H., Christofides, P.D.: Model predictive control for process operational safety: Utilizing Safeness Index-based constraints and control Lyapunov-barrier functions. In: Proceedings of 13th International Symposium on Process Systems Engineering, Computer Aided Chemical Engineering, vol. 44, pp. 505–510. San Diego, California (2018)
215. Wu, Z., Chen, S., Rincon, D., Christofides, P.D.: Post cyber-attack state reconstruction for nonlinear processes using machine learning. Chem. Eng. Res. Des. **159**, 248–261 (2020)
216. Wu, Z., Christofides, P.D.: Economic machine-learning-based predictive control of nonlinear systems. Mathematics **7**, 494 (2019)
217. Wu, Z., Christofides, P.D.: Handling bounded and unbounded unsafe sets in control Lyapunov-barrier function-based model predictive control of nonlinear processes. Chem. Eng. Res. Des. **143**, 140–149 (2019)
218. Wu, Z., Christofides, P.D.: Optimizing process economics and operational safety via economic MPC using barrier functions and recurrent neural network models. Chem. Eng. Res. Des. **152**, 455–465 (2019)
219. Wu, Z., Christofides, P.D.: Control Lyapunov-barrier function-based predictive control of nonlinear processes using machine learning modeling. Comput. Chem. Eng. **134**, 106706 (2020)
220. Wu, Z., Durand, H., Christofides, P.D.: Safe economic model predictive control of nonlinear systems. Syst. Control Lett. **118**, 69–76 (2018)
221. Wu, Z., Durand, H., Christofides, P.D.: Safeness Index-based economic model predictive control of stochastic nonlinear systems. Mathematics **6**, 69 (2018)
222. Wu, Z., Rincon, D., Christofides, P.D.: Process structure-based recurrent neural network modeling for model predictive control of nonlinear processes. J. Process Control **89**, 74–84 (2020)
223. Wu, Z., Rincon, D., Christofides, P.D.: Real-time adaptive machine-learning-based predictive control of nonlinear processes. Ind. Eng. Chem. Res. **59**, 2275–2290 (2020)
224. Wu, Z., Rincon, D., Christofides, P.D.: Real-time machine learning for operational safety of nonlinear processes via barrier-function based predictive control. Chem. Eng. Res. Des. **155**, 88–97 (2020)
225. Wu, Z., Tran, A., Ren, Y.M., Barnes, C.S., Chen, S., Christofides, P.D.: Model predictive control of phthalic anhydride synthesis in a fixed-bed catalytic reactor via machine learning modeling. Chem. Eng. Res. Des. **145**, 173–183 (2019)
226. Wu, Z., Tran, A., Rincon, D., Christofides, P.D.: Machine learning-based predictive control of nonlinear processes. part I: Theory. AIChE J. **65**, e16729 (2019)
227. Wu, Z., Tran, A., Rincon, D., Christofides, P.D.: Machine learning-based predictive control of nonlinear processes. part II: Computational implementation. AIChE J. **65**, e16734 (2019)
228. Xu, X., Tabuada, P., Grizzle, J.W., Ames, A.D.: Robustness of control barrier functions for safety critical control. IFAC-PapersOnLine **48**, 54–61 (2015)
229. Xue, D., El-Farra, N.H.: Actuator fault-tolerant control of networked distributed processes with event-triggered sensor-controller communication. In: Proceedings of the American Control Conference, pp. 1661–1666. Boston, Massachusetts (2016)
230. Ye, N., Zhang, Y., Borror, C.M.: Robustness of the markov-chain model for cyber-attack detection. IEEE Trans. Reliab. **53**, 116–123 (2004)
231. Yeo, K.: Short Note on the Behavior of Recurrent Neural Network for Noisy Dynamical System (2019). arXiv:1904.05158
232. Yin, X., Liu, J.: Subsystem decomposition of process networks for simultaneous distributed state estimation and control. AIChE J. **65**, 904–914 (2019)

233. Zavala, V.M.: A multiobjective optimization perspective on the stability of economic MPC. In: Proceedings of the 9th IFAC Symposium on Advanced Control of Chemical Processes, pp. 975–981. Whistler, Canada (2015)
234. Zhang, C., Ma, Y.: Ensemble Machine Learning: Methods and Applications. Springer, Berlin (2012)
235. Zhang, J., Liu, J.: Distributed moving horizon state estimation for nonlinear systems with bounded uncertainties. J. Process Control **23**, 1281–1295 (2013)
236. Zhang, S., Zhang, S., Wang, B., Habetler, T.G.: Machine Learning and Deep Learning Algorithms for Bearing Fault Diagnostics-a Comprehensive Review (2019). arXiv:1901.08247
237. Zhang, Z., Wu, Z., Durand, H., Albalawi, F., Christofides, P.D.: On integration of feedback control and safety systems: analyzing two chemical process applications. Chem. Eng. Res. Design **132**, 616–626 (2018)
238. Zhang, Z., Wu, Z., Rincon, D., Christofides, P.D.: Operational safety of an ammonia process network via model predictive control. Chem. Eng. Res. Design **146**, 277–289 (2019)
239. Zhang, Z., Wu, Z., Rincon, D., Christofides, P.D.: Operational safety via model predictive control: the torrance refinery accident revisited. Chem. Eng. Res. Design **149**, 138–146 (2019)
240. Zhang, Z., Wu, Z., Rincon, D., Garcia, C., Christofides, P.D.: Operational safety of chemical processes via Safeness-Index based MPC: two large-scale case studies. Comput. Chem. Eng. **125**, 204–215 (2019)

The manufacturer's authorised representative in the EU is Springer Nature Customer Service Centre GmbH, Europaplatz 3, 69115 Heidelberg, Germany. If you have any concerns regarding our products, please contact ProductSafety@springernature.com

Printed and bound by CPI Group (UK) Ltd, Croydon, CR0 4YY
24/07/2026
02175973-0001